Venice in Environmental Peril?

Myth and Reality

Dominic Standish

Foreword by John Eglin

UNIVERSITY PRESS OF AMERICA,® INC.
Lanham • Boulder • New York • Toronto • Plymouth, UK

4501 Forbes Boulevard
Suite 200
Lanham, Maryland 20706
UPA Acquisitions Department (301) 459-3366

Estover Road
Plymouth PL6 7PY
United Kingdom

Printed in the United States of America
British Library Cataloging in Publication Information Available

Library of Congress Control Number: 2011935094
ISBN: 978-0-7618-5664-1 (paperback : alk. paper)
eISBN: 978-0-7618-5665-8

Cover artwork adapted from *Ancient and Modern* by Patrick Hughes

∞™ The paper used in this publication meets the minimum requirements of American National Standard for Information Sciences—Permanence of Paper for Printed Library Materials, ANSI Z39.48-1992

For Laura, who brought me to Venice,
and our sons Riccardo and Steffan.

Contents

PART FOUR: MYTHS ABOUT MODERNIZATION

Illustrations

Foreword

By John Eglin, author of *Venice Transfigured: The Myth of Venice in British Culture, 1660–1797* (2001). October 2010.

I suppose my qualifications for writing this foreword are twofold: first, I have written on the political mythology of Venice (as refracted through the lens of another political culture), and I have lived for the past decade and a half in a part of the world suffused with its own historical myth, a myth with narratives built around the relationship of humans with their environment. One cannot live in the American west as long as this writer has, and not develop a healthy respect for historical myth. The counterfactual character of myth is what makes it mythical, but its connection to deeper truth makes it mythic. We are all of us, in this part of the world, fast in the grip of the great myth of the frontier, of civilization (of a sort, and only as much as we can bear) wrung from the land through the rugged individualism and pioneer self-sufficiency of hardy homesteaders settled in houses of sod, like the mythic proto-Venetians fleeing into the salt marshes carrying the flame of Roman civilization. The frontier myth informs both sides of the environmental debates that are such flashpoints in local and regional politics, as this mythic tradition is a double-sided coin.

So, too, is the myth of Venice, and just as "the tourist Venice *is* Venice," as Mary McCarthy put it,[1] the history of Venice, to a great extent, *is* the history of its myth. Franco Gaeta famously identified three components of this myth: that of the Venetian constitution, the unchanging mixture of monarchy, aristocracy and democracy touted by ancient political theorists as the most stable form of government; that of Venetian independence, dating from the city's foundation by Roman patricians fleeing the barbarian conquest of Italy; and finally that of the "gallant city."[2] To Gaeta's three myths of Venice, we might, however, add a fourth: the imperiled city. The city in peril is mythic, not mythical; how that peril is constructed and narrated depends upon tropes that have been developed over centuries. Venice, born in peril, has never been out of it. The city that emerged from the chaos of late antiquity has for the last half millennium floated precariously between antiquity and modernity. The persistence of external threats to Venice of one kind or another has contributed to the often paranoid quality of its myth. Despotic powers sought to conquer and destroy it, for its government was perfect, naturally so, as its governors were the last true descendants of the Roman patriciate, endowing Venice with an unbroken republican apostolic succession. And naturally the sea, in defiance of which Venice rose on an infinity of wooden piles, and whose subjection was celebrated annually in mock nuptials, might seek to reassert itself somehow.

For centuries, it was feared that the sea would retreat, as it had from Pisa, for the shallow lagoon formed its natural defense against invasion. The fear that Venice would be swallowed not by the sea, but by dry land, was supported by the steady silting up of the lagoon, which proceeded apace from late antiquity. The belief that Venice was *naturally* fortified from the hostility or ambition of neighboring powers concealed the reality that nothing about Venice's situation *was* natural. Only human intervention on a massive scale over the course of centuries prevented Venice from being left high, dry, and vulnerable. As Frederic Lane pointed out forty years ago, such intervention was controversial even in medieval Venice, when as now, "conflicting interests and theories" impeded projects to preserve the city and lagoon.[3] As much as anything else, Venice was a symbol of human mastery over nature. Perhaps nature must ultimately triumph, but it would be well to remember that the more recent danger to Venice is also of human origin.

"Without Napoleon, Venice would not be complete."[4] Napoleon affected the doom of the Republic, finishing what had been begun by the opening of transoceanic trade and the emergence of unitary nation states in the sixteenth century, and allowing Wordsworth and Ruskin to presage the end of the city itself. But as the invasion of 1797 was not the first blow, neither was it the last. The rise of industry reinforced the city's irrelevance; the railroad bridge that finally united Venice to the Terraferma yet another trophy of the victory of modernity, a token that the city's storied insulation no longer mattered. No one knew at the time what Ruskin would have appreciated, had he known that carbon belched into the atmosphere by a coal-fired industrial revolution would cause sea levels to rise, constituting a new danger to Venice. Geopolitical peril has been supplanted by environmental peril.

Dominic Standish finds that just as the political and constitutional myths of Venice focused early modern debates over the direction of the emerging modern state, the myth of an environmentally imperiled Venice has conditioned current debates over the handling of environmental issues in public policy. Now as then, contention centers on the relationship between the city's present and its past. Is Venice a precious heritage site to be preserved at all costs, or is the ecosystem of the lagoon too fragile to be disrupted in the interest of preserving a gaudy tourist amusement park? (Venice, of course, has been both of those things for at least the last three centuries.) The same tension guided internal discussions about adapting, or not, to contemporary infrastructural needs or desires, whether these involved bridging the Grand Canal, reconstructing the Campanile, introducing the railroad, or introducing the automobile, at least over the Lagoon. Myth can blind us to reality, but it can also move us to shape reality -- and it can do both at the same time.

In closing, I turn again to Mary McCarthy's *Venice Observed,* a book that would be profitably read by anyone who writes about Venice or perhaps more profitably avoided, until the task is complete, for some of her observations are warnings, and warning to authors. "'I envy you writing about Venice,' says the newcomer," she writes; "'I pity you,' says the old hand," for there is nothing, according to McCarthy, that can be said of Venice "that has not been said before."[5] Such an admonition might indeed dampen the enthusiasm of a newcomer to the topic, but this particular old hand finds it oddly encouraging. For the corollary to McCarthy's dictum may be that nothing one says about Venice is ever completely wrong - a marvelous tonic against scholarly timidity, that.

NOTES

1. Mary McCarthy, *Venice Observed* (New York: Harcourt Brace, 1963), 8.
2. Franco Gaeta, "Alcune Considerazione sul Mito di Venezia," *Bibliothèque d'Humanisme et Renaissance* 23 (January 1961): 58-75.
3. Frederic Chapin Lane, *Venice: A Maritime Republic* (Baltimore: Johns Hopkins University Press, 1973), 452-453.
4. McCarthy, *Venice Observed,* 5.
5. Ibid., 11-12.

Acknowledgments

I would like to begin by thanking my wife Laura Ceccato. Her impact on this book and my life is hard to put into words and is possibly better illustrated by the first day she brought me to Venice when I proposed to her and we were married there eight months later. Ever since our wedding day in 1997, living in the region of Venice has constantly engaged me with developments in the city. For remarkable patience and continual support during the research and writing of this book, our two sons, Riccardo and Steffan, deserve special praise. Laura also gave her valuable time for editorial assistance and substantial translation help.

Other members of my family have also provided wonderful support. My father, Francis Standish, was once my primary school teacher and headmaster and rediscovered the energy to improve my English again in most of the manuscript. One of my brothers, Alex Standish, who is a Geography Assistant Professor, resolved many of my geographical weaknesses and created both this book's maps.

Jennie Bristow did a wonderful job of editing the whole manuscript twicc, improved my English immensely and made important recommendations. At the University Press of America, Lindsay Macdonald gave useful editorial assistance, as did Brooke Bascietto of the Rowman and Littlefield Publishing Group. My friends Carla Picardi and Kathleen Graham both offered vital comments on two chapters. Stephen Crocker helped with one chapter. Although I am lucky to have received tremendous editorial assistance, any errors in the text are my responsibility.

For constant intellectual inspiration and guidance, I would like to thank Professor Frank Furedi of Kent University, UK. Frank opened my mind as an undergraduate student and provided invaluable direction while supervising my sociology Ph.D about Venice and environmental risk. At Kent University,

Dr Adam Burgess and Dr Jan Macvarish helped refine ideas for this project. Professor Molly Rapert of Arkansas University also gave me important methodological advice that contributed towards my Ph.D and this book. Alan Hudson and Dr James Panton of Oxford University offered instructive comments. Dr Al Ringleb, Cristina Turchet, Anna Fiumicetti and staff at Iowa University and the CIMBA Consortium of Universities for International Studies have supported my teaching and a series of seminars about Venice that has explored ideas for this book. Many of my students have made great suggestions and I thank them all.

For the eloquent Foreword to this book and his own enlightening book about Venice, I thank Professor John Eglin of Montana University. Patrick Hughes generously donated his fascinating painting 'Ancient and Modern' for the book's front cover. For the reproduction of graphs and tables, thanks go to the Istituzione Centro Previsioni e Segnalazioni Maree (Venice) and Venice City Council.

This book would not have been possible without the comments of environmental campaigners I interviewed in Venice. I thank them for their time and insights. Jane Da Mosto also offered valuable observations. Finally, it is with great sadness that I end by noting the death of one interviewee, Maurizio Zanetto, in June 2010.

Introduction: The Venice Problem and a Proposed Solution

> "Since a particularly devastating flood in 1966, an effective defense of Venice has been undermined by a bitter debate over whether to keep out the sea by building 79 huge, mechanized steel barriers at the three entrances to the Venetian lagoon. Engineers backed the barriers, but they were opposed by environmentalists who argued for restoring the lagoon to its natural condition and reversing the effects of what they saw as reckless industrialization."
>
> Barry James, *International Herald Tribune,* 22 March 2002.

The above quote accurately represents the popular perception of the Venice problem and the debate about protecting the city with mobile dams. This book questions this perception and reveals how understanding debates about Venice's problems requires demystifying a delicate interplay between myth and reality. Contemporary debate about Venice's environment is influenced by mythical traditions established during the Venetian Republic from 727 to 1797, when myths were cultivated to provide a system of meaning for domestic cohesion and to impress other states. Despite experiencing devastating floods and environmental disasters, myths and ritual ceremonies were devised to celebrate domination over the seas. The ancient Venetians courageously intervened to curb environmental hazards. By contrast, today's discussion of Venice is pervaded by fear of the sea and a preoccupation with environmental threats; and claims about sinking, rising sea levels, and the destructive impact of tourism, present Venice as a city in peril. We deconstruct these claims to show that they reveal more about how environmentalism permeates our culture than about objective threats. Contemporary claims about Venice in peril can be compared with ancient myths that the Venetian Republic constructed to give the city wider meaning, and provide a way to understand the relationship between the city and the sea that resonates in a Western culture saturated with environmental fears. Interviews with Venetian environmental campaigners, upon which this book is based, help to unlock the relationship between myth and reality in Venice.[1]

Venice has acquired a reputation as an environmentally-threatened retreat from modernization. The reality is that Venice is not in peril from sinking, rising sea levels or tourism. But there is a problem in Venice, arising from the weakening of human ambition since the fall of the Venetian Republic in 1797. Reviving the courage and resilience to intervene to neutralize environmental threats is vital to improve the city (Standish 2002). Rather than restricting intervention in search of a mythical harmony with nature, this book proposes that we should strive for new levels of ingenuity to modernize Venice. To this end we offer, in Chapter Nine, a ten-point proposal for how the city could be developed today, building on the city's unique history and attraction, and the engineering solutions to flooding that are already within our grasp.

INTRODUCING THE PROBLEM: FLOODING AND SINKING IN VENICE

The controversy about the mobile dams is the most prominent example of how current debates about Venice are moulded by ancient myths. Both advocates and opponents of the mobile dams have constructed cultural narratives about the dams and traditions founded during the Venetian Republic to support their claims. To understand these claims, we need to separate imagery and reality. This is notoriously difficult in Venice, although as the historian James Grubb (1986) affirms, it "need not be insuperable, since Venice provides its own interpretative keys" (Grubb 1986, 43).

The debate about the Venice problem emerged after the city was completely flooded in November 1966. Flooding and sinking did not originate as historically unique problems in 1966, and these floods were not peculiar to Venice: their physical impact cannot explain why the city subsequently received so much attention. Rather, the Venice problem was constructed through the responses to the 1966 floods by various players, including the United Nations Educational, Scientific and Cultural Organization (UNESCO), successive Italian governments, and claims-making organizations. These responses politicized the Venice problem, and helped to politicize proposed solutions, especially the mobile dams.

Venice is a cluster of 117 islands among other islands scattered across marshes and open water in a lagoon. The lagoon is approximately 56 kilometers (35 miles) long and 11 kilometers (7 miles) wide. It is protected from the sea by littoral islands (*lidi*), apart from three inlets between the lagoon and the sea (see Figure 1.1): these allow tidal waters to flow between the lagoon and the sea, meaning that high tides can flood the islands.

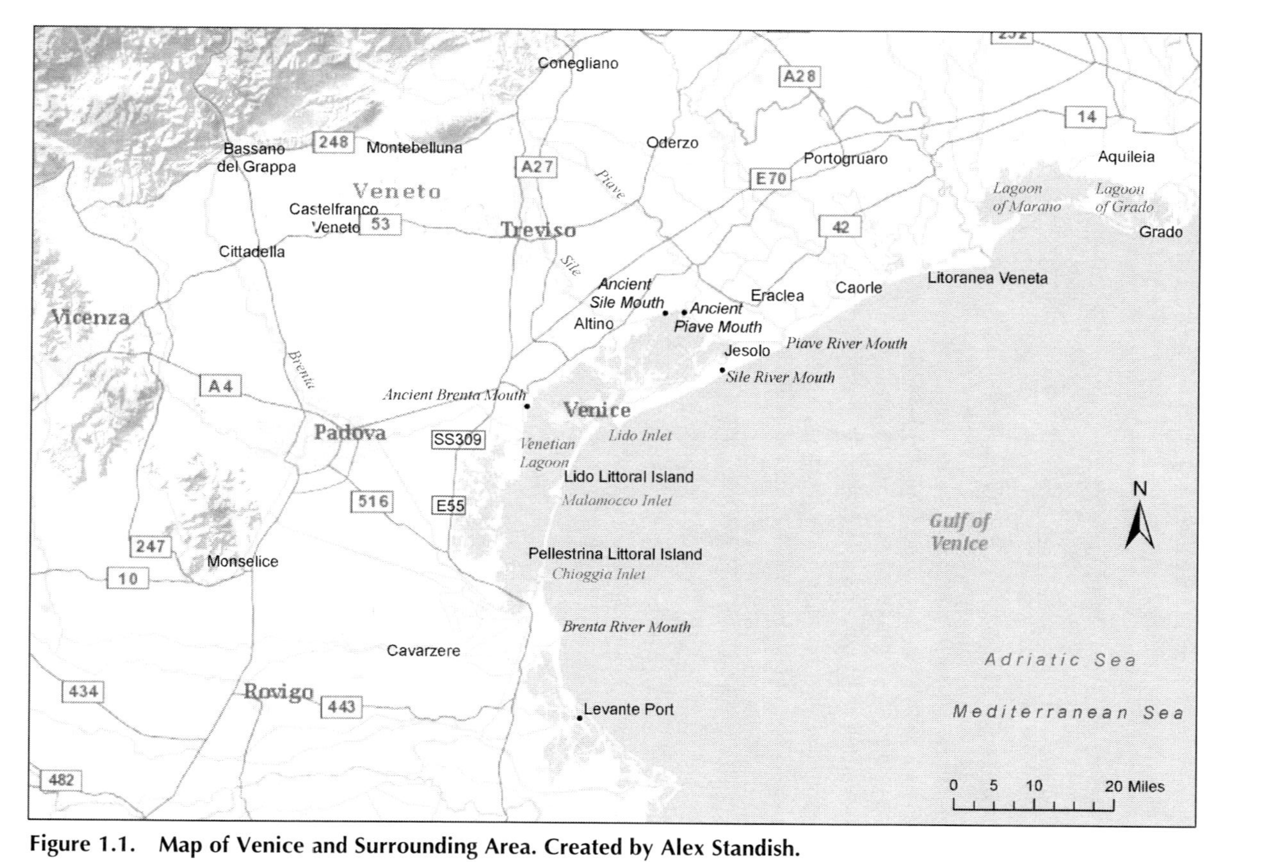

Figure 1.1. **Map of Venice and Surrounding Area. Created by Alex Standish.**

There is little doubt that Venice needs protection against increased flooding. Environmental campaigner Marco Favaro stated, during an interview with the author in 2005, that over the past century the level of Venice with respect to the sea has lowered by 27-28 centimeters[2] (Favaro 2005). This suggests a slightly worse situation than that indicated in the data provided on the Venice City Council website in 2009, which listed the total increase in sea level in Venice between 1897 and 2009 as approximately 25cm (Istituzione Centro Previsioni e Segnalazioni Maree 2009). Venice's annual mean sea level is calculated by averaging all the maximum and minimum tides in a year. The variation in annual mean sea level between 1872 and 2007 is represented in centimeters in Table 1.1: the latest graph available at the time of writing.

The Organisation for Economic Cooperation and Development (OECD) provides useful flooding data for Venice in 2010: "Average water levels are now almost 30 centimetres above levels in the 1880s, and the frequency of high-water events has increased more than tenfold when considered on a decadal basis from the 1880s" (OECD 2010, 84). High-water events are classified as those reaching over 110cm above the zero reference point established in 1897 at the Punta della Salute entrance into the Grand Canal. There is significant variation in flooding from year to year. There were none

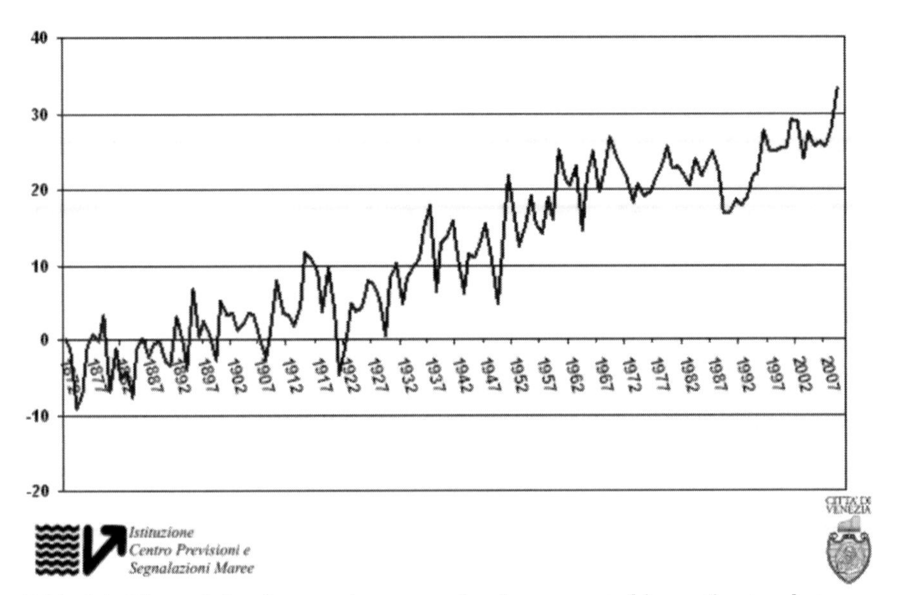

Table 1.1. The variation in annual mean sea level, represented in centimeters between 1872 and 2007. Source: Istituzione Centro Previsioni e Segnalazioni Maree, Venice City Council.

of these high-water events in 2003 or 2007, although there were 10 between 14 November and 8 December 2002. On 1 December 2008, Venice was swamped by its worst floods for 22 years (Standish 2008). The annual frequency of high-water events above 110cm is illustrated in Table 1.2.

To understand what causes high-water events in Venice we need to distinguish between long-term trends and short-term weather factors. Over the long term, there has been an increase in the relative sea level (RSL) in Venice due to subsidence (a lowering of the land level: sinking), and higher sea levels. Regular subsidence can be predicted accurately, as described by the scientists Caroline Fletcher and Jane Da Mosto (2004):

> The land itself is subsiding naturally as the ancient sediments of the coast settle. Also, the movement of the Earth's crust (on a geological timescale), is driving this part of Italy down under the Alps. Together these processes cause Venice to sink by about 0.5mm each year, although the exact amount varies over time, from place to place – the northern and southern parts of the lagoon have registered rates of subsidence of a few mm/year. (Fletcher and Da Mosto 2004, 34)

To add to regular subsidence, there was exceptional subsidence in Venice during the twentieth century. This was principally due to the extraction of groundwater from the aquifers (natural underground reservoirs) beneath the

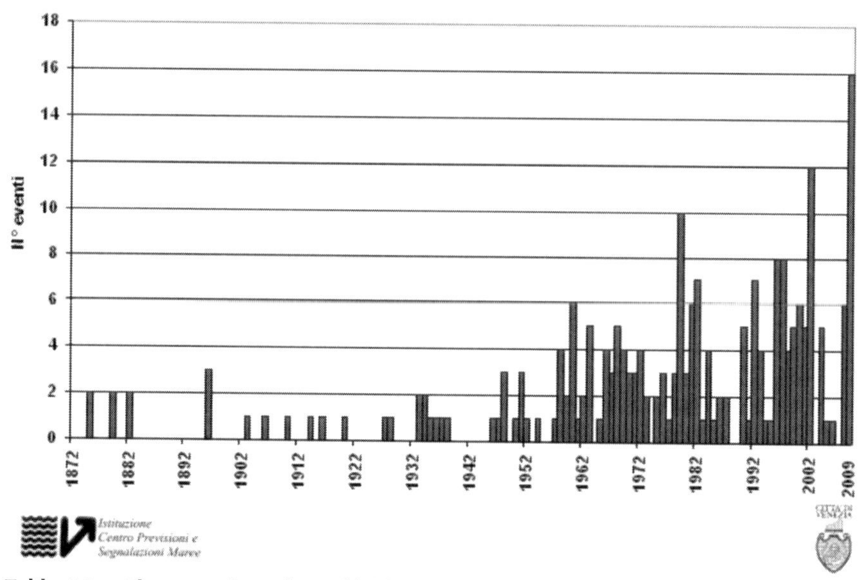

Table 1.2. The annual number of high-water events more than or equal to 110 centimeters above zero on the Punta della Salute tide gauge. Source: Istituzione Centro Previsioni e Segnalazioni Maree, Venice City Council.

lagoon. Most of the groundwater was pumped to industries at the Marghera industrial zones bordering the city (see Figure 1.2), although it was also used for irrigation and Venice's municipal water supply. Electrical pumps were introduced in the 1930s at Marghera to make groundwater pumping more rapid. Padua University's Professor Augusto Ghetti (1988) explains that Venice's natural geological rate of subsidence accelerated from 1.8mm a year in 1930 to 8mm a year by 1950, and became even worse between 1968 and 1969 when there was sinking of 17mm at Marghera and 12mm in Venice[3] (Ghetti 1988, 26-27). The Italian government closed the wells extracting this groundwater from 1971, as noted in paper from the Italian Institute of Marine Sciences by Laura Carbognin et al (2009, 3). A small rebound (2cm) was recorded (Fletcher and Da Mosto 2004, 35) as the aquifers partly refilled; however, this did not compensate for the significant increases in subsidence caused by groundwater extraction. Between 1897 and 1983, the RSL in Venice rose by 23cm, with 12cm of the 23cm RSL rise due to subsidence and 11cm caused by rising sea levels (Cecconi 1997a, 1).

Another long-term concern regarding the rise in RSL is the extraction of methane gas from under the Adriatic Sea, which could exaggerate irregular subsidence. Environmental organizations including Italia Nostra started campaigning in the 1960s to prevent the extraction of natural gas near Venice. Gas extraction in other parts of the Adriatic Sea is occurring, and in 2009, the former Minister for Productive Activity Claudio Scajola suggested that large gas reserves near Venice should also be tapped (*La Repubblica*, 9 January 2009). It remains to be seen whether gas extraction in the upper part of the Adriatic will restart irregular subsidence in Venice: if it does not, we can expect limited subsidence of 0.5mm each year.

When the rate of subsidence in Venice slowed after the extraction of groundwater stopped, more attention focused on the contribution of rising sea levels to flooding. The Intergovernmental Panel on Climate Change (IPCC) established that there was a global temperature rise of 0.74° C[4] between 1906 and 2005, which added to global sea levels rising (eustatic change) by an average rate of 1.8mm/year from 1961 (IPCC 2007, 2). Yet the impact of higher temperatures and sea levels varies in different parts of the world. "It is important to point out that long tide-gauge records around the world show that the eustatic rate over the last century in the Northern Adriatic Sea, along with the whole Mediterranean Sea, is consistently lower (approximately 35%) than the global mean," note Carbognin et al (2009, 6), referring to research by sea level scientist Simon Holgate (2007). These authors recorded that this difference is due to changes in atmospheric pressure, temperatures, salinity and the specific features of the almost-closed Adriatic Sea (Carbognin et al. 2009, 6). Nevertheless, global sea level rise is one element exaggerating

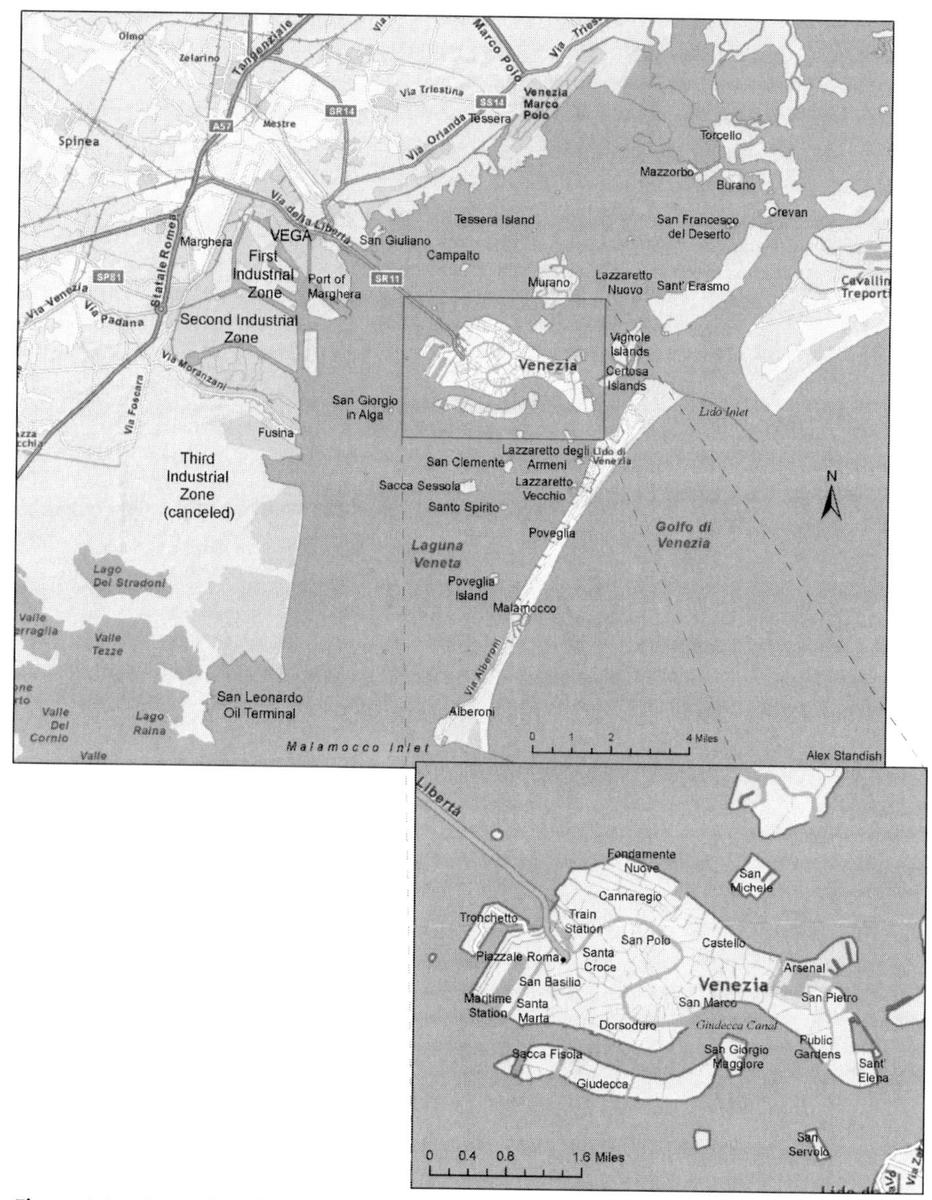

Figure 1.2. Map of Venice, the Venetian Lagoon and Islands. Created by Alex Standish.

general flooding in Venice. It has been estimated that eustasy increased the sea level in Venice by 9cm between the start of the twentieth century and 1970 and at a similar rate by 5cm between 1970 and 2005 (Istituzione Centro Previsioni e Segnalazioni Maree 2009). Another cause of higher water levels in the lagoon is silt, which is deposited by rivers flowing into the lagoon: additional silt makes the lagoon bed rise, pushing the water level up. Subsidence, eustasy and silting have led to a long-term rise in the relative sea level, and combine with the astronomical tide and short-term weather conditions to create high-water events.

Heavy rain falls directly into the lagoon and increases the amount of water from rivers flowing into the lagoon. Storm surges exaggerate flooding in Venice even more than rain. Overall, "the main influence comes from low pressure weather systems and associated winds, which create storm surges that effectively push more water into the lagoon" (Fletcher and Da Mosto 2004, 32). Winds known as the *sirocco* from the south-east and the *bora* from the north-east are famous for contributing to these storm surges. Alterations to the physical structure of the Venetian lagoon have also added to the impact of winds:

> Changes in the shape of the inlets, loss of saltmarshes and deep navigation channels have contributed to a greater volume of water and stronger current entering the lagoon when pushed by the tide and winds. A small effect is also attributed to the reduction in total area of the lagoon, due to land reclamation and other interventions, so the greater volume of water has less area in which to spread itself. (Fletcher and Da Mosto 2004, 40)

In summary, subsidence, a variety of physical changes to the lagoon, and a rise in sea levels, mean that Venice has become more susceptible to flooding. It is the combination of these longer-term factors increasing the relative sea level and an atypical mixture of local weather that causes Venice's high-water events. When storm surges, wind and rain increase the volume of water in the Venetian lagoon, high floods are the outcome. Such bad weather, along with an earth tremor on the Adriatic Sea bed creating tidal waves, produced the high floods of 1966. This happened after significant subsidence since the 1930s. In November 1966, water levels in Venice were nearly two meters[5] higher than usual. But did the floods in 1966 justify defining the Venice problem as unique?

The Unexceptional Floods of 1966

On 3 and 4 November 1966, parts of Venice experienced water depths 1.94m (6 feet and 4.5 inches) above mean sea level. Precise historical measurements of high tides in Venice have only been available since the first tide gauge was placed in 1897 at Punta della Salute. Measurements using this tide gauge show that the level of the 1966 floods was exceptional for the past century: yet when a variety of sources are examined, there is evidence of much more devastating floods in Venice's history.

Listed chronicles show flooding has been plaguing Venice since the sixth century (Zucchetta 2000, 88-109). One of the worst floods that damaged Venice and its lagoon islands occurred in 1106, when, as the eminent Venice historian John Julius Norwich has documented, severe flooding of the Venetian old town Malamocco swept away the entire community and left not one building standing (Norwich 1983, 82). A previous settlement at Malamocco was also destroyed by flooding in the year 527. The historian Peter Ackroyd (2010) writes: "[I]n 1250 the water rose steadily for four hours and, in the testimony of a contemporary, 'many were drowned in their houses or died of the cold'" (Ackroyd 2010, 31). Tintoretto's paintings in Venice's Church of Madonna dell'Orto show palaces and churches drowning in floods. In 1852, Doctor Thomas Burgess, a famous climatotherapist, predicted that Venice was "doomed... to sink under the waves of the Adriatic some sixty years hence, and leave no trace behind" (cited in Pemble 1996, 114). Lagoon islands experienced significant inundations in 782, 840, 875, 1102, 1240, 1268, and 1794 (Keahey 2002, 98). The medieval historian Elisabeth Crouzet-Pavan (2002) describes how the sea frequently overcame defenses during the Middle Ages, and Venetians' fears of destruction prompted the creation of myths about their relationship with the sea:

> Every time an exceptional flood tide opened a breach in the littoral islands that protected the city from the sea, part of the shoreline was destroyed and low-lying land was submerged, thus threatening the community with death by drowning. The risk of sinking into the sea inspired one of the central scenes in lagoon mythology. On 15 February 1340 a flood tide brought such high water that Venice seemed threatened with annihilation. (Crouzet-Pavan 2002, 50)

To compare these floods to 1966, we should concentrate on their impact. We cannot compare the relative water levels because accurate data prior to 1897 is unavailable. Moreover, comparing floods by isolating water levels would be unsatisfactory even if we had this data: the seriousness of a flood is not solely determined by the water level, but also by the ability of the community to protect itself and respond. Less developed infrastructure and transport, the absence of sea walls, and weaker building structures and defenses, meant that past Venetian communities were often devastated by flooding. They suffered much more from flooding than Venetians did in 1966 or do today.

During the November 1966 storm surges, the sea walls built in the eighteenth century on the Pellestrina Littoral Island initially held back the pounding seas. They gave some residents time to fortify their homes or evacuate until the walls eventually crumbled. Between 3,000 and 5,000 people were made homeless by the 1966 floods, mainly from the islands of Pellestrina and Saint'Erasmo. Yet the higher level of development in 1966 compared with the situation during earlier serious floods meant that Venetians could respond more effectively. Motor boats were used to rescue the homeless and transport them to safe shelter in ships, hospitals and military barracks. Properties and furniture throughout the city were badly damaged by flooding, but unlike in 527 or 1106, whole communities of buildings were not swept away. Nobody in the historical center of the city died during the 1966 floods. By way of contrast, a report by the Italian news agency *ANSA* (1997) described deaths from Venice's floods in 1268 when "the water rose from eight o'clock until midday. Many were drowned inside their houses or simply died of the cold." Venice in 1966 was left without functioning electricity and telephones for a week, but Venetians could still hear tide forecasts on battery operated radios that helped them plan how to respond. Rubbish, dead rats and pigeons floated in alleyways: conditions that were commonplace or much worse in previous floods.

Floods have been ravaging Venice and Venetians have been debating the causes for centuries. Marino Folin and Mario Spinelli (2005), scholars from the Venice University Institute of Architecture (IUAV), document that one of the more heated debates about the reasons for flooding in Venice was held during the sixteenth century (Folin and Spinelli 2005, 148). There was an awareness of the problems related to rising sea levels in the fifteenth century and rudimentary measurements began in 1440, as recorded by marine scientist Roberto Frassetto (2005, 36). Concern that the city would be engulfed by the sea provoked English art critic and poet John Ruskin in 1860 to detail the city stone by stone for the future generations who might never see it (Ruskin 1960 [1860]). Thus flooding and sinking did not emerge as historically unique problems in 1966.

Nor were the November 1966 floods unique to Venice: they were much worse in many other parts of Italy. By 19 November 1966, the official final death toll from the floods across northern and central Italy was 110 people, with an additional six missing; there were no flood-related deaths in Venice's historical center, although at nearby Chioggia two people died from heart attacks (*ANSA* 1997). In Florence, 90 people died due to the onslaught of high rivers after more than a month of rain (Berendt 2005, 260). Damage to the monuments, historical buildings and artefacts that make up Venice's heritage cannot explain the framing[6] of Venice's 1966 floods as an exceptional problem. In Florence, treasured paintings from leading art galleries were washed out into the rivers, and low-lying churches and buildings filled up with water, mud and debris to heights of nearly two meters. The British journalists Stephen Fay and Phillip Knightley (1976) report that the flood damage in Florence was so serious that Venice was widely ignored in the immediate aftermath:

> At the same time as the lagoon overwhelmed Venice, further south the river Arno, swollen by the opening of sluice gates to save an upstream dam from bursting, broke its banks and swirled through Florence. The Uffizi, the National Library, the State Archives, the Bargello, the Cathedral, the Baptistry, Santa Maria Novella, Santa Croce, and countless other churches, museums and historic buildings were filled with water, mud, fuel oil and sewage. In the rush to save the immense artistic wealth of Florence, Venice was virtually forgotten. Also, Venice appeared to recover rapidly. Within three weeks, Florians – the oldest surviving café in St Mark's Square – had reopened. (Fay and Knightley 1976, 27-28)

Although the 1966 Venice floods were disruptive, Venetians proved capable of getting their city back to business quickly. The American historian Peter Lauritzen (1986) writes:

> Eventually the flood waters receded and by nine o'clock that night, the Venetians could walk about dryshod to inspect the damage. They found rubbish and detritus scattered everywhere and a slimy coating of black fuel oil stained walls and pavements, although much had floated out to sea on the ebb. Electricity was still out of order and the waters had ruined central heating units installed in ground-floor rooms. But no one had died in the flood and no work of art was damaged. (Lauritzen 1986, 26)

The initial international response to the 1966 floods therefore focused on Florence. "Art lovers around the world formed committees to send aid and assistance. In the United States, the Committee to Rescue Italian Art was established, with Jacqueline Kennedy as its honorary president," writes the American author John Berendt (2005, 260). Attention turned to Venice in the following year:

> It was not until the following summer that it gradually became apparent that the damage to Venice had been much more profound, much more permanent, much more tragic than anyone had imagined. 'On that single night of the floods,' the Mayor said, 'our city aged fifty years.' The 1966 floods finally brought home to the world what had been happening in Venice: the catalogue of its ailments was alarming. The major danger was from the city's age-old enemy, the sea. (Fay and Knightley 1976, 28)

Venice appeared to attract greater attention than Florence due to its fragile relationship with the sea. Long-term perceptions of this relationship have been changed by past and present Venetian myths, as explored in Chapters One, Two and Seven. The 1966 floods was a key moment in shifting perceptions, becoming central to myths about Venice's vulnerability. As time passed, responses to the 1966 floods increasingly concentrated on Venice, despite the continued threat to Florence from the river Arno. During his first administration as the Mayor of Venice (1993-2000), Massimo Cacciari complained: "A sword of Damocles hangs over [Florence] and yet everybody continues to agitate about Venice" (cited in Betts 1999). A series of special laws were enacted: according to Michele Vianello, the former Vice-Mayor of Venice, Special Law 171, passed in 1973, was the first law to declare that Venice and its lagoon were "a problem of essential national interest" (Vianello 2004, 150), and further Special Laws passed by Italian governments in 1984 and 1992 helped to single out the city as a priority and defined the parameters of the Venice problem. "The 'Venice problem' was quickly declared to be of national interest and a priority for action. Italy has since enacted several laws to safeguard the city and lagoon, and a great deal of restoration and protection work against flooding has, and is, being done," note Fletcher and Da Mosto (2004, 9).

Such special measures were often justified by raising the specter of Venice sinking into the sea. Yet the rate at which Venice is sinking does not provide an adequate explanation for the exceptional attention the city has attracted. It has been claimed that Ravenna, the major Adriatic seaport south of Venice, is sinking at a much faster rate: Professor Giuseppe Gambolati of Padua University argues: "In Venice we see natural subsidence of 0.5 millimeter (*sic*) per year, but in Ravenna we have 2.5 millimeters per year" (cited in Keahey 2002, 24). But, like Florence, Ravenna has not been offered the high-priority protection measures contained in special laws for Venice. More recent research on Ravenna by the geologists Werter Bertoni, Carlo Elmi and Francesco Marabini (2005) reveals that during the second half of the last century, "the town and its surroundings have been affected by a severe artificial subsidence of up to fifty times the natural rate, chiefly due to water and gas extraction. The total subsidence reached values of about one meter" (Bertoni, Elmi and Marabini 2005, 23). Although Ravenna's rate of subsidence has slowed down due to the reduction of groundwater use, offshore gas extraction means subsidence has continued as a significant problem. These geologists conclude that at Ravenna "[M]ost coastal areas will probably have to be abandoned to the sea or else continuous defenses will have to be constructed all along the shoreline" (Bertoni, Elmi and Marabini 2005, 31). Given the lack of evidence that the 1966 floods in Venice were unique historically or in comparison with Florence, or that Venice has faced exceptional sinking problems, why was it that the 'Venice problem' became the focus of international political attention?

The Politicization of the 'Venice Problem'

The issue of improving flood defenses is usually a technical matter dealt with at a local level: as shown by the construction of the Thames River flood barrier in London. On 21 September 1969, a storm surge caused flooding down the east coast of England, prompting the London Flood Committee meeting on that day to call for a protective barrier. "That afternoon was so vivid in the decision makers' memory that they decided they must have a barrier for London. Finally things moved quickly. The antibarrier shipping lobby was quieted," writes American journalist John Keahey (2002, 176). A London barrier had been discussed since flooding in the early twentieth century, but the floods in 1953 and storm surges in 1969 assisted the passing of the 1973 Thames Barrier Act. Work on the barrier began in 1974 and it was opened in 1984. The swift and largely technical character of the debate about the Thames barrier is a stark contrast to the discussion of the Venice flooding problem and barrier protection since 1966. The London debate was concentrated within local and national institutions, as well as bodies with a direct industrial interest, such as shipping organizations. In Venice, similar issues attracted the attention of a plethora of local, national and international players. These players included claims-makers from a wide range of organizations, successive Italian governments and international bodies, especially UNESCO. The intervention of various players after the 1966 floods helped construct the city's problems of flooding and sinking as worthy of international political interest.

Claims-makers had an active role in persuading audiences to accept that Venice was a unique case deserving special political attention. "All claims-making is a form of diffusion, in which claimsmakers (transmitters) try to persuade audiences (adopters) by making claims (via channels) about social problems (the object of diffusion)," explains the constructionist scholar Joel Best (2001a, 8). In this regard, we can see that the construction of Venice's problems as unique and deserving of political attention in the aftermath of the 1966 floods provided the object of diffusion. Private committees were established internationally and promoted this claim, including the *Comité Francais pour la Sauvegarde de Venise* in France, *Arbeitskreis Venedig der Deutschen Unesco-Kommission* in Germany and *Pro Venezia* in Sweden. British and American organizations were the most prominent claims-making transmitters of Venice's problems. After the 1966 floods, many Americans had concentrated on restoring Florence through the International Fund for Monuments, but attention soon shifted to Venice: "Recognizing a threat to the very existence of Venice, Colonel Gray established a 'Venice Committee' within his International Fund for Monuments. As the salvage operations in Florence were being completed, Gray enlisted the head of the Committee to Rescue Italian Art, Professor John McAndrew, to be the Venice Committee's chairman" (Berendt 2005, 261). By 1971, the Venice Committee was detached from the International Fund for Monuments and became Save Venice, with McAndrew as chairman. As Berendt (2005) describes, Save Venice became a highly influential promoter of Venice as uniquely deserving attention in the United States and internationally.

Among the others transmitting claims about Venice's problems was Sir Ashley Clarke, the former British Ambassador to Italy. In 1967, Clarke helped to found the Italian Art and Archives Rescue Fund (IAARF) to raise finance for Florence and Venice following the 1966 floods; by the end of July 1967, it was decided that efforts to help Florence should be wound down in order to concentrate on Venice. This contributed to the shift in the response to the floods. In 1971, the IAARF was replaced by a new fundraising organization, The Venice in Peril Fund, the choice of name supporting the claim that Venice is a uniquely fragile city requiring urgent attention. The Venice in Peril Fund has raised substantial finance for restoration projects in the city and has provided channels for the international diffusion of claims through conferences, campaigns and publications. This preservationist association has helped to organize high-profile conferences about Venice's problems, and assisted the publication of influential texts, including *The Science of Saving Venice* (2004), *Flooding and Environmental Challenges for Venice and its Lagoon: State of Knowledge* (2005) and *The Venice Report* (2009).

In the early stages of international claims-diffusion, environmental organizations were minor players. However, they made a significant contribution to this process after 1997, especially through their appeals to the European Commission (Standish 2004): this process is examined in Chapter Five. The most high-profile environmental organization in Venice, Italia Nostra, played an important local part in the formation of the Venice problem: after the Venice branch of Italia Nostra was created in 1959, it compiled a list of endangered Venetian palaces in need of restoration, which was later completed by UNESCO. During his interview with the author, Maurizio Zanetto, the former President and Secretary of Italia Nostra, identified the roles of both Italia Nostra and UNESCO in defining the Venice problem after 1966:

> I knew Italia Nostra because then there were a lot of articles about [the organization] in the newspapers and they used to do a lot of campaigning in those years … I was interested in the problems of safeguarding the environment after the year 1966. I bought and read the publication by UNESCO. So I saw that Italia

> Nostra was the most active association and for some years it was the only association doing something. I remembered the intervention in the building work in the 50s. So I enrolled in Italia Nostra. (Zanetto 2005)

Italia Nostra was influential locally in the Venice problem from its inception, working with UNESCO. It was UNESCO that led the international response to the 1966 floods and facilitated the international diffusion of the Venice problem. UNESCO's impact on this problem is outlined by Howard Moore, the Director of the UNESCO Office in Venice:

> The Campaign for the Safeguarding of Venice had its origins in the immediate reaction by UNESCO to that dramatic event, when an international campaign was launched by the then Director – General of the Organization, René Maheu. In response to the appeal of Maheu, many private bodies (some fifty-nine) were set up around the world to gather funds and use them in the preservation and restoration of buildings, monuments and works of art in Venice. Even today, almost forty years on, there are well over twenty – in 11 countries – that are still active, and UNESCO, through its Office in the city and the Association of Private Committees, is pleased to play a part in their important work. (Moore 2005, xxiii-xxiv)

By bringing together, in 1969, scientists from around the world to discuss the 1966 Venice floods, UNESCO stimulated the momentum behind the international diffusion of the Venice problem. There was some skepticism in Italy about the timing and possible outcomes of UNESCO's intervention: it became interested in the Venice problem in the period just after the Egyptian government had capitulated to UNESCO pressure to intervene and save the temples of Abu Simbel from the Nile. "UNESCO came on the Venice scene only because it was finished at Abu Simbel, and now it would have us become a cross between a Nubian monument and Disneyland," declared Gianni de Michelis, formerly a prominent Italian Socialist Party member (cited in Fay and Knightley 1976, 99).

Italian governments had also self-consciously raised the domestic and international profile of the Venice problem. Emilio Colombo, the Italian Prime Minister in 1971, stated that he sought "[T]o bring the problem of Venice out of the shadows where it seems to have remained grounded until now, and set it on the way to a solution in line with Venetian, Italian, and international expectations" (cited in Fay and Knightley 1976, 61). While Italian governments did not give UNESCO extensive powers of intervention in the Venice problem, UNESCO was assigned a decision-making role through being allocated a seat on the Venice Safeguarding Commission established in 1973. The Venice Safeguarding Commission was granted the authority to advise the government on all building and territorial changes within the Venice lagoon boundary; it was a noteworthy step to invite a representative of an international body to participate in local and national policy decisions. "It was, as Italian politicians never failed to point out, a considerable concession to offer membership to a non-Italian representing an international organization," write Fay and Knightley (1976, 95). Subsequently, UNESCO has actively put pressure on Italian governments: for example, its 1974 General Convention requested that the Italian state guarantee the safeguarding interventions for Venice in the Special Law of 1973. Pressure on Italian governments has also been exerted from 11 countries through the Association of Private Committees, which UNESCO has helped to coordinate, and which since 1997 has held a special status as a non-governmental organization (NGO) in operational relations with UNESCO. Italian NGOs have also attempted to sway government policy on Venice by channeling claims through UNESCO: in 2002, environmental and other organizations in the Save Venice and its Lagoon Committee asked UNESCO to support measures against the Italian government's plans to construct the mobile dams in Venice (Environmental News Service 2002).

UNESCO has played a decisive role in shaping the Venice problem, alongside Italian governments and NGOs from many countries. UNESCO guided the responses to the 1966 floods during the early stages of the Venice problem's construction. Although the 1966 floods in Venice were very disruptive and sinking had accelerated, UNESCO was not simply reacting to these problems: rather, the organization was seeking a new cause, and Venice provided moral attraction that could stimulate international interest. Senior members of UNESCO self-consciously framed the safeguarding of Venice as a moral duty, as indicated in a 1973 document by the organization's International Consultative Committee:

> Firstly, one might observe that the participation of the international community is a *moral obligation* (UNESCO's italics). If Venice really represents, in the eyes of many men and women, a vital common asset, then they must share the burden of its preservation...An enterprise of this kind can represent the 'honour of a lifetime'...or, on the other hand, a lost opportunity. Anyone is free, of course, to leave the task to others, but those with most pride and greatest wisdom will want to say 'I took part.' (Cited in Fay and Knightley 1976, 96)

The members of the UNESCO International Consultative Committee appreciated that the Venice problem was not simply about responding to the 1966 flood damage. Venice is a symbol of Western patrimony, and campaigns about saving Venice combine the city's real problems and its mythical traditions. Presenting Venice's problems as critical and cultivating a role in saving the city thus became an opportunity to raise UNESCO's profile.

Similarly, the shift in focus from Florence to Venice by American and British preservationist organizations indicated an understanding that Venice had greater potential as an international problem. Between 1969 and 1973, the Venice problem became consolidated through the campaign created by UNESCO in 1969, the 1971 establishment of Save Venice and The Venice in Peril Fund, and the 1973 Special Law for Venice. Even though sinking and flooding in Venice increased during the twentieth century, it was the intervention of these various players that turned these difficulties into an internationally-recognized problem. Therefore, it is the *interaction* of the *objective* causes of sinking and flooding and *subjective* political factors that constitutes the Venice problem. But how did the dam project become the most controversial of possible solutions?

INTRODUCING A PROPOSED SOLUTION: VENICE'S MOBILE DAMS

The Venice dam project was proposed in a feasibility study during 1971, legislated for in 1973 and almost 40 years later is still not finished. This timescale is in stark contrast to the London Thames barrier, which was legislated for in 1973 and opened 11 years later. The principal explanation for this contrast is that the London barrier was treated in a technical manner, whereas the Venice barrier became a point of contention between competing political parties and claims-makers. The reasons why the Venice dam project was politicized and how this has affected changing perceptions of the city are explored in the remainder of the book. This introductory overview provides the basic details of how the mobile dam project has evolved.

The mobile dam project is commonly referred to as Project 'MOSE', named after a model of a barrier (*Modulo Sperimentale Elettromeccanico – MoSE* – Electromechanical Experimental Model). Project MOSE is a system of 78 mobile barriers designed to protect the three inlets to the lagoon that surrounds Venice, located in Figure 1.1. The barriers will remain below the surface of the sea until high tides and flooding of the city are predicted, at which point they will fill with air, rise up and block the sea from the lagoon. A simulation film of how the dams will operate can be viewed on the website of the company building them, the New Venice Consortium (*Consorzio Venezia Nuova* – CVN).[7]

Project MOSE's mobile gates have been planned to block high tides reaching over 110cm (43.3070 inches) on the Punta della Salute tide meter when storm surges are anticipated. It is important to note that the mobile dams were originally only designed to block these very high tides, although they have sufficient flexibility to protect against medium tides; but the dam project has never been a solution for all types of flooding in Venice. Footpaths, entrances to buildings, and local defenses have been improved over the last 30 years to limit the impact of medium and low-level flooding. With the mobile dams, a forecasting system will predict when high floods are expected, closing the gates six hours beforehand. High water predictions by Venice's Center for Sea Forecasts have been fairly reliable, although could still be more precise: when Venice was disrupted by its highest flooding for 22 years on 1 December 2008, the forecasts initially underestimated how high the tides would be (Bertasi 2008).

The design and status of the dam project has changed over time. The first proposals for a dam system emerged after the 1966 high floods. In 1969, the Italian government founded the Institute for the Study of the Dynamics of Grand Masses to examine the physical and environmental problems affecting the city and the lagoon, which initiated a 'competition of ideas' in 1970 for proposals to defend Venice from flooding. The competition was won by a design for an underwater mobile barrier system, and Roberto Frassetto, of the Italian National Research Council (CNR) and director of the Institute, produced a feasibility and cost study in 1971. Frassetto claims that this remains the best solution:

> The first gate [for Venice] was presented in 1970…so my institute was the one that proposed and made a meeting and got the [construction] firms to tell us what kind of technical information – geological, meteorological, and so forth – they needed to build this kind of defense. So among the several solutions that emerged there was the gate that lies at the bottom and is hinged at the bottom and is lifted up by compressed air. In thirty years, despite jealous engineers making all attempts possible to ridicule the gates – there is no better solution. This is the only solid and intelligent solution for Venice. (Cited in Keahey 2002, 140-1)

Frassetto's plans to build the dam system were first given legal recommendation in the 1973 Special Law to safeguard Venice. In 1974, the fourth subcommittee of the Grand Committee, which reported to the Director-General of Public Works Ministry, voted in favor of the underwater mobile barrier system that won the 1970-1 'competition of ideas.' The Grand Committee was then dissolved and replaced by the Senior Council of the Public Works Ministry, which in 1975 called for international bids for work to defend the city from high waters using a fixed, rather than a mobile, barrier system. A special adjudication panel decided in 1978 that none of the proposed projects submitted in the international bids was suitable for flood control. In 1980, the Public Works Ministry appropriated all plans from the international bids and set up a technical team to develop a project to protect the lagoon from flooding. By 1981, the technical team had come up with the 'Feasibility study and principal plan' for a major flood control project, known as the *Progettone* (Big Project), which provided a conceptual overview of how a dam system would be part of a general effort to restore hydro-geological and ecological balance to the lagoon. The Senior Council of Public Works approved the Progettone in 1982, as did Venice City Council.

After a period of institutional instability regarding Venice's protection, the government established an institutional framework that has endured. In 1984, another government Special Law for Venice instituted the Committee for Policy, Coordination and Control (*Comitatone* - Big Committee), to decide strategy for all measures to safeguard Venice and the lagoon, especially how to divide the budget. The Comitatone is chaired by the Italian Prime Minister and includes the heads of five government ministries and various local administrators. The creation of this committee confirmed the elevated status of the Venice problem in national Italian politics. The 1984 Special Law also established the New Venice Consortium (*Consorzio Venezia Nuova* - CVN), which became the executive agency of the Ministry for Infrastructure and Transport and the Venice Water Authority (*Magistrato alle Acque* - MAV). Planning and implementing measures to safeguard Venice and its lagoon have been the CVN's direct responsibility since 1984. In 1988, the CVN rolled out an experimental model for the MOSE gate project, which reproduced a single gate in one-to-one scale and was placed at the Arsenal near the Lido inlet to the lagoon. The model was used for research until 1992, when it was declared a success by the CVN and removed. During this period of experimenting with a dam, the CVN and MAV completed the Conceptual Design for the mobile barriers in 1989, which considerably reduced the fixed parts of the barriers proposed in the Progettone. This modification was made to increase water exchange in the lagoon and minimize restrictions on fishing and port activities.

Work on the Preliminary Design for the dam project began in 1991 and concluded in 1992. This design was approved by MAV in 1992 and by an international panel of engineering experts employed by the CVN in 1993. The Public Works Ministry authorized the mobile barrier project to enter the final design phase in 1994 to prepare for construction. In 1995, the Comitatone agreed to subject the mobile dams to an Environmental Impact Assessment (EIA) (*Valutazione d'Impatto Ambientale* - *VIA*) following a request from Venice City Council, and delegated the task of conducting an Environmental Impact Study (EIS) to MAV. The EIS was carried out by various consultancies (Thetis from Venice, Ismes from Bergamo, Technital from Venice-Verona) and assessed in several subsequent reports. A 1997 report by the Ministry of Public Works (Italy), Venice Water Authority and Consorzio Venezia Nuova (1997) concluded:

> [T]he mobile flood barrier interferences are very limited and concentrated locally. They are modest in respect to the benefits that would derive from the elimination of flooding for urban lagoon areas, benefits that should become more and more important as the current mean sea level is foreseen to increase in the next decades due to the combined effect of eustasy and subsidence. (Ministry of Public Works, Venice Water Authority and Consorzio Venezia Nuova 1997)

The EIS was also assessed by other bodies that became involved in the wider EIA process. This delayed the construction of the mobile dams, especially due to conflicting interpretations of the study. The EIS was overseen by a Commission (*Collegio*) of International Experts from Belgium, Holland, the USA and Italy: MAV requested this panel of experts in 1995 and the Comitatone agreed. The Commission of International Experts' report (1998) found that the mobile gate project contained some weaknesses, but was not completely negative on the environmental impact of the project. The report approved the construction of the mobile dams overall, providing some adjustments to the design were made: "The *Collegio* concludes that the mobile gate system is flexible and robust and will be effective for protecting Venice against high water under a wide range of sea-level rise scenarios. However, the project requires some improvements, especially to prevent resonance and unnecessary leakage between gate elements." (Commission of International Experts 1998, 26)

A more negative interpretation of the EIS was published in 1998 by the Environmental Impact Assessment Commission, established by the Ministry of the Environment. At this time, the Environment Minister was Edo Ronchi, a leading member of the Green Party. This Commission's report on the EIS was critical of the mobile dams, although it did not rule them out, instead stating that an improved version of the mobile barrier project "could be reconsidered once basic work to re-establish the general health of the lagoon and the city has been undertaken and its effects on their vulnerability to flooding taken into account" (cited in Keahey 2002, 240).

Federico Antinori, who leads the campaign against the mobile dams for the Italian National Bird Association (LIPU), was interviewed by the author in 2005 at his home on the Lido Littoral Island, where some of the most disruptive construction work for the mobile dams was taking place. Antinori is opposed to Project MOSE, yet his interpretation of the Environment Ministry Commission EIA (VIA) report was that the Project could be modified, rather than rejected: "The VIA Commission said that MOSE was not the right thing and some alternatives should have been presented, at the minimum that

MOSE should be adapted to the situation, how to adjust it, how to make it different" (Antinori 2005).

The Environment Ministry Commission's report called for the completion of work to fortify internal defenses within the city, narrowing the lagoon inlets to reduce water intake and draining potential flood waters from rivers. This report suggested that such measures should be completed before considering the mobile dams or other longer-term technologies, and identified considerable risks with the mobile barrier project: "The Commission has ascertained that the project and the EIS do not cater for even minimally adequate mitigating and/or compensating measures during the executive phase. The Commission has considered that potentially highly critical environmental risks have emerged in the course of the analyses" (Ministry of the Environment - Environmental Impact Assessments Commission 1998). This negative assessment of the EIS was used by Minister Ronchi in 1998 to issue a decree blocking the progression of the mobile dam project, insisting that various steps to improve the defense of the city should be implemented first. Ronchi received support from Venice City Council and its Mayor, Massimo Cacciari. In 1998, Venice City Council included a prominent role for the Green Party: an "excessive influence", according to Project MOSE proponent and former Public Works Minister, Paolo Costa (Owen 1998). Within the government, Ronchi's decree blocking the dams was supported by the Cultural Affairs Minister Giovanna Melandri and it became known as the 'Ronchi-Melandri' decree. Other ministries in the government endorsed progress with the mobile dams: "The barrier has been approved at the ministries of Public Works, Culture, Transport and Research," noted Giovanni Mazzacurati, President of the CVN (cited in Owen 1998).

The Ronchi-Melandri decree was issued despite dominant support for the continuation of the mobile dam project in the Comitatone. Conflict within the government over the dams continued during early 1999 with the Environment Ministry and the Public Works Ministry "in collision" (Possamai 1999). In March 1999, the Comitatone, then chaired by Prime Minister Massimo D'Alema of the Democratic Left (DS), decided that Project MOSE should be "frozen" until the end of the year, so that further investigations into MOSE's environmental impact could be carried out (Testa 1999). The Comitatone passed a resolution in 1999 for MAV to oversee studies to examine the various assessments of the EIS. In April 1999, the European Parliament debated the safeguarding of Venice, and issued a resolution urging that "the Italian government should make a rapid, favourable and definitive judgement in favour of the MOSE project" (European Parliament 1999). MAV submitted documents containing almost 10,000 pages of studies about protecting Venice with dams to the Comitatone on 31 December 1999, yet the Comitatone procrastinated. The stalemate continued into 2000 under the new government led by Giuliano Amato, who had given indications that he was broadly in support of the mobile dams. Amato chaired a meeting of the Council of Ministers in March 2000, which passed a resolution advising that the MOSE dams proceed to the final planning phase. Roberto Biorcio, a leading scholar on the Italian environmental movement, records that the Green Party made abandoning Project MOSE a condition of their participation in the fragile government coalition, where the party held two ministerial posts (Biorcio 2002, 55); and Christopher Emsden, the former editor of the *International Herald Tribune*'s '*Italy Daily*' supplement, reported that "[T]he Green Party has threatened to leave the government if the project is approved" (Emsden 2000).

In July 2000, the Venice Regional Administrative Tribunal (TAR) declared the 1988 Ronchi-Melandri decree blocking Project MOSE canceled due to breaches of procedure and substance. The Environment Minister Willer Bordon responded to the TAR decision by insisting that a proper environmental impact assessment remained necessary for the Project to proceed (Raccanelli and Testa 2000). The Comitatone met to discuss proceeding to the final planning phase of the dam project in July 2000, but conflict between the Environment and Public Works Ministries meant this decision was postponed. A new government led by Prime Minister Silvio Berlusconi was elected in May 2001 and vowed to make several infrastructure projects high-profile initiatives. Project MOSE proceeded to its Executive Phase of final planning in December 2001. On 30 September 2002, the CVN turned in the final design of the dam system, which modified the mobile barriers, most notably by adding navigation locks for ships to pass in and out of the lagoon when the barriers are raised. The Green Party and leading Italian environmental organizations, including Italia Nostra, *Legambiente* (The Environmental League) and WWF Italy, opposed the dams by launching a series of legal appeals to Italian courts and the European Commission (EC): most of these were rejected during 2004. Even though European bodies have criticized Italian governments regarding the dams on several occasions, with limited legal jurisdiction these criticisms did not block the dam project.

In May 2003, Prime Minister Berlusconi came to Venice for the highly publicized placement of the first stone in the construction of the mobile dams. Campaigners, including environmentalists, protested in boats near the location where Berlusconi was laying the first stone (Fornasier 2003). In April 2005, Cacciari was elected as the Mayor of Venice for the second time and his City Council declared war on the mobile dams, calling in the police to stop their construction. Environmentalists and other activists joined together on 22 June 2005 to create a 'No MOSE' permanent assembly organization. Hopes of stopping the construction of the mobile dams were also raised by the election of Romano Prodi in April 2006, replacing Berlusconi as Prime Minister; but the Prodi government backed the continuation of the project in November 2006. The re-election of Berlusconi in April 2008 led to expectations among supporters of the mobile dams that they would be completed in 2011; insufficient funding slowed construction down during 2008 and completion was predicted for 2014 (*Il Gazzettino di Venezia*, 9 October 2008). Completion in 2014 was confirmed in June 2010 by the CVN, which also stated that €3.244 billion (US$4.24905bn)[8] had been assigned of the total €4.678 billion (US$6.12723bn) cost of the MOSE system (Consorzio Venezia Nuova 2010a). Whether the mobile dams are finished in 2014 will depend on continued financial support and the character of interventions by political parties, international bodies and claims-makers.

This overview of how Project MOSE has evolved demonstrates that the causes of the delays to the dam project have changed over time. During the 1970s, institutional instability and doubts about the design created delays, which were largely resolved in the early 1980s. In 1984, the legislative framework to construct the dams was established and the first appeals against the project were launched. After the MOSE model was unveiled in 1988, environmental groups, Venice City Council and the Green Party managed to freeze the project until the 2001 general election. Since then, political opposition has created occasional delays; as has insufficient finance, which has been partly caused by political opposition.

For more than 20 years, the principal reason for the delays is that Venice's mobile dams have been subject to political conflicts and competing claims. This contrasts starkly with the development of London's Thames barrier, and with the Maeslant barrier in Rotterdam's New Waterway in the Netherlands, which started its design phase in 1987, construction in 1991, and was completed in 1997. On the other hand, the stalling of Venice's dam project due to its politicization, institutional instability, the intervention of various players and environmental claims-making has much in common with the St Petersburg flood protection barrier. The Government of the Soviet Union approved the principal design in 1979 and building began in 1980. Public protests stopped work in 1987, highlighting that pollution in the Neva Bay was being aggravated by the barrier's construction, and by 1990 it had become a "highly politicized issue" (Gerritsen et al. 2005, 343). The collapse of the Soviet Union prolonged the lack of progress, and in 1990 foreign players become involved, including international committees of experts and the European Bank for Reconstruction and Development (EBRD). To release funding, the EBRD required an environmental impact assessment and public consultation: these were carried out in 2002 and lively public meetings were well-publicized. The EBRD and Russian government approved further work on the barrier in

December 2002. The European Bank of Investments (BEI) lent €40 million (US$52.3907 million) in 2005 and it was nearing completion in 2010.

Both the St Petersburg and Venice barriers were delayed after they were politicized by intervention from foreign institutions and environmental claims-makers. The policies and language of environmental sustainability were undeveloped when London's barrier was being designed. "In the 1970s and 1980s when the Thames barrier was constructed, the vocabulary of sustainability was not then in use," write David Wilkes and Sarah Lavery (2005, 287), who worked on the Thames barrier at a senior level for the UK Environment Agency. Revisions to the Thames barrier by the Thames Estuary Partnership may be subject to competing claims, as the partnership "brings together over 100 different stakeholder groups to work towards common aims and sustainable development of the Thames Estuary" (Wilkes and Lavery 2005, 292). Environmental claims-making for Venice's barrier project could thus provide insights for the development of the Thames barrier. Furthermore, the examination of how environmentalism and claims-making has affected flood protection in Venice provides a case study for other places. Countries, including Singapore and Bangladesh, considering sea defense measures have investigated the Venice dam project. The combination of extensive engineering projects and growing environmental activism in China means that the Venice case is especially relevant there. Venice's dams were the subject of a political study at National Taiwan University (Tsai 2009). The American city of San Diego has also contemplated Venice's dam system for protection (*Physorg.com*, 24 November 2010).

The impact of environmentalism on claims-making about Venice's mobile dams is explored in Chapters Five, Six and Seven, and the final three chapters provide a broader assessment of how environmentalism has influenced the trends of modernization and sustainability in Venice and elsewhere. A key theme of this book is that policy debates about Venice have been governed by mythology as well as the reality of the city's circumstances; and it is to this discussion that we now turn.

NOTES

1. See Appendix A for interviewee profiles.
2. 2.5 centimeters equals 1 inch.
3. 1 millimeter equals 0.039370 inches.
4. 1 degree Centigrade (Celsius) equals 33.8 degrees Fahrenheit.
5. 1 meter equals 3.28083 feet.
6. 'Framing' is a concept applied throughout this book. It allows claims-makers to redefine problems, as the sociologists Donatella Della Porta and Mario Diani (1999)

explain: "appropriate interpretative frames allow a phenomenon whose origins were previously attributed to natural factors, or which was the responsibility of those already involved, to be transformed into a social or political problem" (Della Porta and Diani 1999, 69).

7. The film can be viewed at http://www.salve.it/uk/soluzioni/acque/mose_uk.htm.
8. All currency conversions were correct on 21 December 2010.

Part One

VENICE'S MYTHICAL TRADITIONS

Chapter One

Founding Myths of the Venetian Republic

"Historians have long recognized the gap between the myth and the reality of Venice. They have been anxious to reveal the contradictions and hypocrisies and to use such exposures to undermine the myth, to deconstruct it and indict it as deliberately misleading fabrication. But it is precisely as a set of 'fictions or half-truths' that myth presents itself as reality, 'forming part of the ideology of a society.' What for the skeptical political or social historian may seem collective self-deception can be viewed more positively and creatively as deliberately conceived self-representation."

Professor David Rosand of Columbia University (2001, 4).

This chapter analyzes ancient myths about Venice that have shaped past and present debates about the city and its environment. Considering Rosand's approach in the above quotation, the aim is not to treat myths as lies that should be eliminated from debates about Venice's environment if they lack historical evidence. Myths reveal how Venetians have depicted their relationship with nature and how this has altered over time.

Venetian myths have gone through a number of historical transitions and their popular impact has varied. Tracing how the city's symbolic representation has changed will assist our understanding of why Venice was and is metaphorically important around the world. The historian Edward Muir (1981) identifies multiple aspects of this representation: "Venice's historical reputation for beauty, religiosity, liberty, peacefulness, and republicanism modern scholars call 'the myth of Venice'" (Muir 1981, 21). The most internationally famous myth about Venice was the political reputation it achieved as an ideal republic of balanced constitutional government. By the early sixteenth century, humanism had transformed the religious myth of Venice protected by Saint Mark into a political ideology (Muir 1981, 21). John Eglin (2001, 42) shows how the political myth resonated around the Anglophone world during the sixteenth century and died with the fall of the Venetian Republic to France in 1797.

The Italian historian Franco Gaeta (1961) distinguishes three myths that Venetians universally accepted: Venice as a free state of liberty; as a republic of mixed government; and as a gallant city. The liberty myth was based on the city's foundation by mainlanders fleeing barbarian invaders and the Republic's independence from foreign powers. The gallant myth built on the city's reputation for frivolity and sensuality, which were most graphically evident during Carnival: the leading event attracting visitors as a primary location on the European Grand Tour. After the French occupation of Venice in 1797, the city's reputation for cultural decadence replaced the political myths of liberty and mixed government.

The changing symbolism of Venice has conditioned environmental debates about the city. We previously discussed that Venice's sinking and flooding problems are political as well as physical issues. This chapter explores imagery that is vital to understand the city's past and present environmental debates, focusing on symbolism that has contributed to the association of Venice with environmental peril.

THE LIBERTY MYTH

The founding of Venice lies at the heart of the problematic relationship between Venetians and nature. Why would anyone build a human settlement on boggy marshland islands in a lagoon? Understanding motivations and interpretations of the city's evolution assists an appreciation of the relationship between Venetians and their environment. Investigating the liberty myth also reveals the gap between the stories and reality of Venetian mythology.

The myth regarding the creation of Venice was based on independence and self-government. The city grew in the northern province of the Roman Empire called Venetia. When Rome fell in 476, the western half of the Roman Empire disintegrated and Venice became part of the eastern Byzantine Empire, ruled from Constantinople. According to the liberty myth, Venice formed its government independently of Roman and Byzantine rule. Liberty was supposedly established by mainlanders fleeing into the lagoon to evade barbarian invaders such as the Goths in 402, Attila the Hun in 452, then the Lombards (fifth to seventh centuries). The legend of fleeing Attila the Hun has been especially durable, as indicated by the popular 1846 opera *Attila* by Giuseppe Verdi.

Mainlanders sought refuge in the lagoon after the Goths attacked Aquileia in 402. These mainlanders had been living in neighboring towns, villages and cities, including Padua (*Padova*) and Aquileia. The islands of the lagoon were difficult to access, which meant they provided an effective hiding place: thus, the challenging lagoon environment was part of its original attraction. Apart from isolated fishermen and fowlers, very few people settled in the lagoon before 452. Although the mythical date for the founding of Venice is 421, this date was constructed later as the mythology of Venice developed. Muir (1981, 70) notes that the myth of Venice's foundation in 421 contradicts the myth that it was created by refugees fleeing Attila the Hun after 452. In fact, the majority of refugees who had temporarily settled in the lagoon during the first half of the fifth century returned to their mainland homes (Norwich 1983, 5). Attila the Hun defeated Aquileia in 452 and the flow of refugees into the lagoon increased, many of whom stayed as conditions in the lagoon improved and became worse on the mainland. It was in 466, not 421, that the earliest evidence of Venice's mythical self-government can be found among the island communities:

> [I]n 466 their respective representatives met together at Grado to work out a rudimentary system of self-government, through tribunes elected annually by each of them. It was a loose association at this stage, and one that was not even confined to the little archipelago which we now know as Venice; Grado itself lies on its own lagoon due south of Aquileia, some sixty miles away. But that distant assembly marks, more accurately than anything else, the beginning of the slow constitutional process from which the Most Serene Republic was ultimately to evolve. (Norwich 1983, 6)

The early settlers' attempt to form governing assemblies was not the same as establishing independent, self-sustained settlements. Early communities on the lagoon islands were economically, politically and religiously dependent on mainland towns, especially Aquileia (Crouzet-Pavan 2002, 4). This dependency had implications for environmental defense. Even the prosperous mainland town of Altino could not afford to help defend the island community six kilometers (3.7 miles) away at Torcello, which at one point was abandoned due to sinking and deposits from the Piave and Sile rivers (see Figure 1.1). Settlers on the lagoon's islands faced a constant battle to build on eroding land that was frequently flooded: "The first houses of the lagoon boatmen were probably wooden huts, raised on stilts to protect them from high tides and floods," surmises the architectural historian Deborah Howard (2005, 4). Settlers consolidated the ground with mats woven of flexible reeds, and by the ninth century they discovered that driving wooden piles (long sharpened poles) into the clay under the water provided support for platforms, which could be extended out across the lagoon. The platforms were initially wooden; later, a layer of stone was added for fortification and to prevent water getting through. The wooden pilings solidified in the absence of contact with the air and still support huge structures: the current Rialto Bridge is built on 10,000 pilings (Fletcher and Da Mosto 2004, 35). These building methods meant that the early settlers were also dependent on the mainland for wood and stone for environmental defense.

The dependency on the mainland and the difficulties of building in the lagoon were two reasons why the islands remained sparsely populated until the late sixth century. Following the Lombard invasions (568) and their defeat of Oderzo (639), wealthier people moved into the lagoon and could afford materials to fortify dwellings. However, as the historian Frederic Lane (1973) writes, "Later chroniclers and genealogists exaggerated the nobility of these immigrants and traced the ancestry of successful Venetians back to distinguished families in the plundered Roman cities" (Lane 1973, 4). This was part of the process of constructing honorable and diplomatically advantageous stories about where Venetians came from. Muir (1981, 66) explains that Venetians entertained legends about their lineage from both the Gauls and Trojans, who allegedly built Venice's mother city, Aquileia. Refugees from Aquileia mainly settled at Grado to the north of the lagoon. From Padua, they established dwellings at Malamocco, as well as Chioggia and Cavarzere at the southern end of the lagoon (see Figure 1.1). The location that we know as Venice was mostly open water with small islands. The main island was called *Rivo Alto* (high bank), which later became Rialto. The first commercial island was Torcello and Rialto did not become the center of Venice until 810.

Far from being independent, the early lagoon communities were largely subject to Roman and Byzantine pressures. Yet the early Venetian historians constructed a mythical liberty narrative that dismissed this reality:

> Andrea Dandolo, a doge and the authoritative fourteenth-century chronicler, ignored the fact that the first doge of Venice was a Byzantine official, created when Venice was part of the Byzantine Empire. He pictures the Venetians gathering in 697 on their own initiative from the various settlements in which they then lived scattered over the lagoons and deciding, nobles and common people together, on the creation of a single leader, the dux or doge, to replace the officials called tribunes by whom their settlements had been hitherto separately governed. (Lane 1973, 87-88)

The first doge did establish a separate military command over the lagoons in 697 (Lane 1973, 4). The Byzantine administrative center was located at Ravenna, until it was captured by Lombard invaders in 751; the doge moved to Malamocco, which was an island settlement bordering the sea on what is now the Lido. After an attack by Pepin, son of the Carolingian Emperor Charlemagne, the doge fled to Rialto in 810. But there is little doubt that the first doge was subject to Byzantine rule:

> This first doge may in fact have been selected by the inhabitants of the lagoons, as later Venetian chroniclers said, maintaining that Venice had been free and independent from its beginnings, but the doge received orders and honors from the Byzantine emperor and Venice was unquestionably considered part of the Byzantine Empire even after the Lombards had captured Ravenna in 751. (Lane 1973, 5)

Crouzet-Pavan (2002, xvii) dates the probable election of the first indigenous doge at around 725/30. Orso Ipato is listed as the first doge by Norwich (2003, 641), ruling from 726 to 737. The independent Venetian Republic was declared in 727, although Venice was confirmed as a semi-independent province of Byzantium in the 811 *Pax Nicephori* between Charlemagne and the Byzantine Empire. Muir (1981, 74) notes that Venetian autonomy was not recognized by Carolingian Emperors until 840. Venice was subject to the Byzantine Empire until 1082 (Crouzet-Pavan 2002, xvii); its independence grew as the Byzantine Empire faded and Venetians remained under the cultural influence of Byzantium for another five centuries (Norwich 2003, 94). The profound hold of Byzantium over Venice was represented by the Greek warrior Theodore, the original patron saint of Venice before St Mark. Independence and religious themes combined in early Venetian mythology: "God inspired men to seek the refuge of the lagoons. God permitted them to remain there safely while Atilla (*sic*) ravaged the West. At that time the community made a pact with God: the Venetians were a chosen people," announced Tomaso Mocenigo, who became Doge of Venice (Crouzet-Pavan 2002, 97-8). The liberty myth was therefore deeply Christian. It was strengthened by constructing the cult of St Mark, which provided a fortified sense of sovereign independence.

THE MYTH OF ST MARK

The Christian roots of Venice were underlined by the claim that Venice was founded at 12 noon on Friday, 25 March, during the year 421. This date is the Feast of the Annunciation to the Virgin. As indicated, historians have questioned the accuracy of this date for Venice's foundation. Conveniently, spring was also supposedly Rome's founding period, with March being the first month of the Roman year. Remarkably, as Muir (1981) points out, "in Venice the founding day of the city was mystically conjoined with the founding of Rome, the beginning of the Christian era, the annual rebirth of nature, and the first day of the calendar year" (Muir 1981, 71). Such was the power of combining various legends into the founding date that it largely superseded the liberty myth based on fleeing Attila the Hun. As Muir writes: "In comparison to the Attila myth, the narrative of the beginning of Venice on March 25 became the more powerful, influential, and popular, not only because of its greater intrinsic symbolism, but also because it was one of the two founding legends incorporated into official ceremony" (Muir 1981, 72).

The founding date of 25 March 421 gave Venice a foothold in Christianity and offered St Mark as an alternative to the Roman St Peter and Byzantine St Theodore. The myth of St Mark therefore helped distinguish the independence of Venice. Norwich (1983) eloquently retells two legends that helped form the myth of St Mark:

> One day – so the story goes – when St Mark was travelling from Aquileia to Rome, his ship chanced to put in at the islands of Rialto. There an angel appeared to him and blessed him with the words *'Pax tibi, Marce, evangelista meus. Hic requiescet corpus tuum'*.[1] The historical evidence for this story is, to say the least, uncertain; the prophecy – since St Mark later became Bishop of Alexandria and remained there till he died – would have seemed improbable; but the legend certainly came in very handy when, in 828 or thereabouts, two Venetian merchants returned from Egypt with a corpse which they claimed to be that of the Evangelist, stolen from his Alexandria tomb. (Norwich 1983, 28)

Muir (1981, 80) establishes that there is no historical evidence St Mark visited Aquileia or anywhere in the region. There is evidence that could support the transfer, or *translatio,* of St Mark's body to Venice in 827 or 828. Norwich (1983, 29) notes that a body was brought to Venice at this time and the doge ordered a chapel to be built for its reception. Although it is impossible to verify whether these were the relics of St Mark (Muir 1981, 82), there is no doubt about the importance of the *translatio* in Venetian mythology, which helped confirm St Mark as the regional symbol in the ninth century. The decision to take St Mark's body to Rialto helped to advance Venice's independence from Byzantium and the Carolingians (Muir 1981, 83), coming shortly after the removal of the Venetian capital from Malamocco to Rialto in 810. Moreover, the acceptance of St Mark's body by the doge transferred political and religious authority from god to the doge, a bond that was celebrated on St Mark's feast day on 25 April. This authority was consummated by the legend of St Mark's apparition, or *inventio*, which appeared during a feast after his relics had been lost. The *inventio* legend was created in the thirteenth century and was required to imitate a pattern established for the afterlife of saints: martyrdom, transfer and rediscovery of relics (*inventio*) (Muir 1981, 87). The *inventio* also supported Venice's struggle for independence by re-asserting St Mark as an alternative to Byzantine protector-saints. The *inventio* celebration day was instituted as 25 Jurre.

The most important legend in the myth of St Mark for the relationship between Venetians and their environment was the evangelist's role in stopping a storm surge to calm the lagoon. During a fierce storm on 15 February 1340, a fisherman was battling to save his boat and equipment from damage. The fisherman was approached by a stranger, who sent him to two other strangers, the last of whom ordered the fisherman to row to the mouth of the lagoon. There he saw a ship of demons, which were causing the storm. The three strangers were revealed as Saint Mark, Saint George and Saint Nicholas. "See," St Mark said to the fisherman, "how high the water has risen in the houses, and how many boats have sunk." He added, "Do you see, this city is going to the bottom and will perish by the waters" (re-told by Crouzet-Pavan 2002, 50). The three saints disposed of the demons and calmed the lagoon by making the sign of a cross. According to Muir (1981, 88), this story was one of the more famous and was highly regarded in official circles, contributing to St Mark becoming the protector of Venetians in battles on land and the defender of the Venetian fleet at sea, complemented by Nicholas, the patron saint of sailors. The legend set a precedent for the creation of myths to help Venetians interpret their relationship with the sea: myths that continue to be constructed to interpret Venice's modern environmental problems. Lane (1973) even compares the legendary calming of the lagoon to a moment during the high floods of 1966: "At the peak of the flood of November, 1966, the suddenness with which a change of wind flattened the waters seemed to me also a miracle" (Lane 1973, 272).

The use of legends included in the myth of St Mark varied as Venice developed. The myth of St Mark served the establishment of the Venetian state, which was symbolized by the construction of St Mark's Church soon after his body arrived in 829. The winged lion of St Mark was displayed on columns, flags, ships, gates and facades. St Mark's lion was an image that protected Venetians on sea and land and represented their amphibious existence: "As befitted the symbol of a maritime republic, the lion was commonly depicted half on land, half on water" (Lane 1973, 431). Domestically, St Mark personified Venetian *communitas,* bonding the community together as a free republic under the stewardship of the doge. As the Venetian Empire grew, the myth of St Mark was integral to colonial policy: when Venice conquered Crete during the thirteenth century, St Mark's Day was enforced in the Cretan calendar. The St Mark myth was flexibly deployed, "at one time legitimating certain political enterprises that supported the city's claims to an overseas empire, at another time presenting Venice as the sole and ultimate rampart of Christianity" (Crouzet-Pavan 2002, 55). Norwich (1983, 30) describes how the image of St Mark's lion incorporated Christian direction, with the lion's paw pointing at the word of God. In the Venetian Republic every important occasion was celebrated with a mass in the church of St Mark, who was credited with authorship of the Gospel.

RELIGIOUS MYTHOLOGY AND THE SEA

Christian mythology was fundamental to Venetians for understanding their watery environment. We have seen how the myth of St Mark, the symbol of St Mark's lion, and the ritual ceremonies of St Mark, dominated Christian mythology for early Venetians and were deployed for maritime uses. Other symbols, rituals and metaphors also featured in Christian mythology about Venice and the sea. In particular, the metaphorical significance of Venice as virgin and 'married' to the sea developed as the Venetian Republic expanded its trading routes across the oceans to become a dominant maritime power. Crouzet-Pavan (2002) suggests that the theme of Venice's dominance over the sea originated during the tenth century and draws on Christian elements within the liberty myth:

> It was probably in the tenth century, the time of Venice's first successful expansion into the Adriatic, that the theme of the city's legitimate sovereignty over the seas began to circulate, completing the more general providential theme. According to the legend of origins, the lagoons were a propitious shelter that God had reserved for the realization of Venice, the site of his privileged pact with the infant community. Logically, the sea was the rightful portion of these 'men born of and nourished by water.' (Crouzet-Pavan 2002, 51)

Christianity also underpinned the construction of virginity as a metaphor for the Venetian Republic. The cult of the Virgin Mary was popular among Venetians (Muir 1981, 139), and the city's impregnability reinforced the virginity metaphor. In 810, Pepin had devastated Pellestrina, which is one of the littoral islands that separate the lagoon and the sea. Resistance at Malamocco helped to weaken his forces, and Hungarian Magyars attempted invasion in 899. They were defeated on the littoral islands. This defeat was partly due to difficulties with navigating the lagoon: a problem that hindered many potential invaders. Although the Genoese in 1379 did manage to capture the city of Chioggia, where the littoral islands meet the mainland, the lagoon was not penetrated. The shallow lagoon with its underwater channels, which only the locals knew how to navigate, created a formidable defense, as invaders became stuck in shallow waters when they strayed out of the channels. This was why the Venetians created palaces with open arcades instead of fortresses: they never felt the need to build the city walls that typically protected cities. The lagoon provided a defensive strength that helped the Venetian Republic to expand. During the twelfth century, the Venetians built on their expanding Empire to sail beyond the Adriatic and Ionian seas and exert their power over the eastern Mediterranean. In 1204, the Venetians helped conquer Constantinople, leading to the break-up of the Byzantine Empire. "Once did she hold the gorgeous East in fee / And was the safeguard of the West," recalled William Wordsworth (Sonnet III.1-2, cited in Lane 1973, 42). The impregnability of the Venetian Empire until the final years of the eighteenth century meant that it became the longest surviving European empire.

Venice was "the only city in Italy never to have been invaded, ravaged or destroyed" (Norwich 1983, xxii), until Napoleon Bonaparte's invasion in 1797. The virgin state generated a mythology of innocence and independence, which was linked by Gaeta (1961) to the historical myth of Venice as a state of liberty. So the myth grew that mainlanders fleeing invaders and seeking their liberty had established an independent state in the virgin territory of the lagoon. Moreover, the virgin state was based on strong feminine imagery, with Venice represented as Venus from 1500 until 1797 (Rosand 2001, 119). Venice's founding date of 25 March was the day Venus was in the ascendant: "Facile and oft-repeated plays on words – Venice as a new Venus, like the goddess born naked and beautiful amid the waves – continued up to the fall of the Republic," writes Crouzet-Pavan (2002, 49).

The female representation of Venice was inverted for the ceremony of *Spozalizio del Mare* (Marriage to the Sea), when the city's male doge married the female sea. The Marriage to the Sea ceremony was held annually on Ascension Day and commemorated an expedition down the Dalmatian coast in approximately the year 1000, which achieved Venetian control of the northern Adriatic Sea. At first, the ceremony included a short and simple prayer; in subsequent years, it was made grander to express the domination of the Republic over the sea. "The essential political point, then, was that in marrying the sea the doge established his legitimate rights of domination over trade routes and over the lands lapped by the waters of the Adriatic," writes Muir (1981, 121). The doge rode out on his spectacular *Bucintoro* (Bucentaur) galley to the Lido lagoon inlet, where he dropped a golden ring overboard, saying "We espouse thee, O Sea, as a sign of true and perpetual dominion" (Muir 1981, 122). According to Christian mythology, the ceremony received the personal blessing of the Pope and this helped to develop the original expedition commemoration. Crouzet-Pavan (2002) explains: "During the thirteenth century, after the conquests of the Fourth Crusade, the rite of wedding the sea was grafted onto it. The doge threw into the sea a gold ring like the one Pope Alexander III was supposed to have given the doge, and by that gesture he wedded the sea and renewed Venetian privileges...later, the doge's wedding to the Adriatic became the central ritual" (Crouzet-Pavan 2002, 48).

Norwich (1983, 116) documents the lack of historical evidence that Pope Alexander III offered this ring. Nevertheless, the myth of the Pope's role cemented the Christian foundations of the ceremony and fortified the image of Venetian maritime superiority. Eglin (2001, 172) refers to Jacopo Sannazaro's comments on the "god-built" city of Venice, standing in the Adriatic waves and giving law to the whole sea. The Christian character of the ceremony was also emphasized by its occurrence on whichever day Ascension Day fell (40 days after Easter), and by paying homage to St Nicholas. Norwich (1983, 116) notes that it is reasonably safe to assume that the Pope did attend the ceremony in 1177, alongside German Emperor Frederick Barbarossa, but there is no indication the Pope actively participated in events. This ceremony of 1177 helped establish Venice as the naval center of Christian Europe, with the doge as the peacemaker between the Pope and Barbarossa. "Throughout that memorable summer, [Venice] was the focus of attention of the whole of Europe – the capital, in a very real sense, of Christendom. Her Doge was playing host to the two leaders of the Western world," writes Norwich (1983, 116-7).

By the sixteenth century, the Marriage to the Sea ceremony had become the most significant date in the Venetian state's calendar. It initiated a 15-day fair that attracted foreign visitors and large crowds. In the sixteenth and seventeenth centuries, the ceremony also inaugurated the Venetian theater season that ran into July. However, perceptions of Venice's relationship to the sea changed as other powers challenged its maritime domination. Other port cities performed marriage to the sea rituals, such as the ancient Roman city of Ravenna, just down the coast. From 1446, the Archbishop of Ravenna married the city to the sea at the village of Cervia with his pastoral ring. But it was the Turkish and European powers who confronted Venice's real supremacy over the eastern seas.

In 1499, Venice suffered humiliating military defeats in the waters off Sapienza to the Turks, who sailed on to take the key location of Lepanto. This forced the Venetian Republic's marriage to the sea onto rocky ground, as the Ottoman Empire challenged the Republic's military domination of eastern oceans. In 1500, a message was sent from the Turkish Vizir to the Venetian Signoria governors: "Tell the Signoria that they have done with wedding the sea; it is our turn now" (cited in Norwich 1983, 383). Venice managed to defend its dominance over the Adriatic Sea against Turkish forces, but leading European powers formed the League of Cambrai against the Venetian Republic in 1508, and the entire Venetian mainland was temporarily lost in 1509. The second blow to Venice's maritime superiority came as a result of Vasco da Gama's landing in Lisbon from India via the Cape of Good Hope, also in 1499. The Cape of Good Hope had been rounded 13 years previously, but this was the first time the journey had been made from India to Europe entirely by sea. The new route meant that goods could be transported between Europe and the Orient without stopping in Venice or even passing through the Mediterranean. Lisbon replaced Venice as the international port for eastern trade and Venice went from being the gateway to the east to a backwater.

Venice's decline on land and at sea reduced its power to respond as the Adriatic was increasingly threatened by Uskok, Barbary, Maltese, Florentine, Spanish and English pirates in the last few decades of the sixteenth century. During the seventeenth century, Venice asserted claims that it still ruled the Adriatic against the King of Spain. But in 1702, Venice panicked when the French fleet conducted operations against Austria off its coast, and the Marriage to the Sea ceremony was canceled. The loss of Venetian command of eastern seas in the eighteenth century meant the Marriage to the Sea ceremony could no longer express Venetian maritime dominance.

Although the ceremony of the Marriage to the Sea had primarily developed to symbolize Venice's trading and naval domination over the Adriatic, it also expressed the desire for calm seas. On the morning of Ascension Day, the doge's cavalier responsible for ceremonial preparations judged whether the waters were sufficiently placid for the Marriage to the Sea to take place; and during the ceremony, there was a plea that God would keep the seas generally calm for Venetians and others. "We worthily entreat Thee to grant that this sea be tranquil and quiet for our men and all others who sail upon it," the patriarch repeated three times (Muir 1981, 122). The desire for calm waters in the Marriage to the Sea ritual, also known as the Sensa, illustrated a belief that nature could be understood. In addition, there was a conviction that the negative impact of some natural forces, such as storms, could be modified. Muir (1981) writes: "The Sensa revealed two profound psychological habits of belief: that natural forces could be comprehended by personifying them, and that through understanding these forces one could better control them, or least predict their influences" (Muir 1981, 133). Venice's Marriage to the Sea ceremony expressed an early version of the belief that man could understand and exercise control over nature. We will examine how this belief developed as the driving force behind modernization, and was increasingly criticized by conservationists and environmental campaigners as a manifestation of human arrogance.

The location of Marriage to the Sea ceremony is highly significant. It took place at the Lido inlet between the lagoon and the sea, which is one of the three inlets where the mobile dams are being built. One wonders whether the dams will manage to have a more effective impact on calming the seas than the numerous doges presiding over the Marriage to the Sea ceremony. Or will the dams be as capable as saints who stopped floods at the Lido inlet? This is also where St Mark, St Nicholas and St George halted the storm surge created by demons, according to the legend in the myth of St Mark. Christian myths governed how Venetians initially understood their relationship with the sea, while new myths permeate contemporary perceptions.

THE POLITICAL MYTH OF CONSTITUTIONAL HARMONY

Christian mythology dominated Venetian mythology, rituals and symbolism from Venice's foundation into the Middle Ages. From the sixteenth century, Christian mythology was gradually replaced by political myths about the Venetian Republic. The myth of Venice's constitutional harmony is important to two of the key insights in this book. Firstly, constitutional harmony was based on the notion of equilibrium between different arms of government. This political equilibrium supported the belief that Venice also achieved environmental equilibrium during the fifteenth century ascendancy of the Venetian Republic, which was consummated during the Marriage to the Sea ceremony. The loss of political equilibrium with the fall of the Republic in 1797 is linked to increasing environmental disequilibrium by past and present conservationists. Secondly, the myth of constitutional harmony had the biggest international impact of all Venetian myths, its popularity laying the basis for the prominence of environmental myths about Venice today. The myth of constitutional harmony needs to be situated within the evolution of Venetian mythology to appreciate its relationship to other myths and developing environmental themes.

The political myth of constitutional harmony was created as Christian mythology about Venice started to diminish, from the fifteenth century onwards. As Crouzet-Pavan (2002) notes, the state gradually assumed greater responsibility for community wellbeing:

> At first the image of the city of divine election, the city to which, from its first settlement on a predestined site at the heart of the lagoons, God 'had always lent a protective hand,' triumphed. During the fifteenth century the dream of Venice as a city divinely brought to power and glory began to fade, and it was seen as the task of the Republic – of the state – to assure the prosperity of the community and guarantee it a perpetually happy destiny. (Crouzet-Pavan 2002, 189-90)

The Venetians were still God's chosen people, protected by St Mark. Yet they believed their state was primarily responsible to its people:

> In the Middle Ages, there were two schemes of thought and emotion by which power was made to seem legitimate and right. One was the descending theory

> which held that all rightful power was handed down by God to the pope and emperor and passed on by them to those below. The other was the ascending theory which held that lawmaking and similar political powers resided in the community and could be handed over by it to those it designated. Rightful rulers were in this sense representatives of the community and were responsible to it. Of these two contrasting theories, the Venetians wholeheartedly embraced the second. (Lane 1973, 90)

The Venetian conception of the state depended on sovereign independence, represented by the doge. All power was still ultimately derived from God. The doge received his office from St Mark, even though he was elected. Although ascendant and descendant elements were combined in Venetian government, sovereignty was not directly dependent on the pope. This understanding of government inevitably led to conflicts with the Vatican, and in 1606, the Venetian Republic was served its fourth papal interdict, following interdicts in 1284, 1309 and 1483. The principal difference between the 1606 interdict and earlier ones was that it became a point of controversy throughout the Christian world, in which one's interpretation of Venice's relationship with the Vatican became an indicator of one's position on Catholicism. Venice's defiance of papal authority became an issue on which the leading European nations were forced to take a stance. Eglin (2001) suggests that it was Venice's defiance of the Counter-Reformation papacy that stimulated British political affinity for the city. The 1606 interdict on Venice was lifted in April 1607 and the city never abandoned the See of Rome; but after Venice's defiance of this interdict no pope ever felt he had the authority to impose another (Norwich 1983, 516).

While Venice acquired an international reputation for challenging the Vatican, this did not undermine Christian mythology in the city. The defiance of papal interdicts was a rejection of papal discipline, not Christianity itself. Although the Venetian Republic was strongly secular, Venetian patriotism was based on a deep sense of piety. "The cardinal virtues – Faith, Hope, and Charity – underlay the republican virtues; so obedience to the state was metaphorically obedience to the will of God," writes Muir (1981), describing the opinion of the fifteenth century humanist Giovanni Caldiera (Muir 1981, 16). Christian mythology was subordinated to the secularism of the constitutional myth, rather than destroyed by it.

Similarly, the constitutional myth incorporated the liberty myth rather than eliminating it. The earlier liberty myth was transformed in the fifteenth century: independent self-government had been more mythical than real in the early days of Venice, but during the last 500 years of the Venetian Republic, liberty became more real than mythical as the Republic asserted its independence from Byzantium. Venice also acquired a reputation for diplomatic independence from its unwillingness to form long-lasting alliances with other powers. In the early fifteenth century, the Republic expanded across the mainland and was perceived as a potential liberator for Italian city states, even though this potential was not realized. The myth of Venice as an independent state of liberty was a founding component of the constitutional myth.

As Venice asserted its independence and turned away from Byzantium, Italy and the West provided inspiration. In the fourteenth and early fifteenth centuries, Venice began to portray itself as a 'New Rome', heir of the ancient Roman Republic and Empire. Yet the myth of Venice was incoherent at this point and lacked theoretical exposition. It was the flourishing of Renaissance humanism in Italy that provided the myth with political and cultural meaning. The most prominent Venetian humanist was Bernado Giustiniani, author of *History of the Origin of Venice,* which, according to the Renaissance scholar Patricia Labalme (1969, 2), was published around 1492. Giustiniani developed the liberty myth that as the Roman Empire dissolved, God led his chosen people into the lagoon, making Venetians the heirs of Roman political traditions. Muir (1981) explains: "By the early decades of the sixteenth century, the language of humanism had transformed the traditional myth of a holy city protected by Saint Mark, independent from all foreign powers, into a coherent political ideology that was classical in its derivation. Venetian councils, for instance, were now seen as evolved forms of Roman institutions" (Muir 1981, 25-6).

Renaissance humanists gave Venetian republicanism theoretical expression. The Renaissance came later in Venice than many Italian city states, with the Venetians characteristically being "doers" rather than thinkers (Norwich 1983, 341). Although the center of the Renaissance was Florence, Florentines regarded Venice as an emblem of republicanism. In the fifteenth century, Florentine citizens were encouraged to emulate the Venetian example by Florentine leaders who also sought constitutional inspiration from Venice. Norwich (1983) remarks that by 1510 "Venice had become the intellectual centre of Italy", with Venice's University at Padua having unrivalled prestige in Europe (Norwich 1983, 412 and 594). Venice's leading role in the European Renaissance helped perpetuate the constitutional myth internationally. Renaissance Venetians cultivated the myth of Venice as a place of beauty, harmony, liberty, and religiosity through their art, poetry, music, history and ritual pageantry.

By the sixteenth century, constitutional harmony dominated Venetian mythology. Gaeta (1961) characterizes this as the political myth of Venice's perfectly balanced constitution. The myth was founded on an apparent equilibrium between the three types of domination: monarchy (the doge), democracy (the Great Council) and aristocracy (through the Senate and the Council of Ten). This combination of government mirrored the republic outlined in Plato's *Laws* with elements of one, few and many. The three branches of government were the basis of the constitutional myth that the Republic beautified as an ideally governed state. Rosand (2001) sketches out how the constitutional myth became a series of half-truths and icons that the Republic invented:

> Out of the facts and fictions of its history, the Republic of Venice wove the fabric of propaganda that represents the essence of the myth of itself: an ideally formed state, miraculously uniting in its exemplary structure the best of all governmental types – that is, monarchy, oligarchy, and democracy – and, most significantly institutionalizing this harmonic structure in a constitution that was to inspire other nations for centuries. (Rosand 2001, 3)

Over time, the balance between the different forms of government shifted in correspondence with changing political forces. The fact that the constitution was based on custom, rather than on a law or a written document, gave Venetians the ability to adapt to changing needs, and the flexible constitutional arrangement helped the Venetian state to endure. For instance, the doge was elected in different ways after the ninth century, and the variations and complexities of Venetian electoral systems became vital components of the constitutional myth. The Venetian system of government was founded on a pyramid structure with the General Assembly at its base, followed by the Great Council, then the Forty and the Senate, with the Ducal Council forming the last layer before the doge at the apex. After 1178, there were six Ducal Councilors. They met regularly with the three heads of the Forty and the doge to perform the basic tasks of government. Together, these 10 men were referred to as the Signoria. Although the doge was a princely figure, he enjoyed no dynastic rights and was subject to numerous restrictions, from receiving gifts to personal memorials. As time passed the development of republicanism led to a declining emphasis on the doge's princely charisma, which was perceived as a threat to the Venetian Republic. These arrangements of government provided the Republic with an almost uninterrupted succession of 118 doges, a continuity that "gave substance to the Venetians' claims that theirs was a long-lived republic worthy of imitation" (Muir 1981, 299).

A strong sense of justice, emanating from the doge downwards, was at the heart of the constitutional myth and many explanations for the Republic's endurance. The doge was personally responsible for justice, alongside the other members of the Signoria. This enabled the Doge's Palace to project an image of justice, echoed in many of Shakespeare's plays, and the ceremonial emphasis on justice perpetuated the constitutional myth and may have helped convince commoners that they would be treated justly. Lane (1973, 271) suggests that the justice system was the key factor keeping order among the lower classes. Formally, nobles and commoners had an equal standing in court. Of course, differences in power and wealth created real inequalities and the rich could hire the best attorneys, but the poor at least had the right to an officially appointed attorney. State attorneys were particularly important because they had the responsibility for prosecuting Venetian state officials: all Venetian officeholders were liable to prosecution for abuse of office.

Despite Venice's reputation for justice, Muir questions whether the judicial system was the main source of stability: "A well-ordered system of magistracies…was insufficient to maintain political stability without an auxiliary ethic of political service" (Muir 1981, 21). Instead, he argues that the civic ethic of service to the Republic was the key to stability, requiring meekness, humility, faithfulness and unqualified love for the Republic (Muir 1981, 21). It is true that civil service appointments created a lower-level elite, which consisted of people who had special privileges due to positions in the diplomatic corps, legal services and medicine. This tier below the nobles assisted the maintenance of social order throughout Venetian society, and there are few instances of revolt by the lower classes in the history of the Venetian Republic. The British historian Christopher Hibbert (1988, 8) records that the fourth Candiano Doge provoked revolt and his eventual assassination, in 976, through his extravagances and demands for Venetians to support his personal feuds. As the Republic developed, social peace was encouraged by the relatively plentiful supply of food for Venetians, apart from occasional bread riots during the sixteenth century. Class tensions were also defused through the division of Venice into 60 parish communities including rich and poor. The maintenance of political and social harmony assisted the elite's legitimacy, concealed the force behind social domination, and provided its members with a unifying sense of mission. Social stability and the image of harmony added to the sense of environmental equilibrium. The relative lack of class conflict in the Venetian Republic helped perpetuate the myth of constitutional harmony, although this harmony has been questioned by those who have highlighted significant political crises.

The Italian historian Gaetano Cozzi (1973) explores the conflicts between nobles in different councils within the Republic, and records how the *avogadoria di comun,* a judicial body that guaranteed legal equality to all nobles, gradually lost its jurisdiction to the oligarchic Council of Ten that was established in 1310 (Cozzi 1973, 293-338). The result was a decline in equality before the law, the central principle of Venetian republicanism. Ironically, as the Council of Ten asserted authoritarian rule and subverted legal equality, the myth of political justice and harmony was increasingly broadcast. According to Muir (1981), "the humanist examination of the republic was an attempt to understand frightening social and political changes by recalling the example of a lost past" (Muir 1981, 29). Opposition emerged in 1582 among a group of Senators referred to as the *giovanni*, who opposed the oligarchic clique of the *vecchi* that controlled the Council of Ten. Victory of the *giovanni* over the *vecchi* meant that the power of the Council of Ten declined and political control reverted to the Senate. This shift in power illustrated how the appearance of constitutional harmony was created by a high degree of adaptability in the evolution of state institutions. Norwich (2003, 5) points out that the constitution essentially remained unchanged for over 1,000 years, also surviving earlier threats in 1310 and 1355. There was another internal crisis in 1628-9, prompting Renier Zeno, one of the protagonists, to contrast the myth of Venetian justice and equality before the law with gross corruption. The historian Donald Queller investigates the extent to which Venetian harmony was mythical. Queller (1986) recounts numerous examples of elite corruption, including the *broglio* system whereby poorer nobles sold their votes to richer nobles; and with some justification, criticizes scholars for concentrating their studies on the harmonious Venetian myth more than the reality (Queller 1973, 174).

Inconsistencies and conflicts have led to questions about whether harmony within the constitutional myth corresponded to reality. Yet the comparative stability of Venice maintained the myth until the final days of the Republic. Crouzet-Pavan (2002, 25) reminds us that although we need to be cautious about myths, we should acknowledge the relative calm in the politics of the Venetian Republic compared with the European breakdown of Christian unity and unstable Italian city states. Moreover, in a world dominated by monarchies and aristocracies, Venice stood out as a unique form of democratic government. The Venetian Republic's political institutions were theoretically democratic, even though the vast majority of Venetians were without the franchise. From 1297, the Great Council, consisting of all noblemen over the age of 25, served as the electorate with equal rights, voting to select members of other judicial, administrative and legislative offices that ran the state. Ackroyd (2010, 175) estimates that the patricians who governed the state represented four percent of the Venetian population, and the citizens six percent, leaving 90 percent powerless. There were monarchical and aristocratic tendencies within Venetian government, although its experimentation with democracy provided aspiring democrats with real inspiration. "More vital than the fossilized memory of Athens or Rome or an utopian dream, anti-monarchists in Europe and colonial America found in the living Venice an example that justified their cause," writes Muir (1981, 23).

Renaissance Venice was the only example for republican ideologues of an observable commonwealth. It was this constitutional myth of the Most Serene Republic of Venice, *la Serenissima Repubblica*, which became the leading international myth about Venice. "That collective image – of the self-proclaimed Most Serene Republic is an ideal political entity whose ruling patriciate were selflessly devoted to the commonweal – has come to be known as the 'myth of Venice,'" writes Rosand (2001, 2).

THE INTERNATIONAL IMPACT OF THE CONSTITUTIONAL MYTH

The English also played a conspicuous role in broadcasting the myth of Venice. Although English interest in Venice was well established, political debate about the Venetian constitution was largely prompted by the seventeenth century English *interregnum*. From the execution of King Charles I in 1649 until the restoration of his son to the thrones of England, Scotland and Ireland in 1660, the absence of constitutional agreements provoked consideration of Venice's political arrangements. Like Venice, England did not have a written constitution and sought to adapt government to changing circumstances. English books during this period illustrated the English fascination with Venice: for example, James Howell's *S.P.Q.V., a Survey of the Signorie of Venice, of Her Admired Policy, and Method of Government, &c, with a Cohortation to All Christian Princes to Resent Her Dangerous Condition at Present* (1651) admired Venetian government, its virginal purity and independence from foreign influences:

> Could any State on Earth Immortall be,
> Venice by Her rare Government is she;
> Venice Great Neptunes Minion, still a Mayd,
> Though by the warrlikst Potentats assayed;
> Yet She retaines Her Virgin-waters pure,
> Nor any Forren mixtures can endure.
> (Cited in Muir 1981, 53)

Interest in Venetian politics persisted after the English restoration, with over 10 books published between 1668 and 1672 that addressed themes related to the constitution of Venice. Eglin (2001) explains why the myth of Venice in English culture persisted into the seventeenth and eighteenth centuries:

> The international prestige of Venice, and the force of its myth, was bolstered – especially in Protestant Europe – by its defiance of a papal interdict a century after the League of Cambrai, by its survival in the maelstrom of the Italian wars, and by its holding out against the ambitions of ultramontane powers, Spain in particular. For these reasons, the myth of Venice resonated in England, a nation that had also defied Rome, and that had also defended itself from being ground under the juggernaut of Spanish ambition. (Eglin 2001, 14)

The myth of Venice became a statement about attitudes towards liberty in debates between Whigs and Tories in England. Thus this myth not only reflected interpretations of what was happening in the city, but was used as a statement about one's own values and what was happening in one's own country. "[T]he mythologized Venice was often replaced by a correspondingly idealized Britain, with identical attributes described in identical language, as if the British had learned from the Venetians what to think of themselves" writes Eglin (2001, 6). The Venetian model of government in the constitutional myth resonated prominently in European, American and Anglophone political thought. "This model of Venetian government significantly enlightened the political discourse of Europe: the nature of man could not be changed, but laws and institutions could be copied, guaranteeing, many hoped, a polity guided by reason," writes Muir (1981, 31). Furthermore, the Venetian constitutional model generated practical political innovations. *The Commonwealth of Oceana* by John Harrington was published in 1656 and widely read in England and America in the century that followed. Harrington praised Venice for its rotation of offices and secret ballot, which limited the power of the rich while preserving rule by an enlightened patriciate. American William Penn, impressed by Harrington, adopted the secret ballot in draft plans for the colonial constitutions of Pennsylvania and New Jersey. Similarly, the 1762 publication of the tenth volume of *The Modern Part of the Universal History* dealt with the history of Venice. This book was reported to have influenced framers the United States Constitution, who incorporated elements from the Venetian constitution (Eglin 2001, 173).

The survival of the Republic until it was conquered in 1797 meant it achieved an idealized status in the other nations. Eco Haitsma Mulier (1980) discusses the importance of Venetian mythology in Dutch constitutional history of the seventeenth century. Venetian governance became an icon of republicanism in Britain, Europe and throughout the English-speaking world: "More than any other political entity of the early modern period, the Republic of Venice shaped the visual imagination of political thought; just as she instructed Europe – and, ultimately, the independent colonies of America – in the idea of statehood, so she taught how to give that idea eloquent pictorial form, especially through the figuration of the state" (Rosand 2001, 1).

Alongside international admiration of the constitutional myth, a counter-myth emerged that characterized the Venetian regime as tyrannical, repressive and militarily weak. This developed from the repressive and clandestine activities of the Council of Ten, created with emergency powers after a period of unrest in 1310. By 1334, this Council was permanent. The counter-myth of Venice was promoted by Machiavelli, although its origins have been traced back to Giovanni Conversino's work in Padua in 1390-1405 (Muir 1981, 49).

Venetian territorial expansion across the Italian mainland during the fifteenth century created resentment and bolstered the counter-myth. The myth of Venice as a state of constitutional harmony was widely questioned from the sixteenth century when rule was increasingly enforced through government decree. During the late sixteenth century, the Council of Ten established a committee of Inquisitors, who violated Venice's traditions of justice by passing judgments without the presence of attorneys. The counter-myth was particularly espoused by Catholic sympathizers in response to the papal interdicts against Venice, and flourished in countries that had major conflicts with Venice, especially Catholic, monarchical France. Frenchman Jean Bodin was the most coherent proponent of the Venetian counter-myth in the sixteenth century, believing that Venice's mixed state contravened the indivisible idea of sovereignty, which could only be concentrated in one institution: "Bodin accordingly devised a new metaphor for Venetian government, a series of concentric circles with gradually increasing diameters: in the central circle was the doge; at the outside, the holder of supreme power – the Great Council" (Muir 1981, 50).

Shady diplomacy was another component of Venetian counter-myth. Venice was the first European state to maintain permanent embassies in foreign countries, and by the fifteenth century, its intelligence and diplomatic spying services were internationally unparalleled. Venice's tight-knit society of narrow alleyways and state informer networks linked to the Council of Ten meant the city came to be regarded as Europe's primary location for spying. The Venetian counter-myth of repression and clandestine activities was amplified by Venice's enemies, and some believe it to have contributed to the downfall of the Venetian Republic:

> It pictured Venetian government as a tyrannical oligarchy maintained by a terrifying efficiency in the use of spies, tortures, and poisons. This countermyth (*sic*) began among Venice's Spanish enemies during the Counter Reformation, but it matured in the eighteenth century, when there was some truth in it, and reached full flower in the propaganda by which the Jacobins and Napoleon justified their destruction of the Republic. (Lane 1973, 89)

The anti-democratic tendencies in Venetian government were noticed by American democrats. As democratic revolutionary spirit swept Europe and America in the eighteenth century, Venice's political institutions remained hostile to full democratic rights. While early American colonists considered Venice a model of republican government, by the time American independence was declared the Venetian Republic was largely identified as a metaphor for oligarchy and decay: to the point that Thomas Jefferson cited Venice as an example of constitutional difficulties resulting from a lack of separation of powers. "American thinkers had no reservations when it came to comparing the mother country to Venice and a number of other states that had once been free but had since declined into despotism," records Eglin (2001, 181). For Americans and the British, the 'fate of Venice' also became a reference point for the dangers of diplomatic neutrality. The refusal of the Venetian Republic to form protective alliances as Napoleon's armies advanced across Italy towards Venice was perceived as contributing to the fall of the Republic to Napoleon (Eglin 2001, 195).

VENETIAN MYTHS AND THE END OF THE VENETIAN REPUBLIC

After the fall of the Republic in 1797, Venice was internationally used less as a mirror that reflected good governance. Eglin (2001) notes "the almost universal substitution of England for Venice as the paragon of wise government" in the eighteenth century (Eglin 2001, 34). As well as destroying Venice's myth of constitutional harmony, Napoleon's victory over the Venetian Republic had a dramatic impact on Venetian myths about liberty, St Mark, virginity and other Christian mythology. In turn, the changes to these myths assisted the construction of the myth that Napoleon introduced environmental disequilibrium in Venice.

The French occupation of Venice in 1797 graphically affected the myth of St Mark. Symbols that were fundamental to this myth were removed following orders from the city's Municipal government installed by the French, and on 7 December 1797, the four horses of St Mark were taken away from above the west portal of St Mark's Church, where they had been for more than five centuries. The lion of St Mark had stood for even longer on the eastern column of the Piazzetta and, like the horses, was transferred to Paris. The French Municipal government in Venice did continue to use the emblem of St Mark's lion on its edicts, but the message promoted by the lion was changed to include a French definition of liberty. On the image of the book held in the lion's paw the words *Diritti e doveri dell' uomo e del cittadino* (Rights and duties of man and citizen) replaced *Pax tibi Marce evangelista meus* (Peace unto you Mark my evangelist). The French championed their revolutionary conception of liberty to redefine the Venetian myth of liberty. This was symbolized by the placement of the French Tree of Liberty in St Mark's Square for the French-inspired Fair of Liberty, held to usher in the post-Venetian Republic Municipal government. Universal suffrage was advocated in plays during French rule at Venice's Fenice Theater, where workers were invited to attend free of charge. Nonetheless, there was little Venetian popular enthusiasm for either the Fair or for French notions of liberty, as the art historian Margaret Plant notes (Plant 2002, 30). Weak popular enthusiasm for French governance was unsurprising: 60 members of the Municipal government, which ruled for eight months after the end of the Republic, were appointed by the French without popular mandate (Norwich 2003, 10).

Religious mythology was undermined by the French defeat of the Venetian Republic. Practically, the Municipal government curtailed support to the clergy in Venice (Plant 2002, 30). Symbolically, the invasion of Venice by Napoleon's army represented the annihilation of the virginal Venice metaphor. In 1797, Napoleon's forces penetrated the previously impregnable Venetian Republic, which had resisted all invasions during its 13 centuries of existence. The complicity of the Venetian government in France's victory did not soften the destruction of the virginal myth. The Grand Council had voted itself and the Republic out of office on 12 May 1797, and the Venetian government even sent boats across the lagoon to aid the passage of French invaders. Although the Venetian Republic had been in decline for over 200 years, "[t]he fall of the Venetian Republic in 1797 was nevertheless a tremendous shock. It represented the final destruction of the myth of Venice, which had long been regarded as a 'virgin state' because of its legendary independence and successful resistance to the efforts of hostile powers" (Eglin 2001, 8). The betrayal of the virginal city by the infidelity of the Adriatic changed the perception of the relationship between the sea and the city. The *Bucintoro* barge, from which the doge had performed the Marriage to the Sea, was ceremoniously burned by the French over three days in full view of Venetians; after which the French temporarily dismantled Venice's mighty shipbuilding industry at the Arsenal. Symbolically and practically, Venice was no longer married to the sea.

The divorce of this previously mythical relationship of balance between Venice and the sea led to the notion that environmental disequilibrium began with Napoleon's reign. The destruction of the virginal metaphor of Venice along with the end of the Marriage to the Sea contributed to the belief that environmental disequilibrium emerged between the sea and the city. These beliefs are discussed in detail in Chapters Seven and Eight. Napoleon's reign did bring physical changes to the city's environment, including the destruction of numerous ecclesiastical buildings and the creation of gardens, and these were incorporated into narratives about Venice's environmental disequilibrium. "In the broader history of the environment, Venice was perhaps a pioneer in a common search for an equilibrium between human life and the environment's time scale," writes Crouzet-Pavan (2002, 45).

The belief that Napoleon's reign marked the start of Venice's environmental disequilibrium also built on the downfall of the constitutional myth, which contributed to a perception of social harmony during the Republic. "The political myth that followed the earlier mythic motif of the providential city turns out to be essential to an understanding of the very fact of the much-praised Venetian equilibrium. It set the city within an imaginary of happy times and peaceful continuance in a world in which everything else seemed swept away by swiftly moving events," explains Crouzet-Pavan (2002, 191). After the Venetian Republic fell, Venice entered a period of unstable government and discord. The city was passed to Austria in 1798, before France regained control in 1806; Austria recovered Venice in 1814, only to face revolution from Venetians in 1848-9. Political disequilibrium accompanied sentiments of environmental disequilibrium.

Venice largely dropped off the political agenda after 1797. The Venetian metaphor was eclipsed in international political debates by the late eighteenth century French and American revolutions, and Venice had become associated with disequilibrium, repression, decay, and intrigue. Following his visit to Venice during the 1820s, the American novelist James Cooper concluded that America had taken up the mantle of democracy cast aside by the degraded politics of intrigue in Venice and Europe, and his 1831 novel *The Bravo* depicts a ruthless oligarchy operating in the shadows of Venetian government. This was indicative of the negative cultural imagery that dominated interpretations of Venice from the early nineteenth century.

NOTE

1. "Peace be unto you, Mark, my evangelist. On this spot shall your body rest." These words are usually inscribed on the open book held in the paw of St Mark's lion.

Chapter Two

Cultural Myths

"[I]n the fall Of Venice think of thine."

Lord George Gordon Byron, *Childe Harold's Pilgrimage,*
Canto IV, (2008 [1812], 153).

Even as Lord Byron warned that England could suffer a similar fate to the Venetian Republic, he understood that the end of the Republic was a metaphor for a human moral fall from grace. Byron typified the sentiments of cultural romantics who lamented a decaying Venice as a focus for their pessimism about the state of humanity. Yet the cultural romantics also captured a growing cultural fascination with Venice. After the political myth died with the defeat of the Republic in 1797, Venice's cultural influence became paramount. From the seventeenth century, the world's elite had flocked to the city on the Grand Tour, which became especially popular during the eighteenth century; in the nineteenth century, Venice attracted artists and intellectuals associated with the Romantic Movement. At the Olympia exhibition hall in London, parts of Venice were reconstructed in a show titled 'Venice the Bride of the Sea,' visited by over 30,000 people on one day alone in June 1892. During the twentieth century, the international intelligentsia made Venice and its Lido beaches the center of their social season.

Cultural interpretations of Venice have varied widely since 1797. Three themes have emerged that are fundamental for environmental considerations: decadence, decay and death. Cultural works including these themes became more frequent as the Venetian Republic went into decline. The Republic was renowned for its decadence during its final century; death and decay became prominent themes during the nineteenth and twentieth centuries. These themes added to fears that Venice was sinking, as the city deteriorated from its position as the high point of Western civilization, commanding the seas, and became widely associated with moral decline and human weakness. Culturally, Venice was typically depicted during the nineteenth and twentieth centuries as a place of frailty and failure, unable to combat the challenges presented by nature.

VENICE AS A CULTURAL CENTER

Venice started to become the artistic center of Europe in the mid-sixteenth century, largely due to the paintings of Titian, Veronese and Tintoretto. Foreigners flocked to the city to view and purchase its art treasures. Andrea Palladio (1508-80) became associated with Venice and was the foremost influence on Western architecture for centuries. Giovanni Antonio Canal, known as Canaletto, specialized in painting Venice as a subject rather than a mere background from the 1720s, developing the styles established by Gentile Bellini and Vittore Carpaccio. Venice appealed so much to foreigners that Canaletto moved to London for 10 years so he could meet demand more effectively. Antonio Canova, who hailed from the Venetian mainland and worked in Venice, was the most famous European sculptor during the last decades of the eighteenth century. Writers from Venice had a huge impact in Europe's capitals, including Giacomo Casanova and Carlo Goldoni in Paris. The city's reputation as a center for music, theater and opera thrived in the seventeenth century: Venetian myths were incorporated into Shakespeare's plays *Othello* and *The Merchant of Venice*, and by the end of the seventeenth century, Venice was a city of 140,000 inhabitants with seventeen theaters giving four opera performances simultaneously every season. In the eighteenth century, Antonio Vivaldi, Venice's most distinguished composer, composed 44 operas. Venice featured prominently in the work of Mozart in Vienna.

During the eighteenth century, the Venice Carnival became an international cultural icon. The first two Venetian myths identified by Gaeta (1961) - Venice as a state of liberty, and as a mixed constitutional state - were replaced by the gallant cultural myth when the Venice Carnival displaced the political and economic power of the Venetian Empire as a symbol of the city. "Republic heretofore the most considerable in Europe, for her Naval Force and the extent of her Commerce; now illustrious for her Carnivals," wrote the English poet Alexander Pope in the early eighteenth century (cited in Eglin 2001, 179). The gallant cultural myth of Venice expressed a conflict between virtue and vice, which came to the fore during Carnival. It raised questions about social order and etiquette for those who took part, and through its existence as a model for masquerades elsewhere. Attempts were made to replicate

Venetian experimentation with social norms through masquerades in London and secret societies, such as the Society of Dilettanti, and the Carnival became a source of European cultural instruction. Even though Venice's Carnival was suspended by Napoleonic rule, it was reinstated in a less extravagant manner after Austria regained control of the city in 1814, featuring as a theme in Johann Strauss' opera *Eine Nacht in Venedig* (A Night in Venice), which premiered in Vienna and Berlin in 1883, and was scheduled to tour in 16 European capitals and New York.

The flowering of Venice's gallant cultural myth was assisted by growing numbers of foreign visitors to the city, often captivated by its reputation for artistic and social innovation. The city was attracting vast numbers of merchants, pilgrims and curious travelers in the fifteenth century. By 1600, the Renaissance encouraged the educational virtues of foreign travel that allowed Venice to promote itself as a prime location for learning. In the eighteenth century, Thomas Nugent's (1756) book *The Grand Tour* described visiting Venice as vital for Europe's elite. Norwich (2003, 2) notes that in the early eighteenth century Venice "assumed the mantle of pleasure capital of Europe." As Venice became a standard stop on the well-trodden Grand Tour, the cultural myth of the city as a place of pleasure replaced its seventeenth-century reputation for repression:

> It was the age of the Grand Tour—that age, indeed, when tourism might be said to have been invented, when not only young English nobleman but the whole aristocracy of Europe was, at some time or another, to be seen in the loveliest and most magical of all cities. It was the age, too, of the carnival—still the most protracted and abandoned in Europe, the mandatory masks providing all the anonymity that could be desired. The Council of Ten, the Inquisitors of State and the secret police were no longer feared as they had been a century before. To give but one example, in 1718 the Inquisitors employed a staff of only three. (Norwich 1983, 585)

Despite the eighteenth-century elevation of Venice's cultural attractions over its degraded political reputation, Venice's constitutional myth was not entirely subsumed by culture. Acknowledging this, Eglin (2001) further notes that eighteenth-century travel and travel writing perpetuated political myths about the Serene Republic: "The Venetian system of government, which had been a ground of contention among political thinkers, survived as a model constitution, and this, along with the city's reputation as a center of painting and music, secured its position on travel itineraries" (Eglin 2001, 103). The Venetian Republic's political system continued to be of interest to many Venetians, constitutional scholars, and visitors, but this political interest in Venice dwindled after the Republic's fall. Venice featured as an icon in the nineteenth century cultural Romantic Movement, with the work of Lord Byron, Wordsworth, J.M.W. Turner and Ruskin tending to depict less gallant images of the city.

DECADENCE, DECAY AND DEATH

Decadence, decay and death were emergent themes in representations of Venice before 1797, which became prominent after the fall of the Republic. Prior to the 1500s, Venetians typically retained a belief that people could overcome the problems that nature presented, such as disastrous flooding and chronic epidemics of disease. Similarly, they faced military defeats and the religious condemnation of papal interdicts with courage. The Venetian Republic's leaders had faith in their capacity to solve these problems without the help of others and maintained a tradition of diplomatic isolation. Yet, as the Republic's decline set in during the eighteenth century, this resilient spirit waned. Venetians seemed to become lazy, enjoying their wealth while not investing in defending and expanding it. Decadence replaced courage. When faced with the threat of Napoleon, Venetians failed to put up a fight to defend the most enduring Republic the world had ever known; and they appeared increasingly incapable of protecting themselves against adversity. Prominent cultural figures such as Byron anticipated the sinking of the city during the nineteenth century, and for Friedrich Nietzsche, Venice and its lagoon became a mirror for a general sense of human melancholy (Nietzsche 1982, 494). Following the end of the Republic and repeated predictions that the city would sink, death became the dominant cultural metaphor. In the twentieth century, this was illustrated by Thomas Mann's 1998 [1912] book *Death in Venice*, and by Luigi Visconti's film (2004 [1971]) and Benjamin Britten's opera (1973) of the same title. By the early twenty-first century, Venice had been transformed from a metaphor for human dominance to one of decay (Berendt 2005, 36). Here we discuss how the transformation from a robust human outlook to a predominant self-image of frailty affected cultural perceptions of Venice and its inhabitants' ability to confront environmental challenges.

Venice has experienced many dark periods in its history, when it faced considerable natural and human threats. Norwich (1983) recalls the troubling period preceding the death of Doge Dandolo in 1289: within 20 years, Venice suffered an earthquake, terrible flooding, revolt in the principal Venetian colony of Crete, military defeats on land and sea, and a papal interdict (Norwich 1983, 172-3). Venetians pulled through this period and did not allow a pessimistic sense of decline to become established. Crouzet-Pavan (2002, 39) records the confident, optimistic outlook that dominated in historical documents before the fourteenth century. Norwich (1983) comments upon the courage of Doge Gradenigo and those around him when faced with economic ruin and papal excommunication in 1309 (Norwich 1983, 188). Venetians also had the courage to confront formidable environmental challenges. Silting from the three main rivers flowing into the lagoon threatened to turn it into dry land, removing Venice's defensive waters: so in 1324, the Venetian state began the huge engineering feat of diverting these rivers away from flowing into the lagoon. They had the confidence that they could combat the forces of nature, although this extremely difficult project was not completed until 1683.

This spirit of optimism was affected by repeated waves of plague from the mid-fourteenth until the first third of the fifteenth century. The Black Death caused an estimated 600 deaths a day at its height during the summer of 1348, wiping out three-fifths of Venice's population before it abated. The densely populated city was frequently threatened by public health problems, poverty and crime. Undeterred by these difficulties, Venice recovered and returned to prosperity, entering what Norwich (1983, 279) refers to as its "Golden Age" from 1405. In the fifteenth century, Venice was the richest city in Italy, with an annual budget comparable to Spain or England (Ackroyd 2010, 215). While Crouzet-Pavan (2002, 39) acknowledges declining optimism during the fifteenth century, she notes that Venice retained a sense of life despite the threat of death.

By 1509, Venice's Empire had been reduced to a few toeholds in an Ottoman world and "[n]o longer was Venice mistress of the Eastern Mediterranean" (Norwich 1983, 400). Norwich adds that, in the late 1530s, a "general breakdown of morale seems to have infected the whole population. The old public spirit had evaporated. The Venetians were growing soft. Wealth had led to luxury, luxury to idleness, and idleness to inertia, even when the state itself was threatened" (Norwich 1983, 455). Another epidemic of plague struck the city in 1576 to 1577, killing some 51,000 Venetians, and two fires in 1574 and 1577 destroyed most of the Doge's Palace. In contrast to difficult periods in the past, decline and death were met with a sentiment of decadence. Although the Venetian Republic was experiencing a loss of Empire and collapse of commerce, the Venetians held a sumptuous Carnival in 1510. During the late sixteenth century, conspicuous cultural extravagance was self-consciously used to impress foreign visitors as the Republic suffered political, military and commercial decline.

The decline of Venice's political power and reputation continued during the seventeenth century. A quarter of this century (1645-1669) was dominated by Venice's defense of its most valuable colony, Crete, against the Ottoman Empire. The loss of Crete was a devastating blow to Venice's fortunes and power over the seas. "Crete had been her last major possession outside the Adriatic, and with its loss not only her power but even her effective presence in the Eastern Mediterranean was dead for ever," remarks Norwich (1983, 557). During the seventeenth century, domestic politics were governed by repression and intrigue. The activities of the Council of Ten and its network of spies created the impression that Venice was a police state. This climate of intrigue was also furnished by rumors that Spain had sponsored conspirators to defeat Venice in 1618, which built on the real power of the Spanish Empire and its close relationship with the Vatican. The Spanish Embassy became the focus of intrigue and hostile demonstrations were held outside it. Two suspected conspirators were found publicly hanged on 18 May 1618, and rumors spread that their accomplices had been drowned the night before in canals. The 'Spanish Conspiracy' added to Venice's notoriety for spying and disreputable politics. It became the subject of Thomas Otoway's play *Venice Preserved*, which was performed almost 400 times in London between 1682 and 1795, when it was banned.

Culturally, Venice was depicted as degraded. By the eighteenth century, Venetians' reputation for decadence surpassed their earlier reputation for courage: "a people famous for centuries as the most skilful seamen, the shrewdest and most courageous merchant adventurers of their time, were now better known for their prowess as cheapskates and intriguers, gamblers and pimps" (Norwich 1983, 583). Although foreign visitors were attracted to Venice by its history, theaters and cultural wealth, they also came for its gambling halls and vice. Eighteenth century Venice appeared to be one long, meaningless party:

> What distinguished the vice of eighteenth-century Venice was not the depths of its iniquity but the pervasiveness of its frivolity. Eighteenth-century Venice had the reputation of being the gayest and most inconsequential of European capitals. The carnivals in which men and women went masked and indulged the liberties of make-believe created a spirit which lasted the year around, a pervading spirit of festivity. (Lane 1973, 434)

Venice's decline was also experienced in relative economic terms. "Venice did not have as large a share of Europe's trade or wealth in 1797 as in 1423," notes Lane (1973, 424). This was not due to absolute Venetian decline: although Europe was industrializing and developing at a faster rate than Venice, during the eighteenth century the Venetian Republic grew economically, mostly through Venetian control over parts of mainland Italy. Agricultural production, especially of maize, expanded rapidly in the Veneto mainland countryside. Textile, metal and silk manufacturing also augmented significantly during the eighteenth century in Brescia, Padua, Bergamo, Vicenza and Friuli. In the city of Venice, the textile industry declined and was only partially compensated for by the expansion of fishing, marine insurance, naval and port trade. These trends were reflected in population shifts in Venice compared with the mainland. "In 1764-66, the census reported 141,056; in 1790, 137,603 inhabitants. The Mainland domain, which had a population of only 1,500,000 in the mid-sixteenth century, grew to more than 2,000,000 by 1770," chronicles Lane (1973, 424). As Venice's economy declined relative to Europe and the Italian mainland, Venetian nobles lived the high life from investments in land and finance instead of investing in manufacturing or trade, provoking some critical responses:

> The most powerful spokesman of the ruling circle during the second half of the century was Andrea Tron, son of the Nicolò Tron who was one of the few Venetian nobles to finance an industrial development. In a celebrated speech to the Senate in 1784 surveying the state of the economy, Andrea called on rich fellow nobles who had their wealth in land or public funds, or who were wasting it in conspicuous consumption, to turn to commercial maritime enterprises as their ancestors had done. (Lane 1973, 428)

Almost inevitably, decline during the eighteenth century entrenched the sense of decay that culminated in the Republic's death in 1797. During the final decade of the Republic, there was already a palpable sense that it was politically dead:

> Venice was utterly demoralized. It was so long since she had been obliged to make a serious military effort that she had lost the will that makes such efforts possible. Peace, the pursuit of pleasure, the love of luxury, the whole spirit of *dolce far niente* had sapped her strength. She was old and tired; she was also spoilt. Even her much-vaunted constitution, once the envy of all her neighbours, seemed to be crumbling: votes were bought and sold, the effective oligarchy was shrinking steadily, the Senate was reduced to little more than a rubber stamp. In this last decade of her existence as a state, almost every political decision she made seemed calculated to hasten her end. Did she, one wonders, have a death wish? (Norwich 1983, 609)

The end of the Republic was worse than death by conquest: it was more of a suicide. Venice had lost the will to defend itself. Instead of negotiating new diplomatic alliances, as Austria and France flexed their muscles Venice relied on its traditional isolation. "Passively it refused to ally with either and rejected proposals for a defense league of Italian states. The traditional policy of unallied neutrality took less effort," writes Lane (1973, 434). When Napoleon's armies defeated the Austrians in Milan and marched towards Venice, the city's leaders could not be bothered to put up a fight: "There was evidence of a will to fight among lower ranks, especially among the Slavic militia brought from Dalmatia, but there was no fighting spirit among the nobles in charge" (Lane 1973, 434). Napoleon did not even feel the need to set foot in Venice in 1797; he sent an understudy. Gone was the heroic spirit of resistance demonstrated by Venetians during their defense of Constantinople in 1453. By 1797, there was no sense of honor in the death of the Republic. "The real tragedy of Venice was not her death; it was the manner in which she died," writes Norwich (1983, 635).

The death of the Venetian Republic generated an international reaction. Cultural figures and intellectuals led the chorus of regret. "The fallen Republic and its perceived hubris fascinated Europe and the New World and provoked operas, paintings, plays, poems and novels," writes Plant (2002, 87). One of the most prominent examples of despair at the fall of the Republic came from Wordsworth, whose 1802 poem *On the Extinction of the Venetian Republic* lamented Venice's loss of power over the seas:

> Once did she hold the gorgeous East in fee;
> And was the safeguard of the West: the worth
> Of Venice did not fall below her birth,
> Venice, the eldest Child of Liberty.
> She was a maiden City, bright and free;
> No guile seduced, no force could violate;
> And, when she took unto herself a Mate,
> She must espouse the everlasting Sea.
> And what if she had seen those glories fade,
> Those titles vanish, and that strength decay;
> Yet shall some tribute of regret be paid
> When her long life hath reach'd its final day:
> Men are we, and must grieve when even the Shade
> Of that which once was great is pass'd away.
>
> (Wordsworth 2000, 268)

In a similar vein, Byron wrote:

> The spouseless Adriatic mourns her lord;
> And, annual marriage now no more renewed,
> The Bucentaur lies rotting unrestored,
> Neglected garment of her widowhood!
>
> (Byron 2008, 151)

The fall of the Venetian Republic was followed by further decline. Initially, this was largely due to trade restrictions during Austrian rule of Venice between 1798 and 1806, but the return of Venice to French rule in 1806 led to the consolidation of decay. From 1809, Napoleonic territories were subject to blockades by the English at sea and the Austrians over land. These blockades isolated Venice from trade and importing basic foods and building materials. Although the city was retaken by Austria in 1814, the consequences of blockades and poor harvests caused a severe depression in Venice between 1814 and 1818. Austrian bureaucrats did little to alleviate starvation, disease, unemployment, malnutrition and migration, and visitors to the city were shocked by impoverishment and destitution. English writer Charles Dickens described his disgust with Venice's stinking canals during his 1844 visit to the city, and his compatriot, Byron, had been struck by the city's dilapidation in 1816:

> In 1815, by the time of Waterloo, Venice was the place of ruin that would soon be recognisable in the poetry of Byron: it was a fallen city that had passed through three periods of foreign rule, with depletion of population and personal losses to its subjects and minimal commercial viability. Its patrimony was diminished, buildings had been demolished and works of art and manuscripts dispersed. Venice's long history was now harshly interpreted not as a tradition of liberty, but of servitude and inquisition. (Plant 2002, 48)

Byron understood that the end of the Venetian Republic represented human moral atrophy as much as physical downfall. "The city's demise was, again, as much moral as physical. Venice was a perfect allegory of decline, which could be sustained as a personal as well as a national admonition," writes Plant (2002, 89-90). This sense of Venetian moral decline into decadence was exaggerated by the emergence of a culturally degenerate definition of liberty, replacing the political liberty myth discussed in Chapter One. Instead of liberty representing political freedom, it became a lifestyle celebration of defying religious and sexual conventions. "[N]egation of the political and historical myths of Venice allowed Venetian liberty to be trivialized as having given way to libertinism," notes Eglin (2001, 56).

The degradation of political life in Venice became part of the city's appeal. Byron went to Venice after failing to fulfil his political ambitions in England (Jerome McGann 2008, xix). Deposed royals escaping democratic political movements adopted Venice as their asylum, especially during the second half of the nineteenth century. The British historian John Pemble (1996) documents this trend:

> Royal refugees accumulated. Following the Bourbons into Venetian exile came a new generation of losers from the hectic casino of dynastic Europe. They included Princess Darinka, widow of the assassinated Danielo II of Montenegro;

> Don Carlos, prince of the Spanish Bourbons, whose bid for the Spanish throne had ended in military failure; and Prince Augustin Iturbide, adopted son of the assassinated Emperor Maximilian of Mexico. (Pemble 1996, 30)

Despite Venice's nineteenth century reputation as a place to escape from democracy, there was a brief revival of Venetian political liberty in the mid-nineteenth century. After suffering political subjugation to France and Austria since 1797, the Venetian revolution against Austrian rule in 1848-9, led by Daniele Manin, generated international admiration for the Venetian spirit of liberty in the face of adversity: "In contrast to the diagnosis of decadence and inertia in 1797, there was only praise for the fortitude, resolution and zeal shown by the Venetians in 1848 and 1849 in the face of attack, famine and cholera" (Plant 2002, 141-2).

Defeat of the revolution, reassertion of Austrian rule, and repression during the 1850s dampened this revival of political liberty, as did the city's exclusion from the initial unification of the Kingdom of Italy in 1861. Eventual unification of Venice with Italy happened after negotiations led to the Treaty of Vienna in 1866. Unification raised expectations that Venice's decline could be reversed and the city did experience an upturn with modernization projects and the reopening of theaters and Carnival; but the cultural definition of liberty as defying religious and sexual convention prevailed. This was illustrated in cultural representations of love and sex during the nineteenth century. Venice became an illustrious metaphor for love, assisted by the publication of Casanova's memoirs in 1798 - which were published in French between 1826 and 1838, and in English in 1894. The international publication of Casanova's memoirs added to the city's reputation for loose morality. By the end of the nineteenth century, the cultural representation of love in relation to Venice was often depicted in dangerous and degraded forms: in Verdi's penultimate opera *Otello,* first performed at Milan's *La Scala* in 1887, Venice is the setting for the dangerous consummation of love in the duet sung by Otello and Desdemona. For the American author Henry James (1909), love in Venice was degraded by being prostituted: the city was a "battered peep-show and bazaar" with St Mark's Square "the biggest booth" in Venice (James 1909, 7-8). After James made his first visit to Venice in 1869, he concluded that the degradation and decay of the city was part of its enchantment: "The misery of Venice stands there for all the world to see; it is part of the spectacle – a thoroughgoing devotee of local colour might consistently say it is part of the pleasure" (James 1909, 1).

Degradation in nineteenth century Venice was also perceived by German composer Richard Wagner, who had a revolutionary impact on Western music. It was in 1858 that Wagner made his first visit to Venice, beholding the "oppressed and degenerate life of the Venetian populace" (Wagner 1983, 577). Yet Wagner worked on his final opera, *Parsifal*, in Italy and Venice, and died in Venice in 1883. The talented English poet Robert Browning died in Venice in 1889. Death became the most prominent Venetian metaphor for the *fin-de-siècle.* It was the theme for Gabriele D'Annunzio's Venetian novel *Il Fuoco (The Flame)* 1991 [1900], Maurice Barrès' *La Mort de Venise* 1990 [1903] and Mann's *Death in Venice* 1998 [1912]. When the French novelist Marcel Proust first visited Venice in 1900 he wrote about the city's reputation for mortality and anticipated his own death (Proust 1994 [1971], 191-2). The Venetian island cemetery of San Michele became popular as a desirable place to be buried. Between 1872 and 1881 the cemetery was expanded, and each year on All Saints' Day large numbers of Venetians made their way to the cemetery to pay their respects across a temporary bridge built on boats. This became a highly visual reminder of the importance of death in Venice. San Michele was the preferred final resting place for famous twentieth century cultural figures including Serge Diaghilev, Ezra Pound, Igor Stravinsky and Joseph Brodsky. By the end of the twentieth century, the renowned literary critic Edward Said pronounced that Venice "is a place where one finds a quite special finality" (Said 1999, 46).

During the twentieth century, death in Venice became a virtual cliché. When Austria dropped approximately 1000 bombs on Venice in the First World War, the myth of Venice dying had the potential to be fulfilled: although human casualties were relatively minimal, with 60 dead. "If the danger to Venice had not been real, it would have been necessary to invent it, because plight and peril had become essential to its post-historical existence. In assaulting the city that had become a paragon, the Austrians were conforming to a myth that had flourished, not failed, amid the rank disarray of the nineteenth-century mind," notes Pemble (1996, 184).

Death in Venice had acquired meaning beyond a physical place of mortality. Léonard Sismonde de Sismondi's *Histoire des républiques italiennes du moyen age,* published in 16 volumes between 1809 and 1818, depicted the history of Venice as the tragedy of people who had gone from being strong to weak (Pemble 1996, 89). Nietzsche believed the Venetian lagoon reflected a wider condition of nineteenth century human melancholy: "all is now motionless, flat, dejected, gloomy, like the lagoon of Venice" (Nietzsche 1982, 494). Similarly, Ruskin (1960 [1860]) pondered why the two principal corner statues in the Doge's Palace portraying the Fall of Man and the Drunkenness of Noah were "representations of human weakness" (cited in Norwich 2003, 75). During the nineteenth and twentieth centuries, Venice was interpreted as a metaphor for fallen humanity. The German sociologist George Simmel (1922, 71) acknowledged that Venice was "a unique order of the form of world consciousness," but described the city's decay as "the tragedy of Venice." English author Virginia Woolf perceived Venice as "the playground of all that was gay, mysterious, and irresponsible" in the inter-war years (cited in Pemble 1996, 109). Venice and its

Lido beaches continued to attract the decadent Western cultural intelligentsia in the 1960s. From the 1970s, Luigi Visconti's film *Death in Venice* (2004 [1971]) became a metaphorical expression of decadence, sickness and fallen humanity. According to Plant (2002), in this film "[n]ot only the characters are in a state fallen from grace and integrity, Venice itself is continuously targeted as a metaphorical expression of the Fall" (Plant 2002, 334).

Human weakness and death in Venice were also linked in the 1976 book *The Death of Venice*, in which Fay and Knightley report on the industrial pollution and corruption in Venice and summarize: "Now Venice is dying and there is no hope of saving her. This book is an account of that death, and an indictment of those responsible for it" (Fay and Knightley 1976, 15). Corruption was another sign of human frailty in Venice, becoming a prominent theme as scandals swept Italy and Venice during the 1970s and 1980s. A report by the journalist campaigner Indro Montanelli, published in the Italian national newspaper *Corriere della Sera* in 1970, identified corruption as a problem in the funding allocated for Venetian restoration projects (Mencini 1996, 16). Rumors of corruption have persisted into the twenty-first century, expressing the perception that the overriding problem in Venice is not controlling nature or restoring ancient buildings, but human weakness. "Not only the elements and the long history of its buildings make the city vulnerable; human management was at fault," writes Plant (2002, 358).

The cultural themes of decadence, frailty and death have helped transform Venice from a metaphor for human dominance to one of decay. "It is this Venice that we have come to know – not the triumphant and arrogant conqueror but the humbled and crumbling ruin," declares Berendt (2005, 36). In contrast to the ancient myths about Venetians dominating the seas, cultural scripts since 1797 cast Venetians as threatened by their watery environment.

CULTURAL MYTHS ABOUT VENICE SINKING

After the fall of the Venetian Republic, Byron felt the sinking of the city would bring Venice a welcome rest from its enemies:

> Venice, lost and won,
> Her thirteen hundred years of freedom done,
> Sinks, like a sea-weed, into whence she rose!
> Better be whelm'd beneath the waves, and shun,
> Even in destruction's depth, her foreign foes,
> From whom submission wrings an infamous repose.
>
> (Byron 2008, 152)

The sentiment that the fall of the Republic would be followed by the city sinking into the sea was taken one step further by Ruskin, whose 1960 [1860] book provided details of the city's stones for the generations he feared would never see it: "I would endeavour to trace the lines of this image before it be for ever lost, and to record, as far as I may, the warning which seems to me to be uttered by every one of the fast-gaining waves, that beat, like passing bells, against the STONES OF VENICE," writes Ruskin (1960 [1860], 1). Similarly, Lauritzen (1986) refers to "the memoirs of English delegates at the Congress of Vienna who had written in 1814 that Venice would not last another fifty years" (Lauritzen 1986, 30). Prominent writers and intellectuals responded to the fall of the Republic in a manner that emphasized the decline of the city into the advancing sea. "Byron, Shelley, Samuel Rogers, and Tom Moore had all helped to popularize the idea of a watery grave, and the Italian historian Carlo Botta predicted in 1826 that Venice would soon be a heap of ruins half hidden by the sea," writes Pemble (1996, 113). Dickens observed that the water of the lagoon was "waiting for the time... when people should look down into its depths for any stone of the old city that had claimed to be its mistress" (cited in Pemble 1996, 113-4). These pessimistic interpretations of Venice's fate contrasted with the optimism of German author Johann Wolfgang Goethe, who visited the city in 1786 before the Republic fell, and concluded: "the Venetians have little to worry about: the slowness with which the sea is receding guarantees them security for millennia, and, by intelligently improving their system of dredged channels, they will do their best to keep their possessions intact" (cited in Bull 1981, 113). The writings of Byron and Ruskin particularly contrasted with earlier depictions of the sea as a means for realizing the Venetian Republic's power:

> [T]he . . . romanticization of Venice stemmed from the apprehension that the sea once dominated by the city would eventually prevail, and that the towers that once seemed to float on the waves would be sunk beneath them. Once the ancient Republic had vanished, conventional wisdom held that the city itself could not be far behind. It is difficult to believe, reading Byron and Ruskin, that at one time observers feared that the sea's retreat would leave the Virgin Republic vulnerable to potential ravishers. (Eglin 2001, 200)

Yet nineteenth century fears about threats to Venice included the resurrection of the belief that it would be left 'high and dry.' In a guidebook published in 1812, the Reverend John Eustace predicted that sedimentation would mean Venice would soon be surrounded by sand (Pemble 1996, 115): and indeed, serious sedimentation problems occurred from 1840. In that year, Venice's Austrian rulers diverted the River Brenta into the Venetian lagoon to protect the province of Padua from flooding, causing heavy sedimentation in the Venetian lagoon near Chioggia. By 1875, *The Times* (London) warned that "the day must come...in which the sands of the sea and the deposits of the rivers shall choke up the lagoons, obstruct the canals..., swamp Venice, and either kill by wholesale or compel to flight its fever-stricken population" (cited in Pemble 1996, 115). Meanwhile, Ruskin and others were predicting that Venice would be engulfed by water.

Sinking theories were supported by evidence of exaggerated subsidence, despite the absence of standardized subsidence measurements before 1897. In 1810, French engineers excavated the patricians' promenade in the lower arcade of the Ducal Palace. They found that the missing bases of columns were below the level of the pavement, which had been raised by about 38cm (almost 15 inches) over 500 years to compensate for soil subsidence. This discovery created widespread concern about sinking (Pemble 1996, 114). How was it possible that fears about Venice sinking and the lagoon becoming dry land could exist simultaneously during the nineteenth century?

Although there was real evidence of subsidence and sedimentation, nineteenth century predictions of Venice being destroyed also expressed cultural pessimism. The fall of the Venetian Republic meant that there was already a predisposition to embrace any sign of Venice's downfall. This willingness to fear the worst was illustrated by the response to the collapse of the emblematic bell tower on St Mark's Square, on 14 July 1902: an event that was interpreted by the leading Viennese architect Otto Wagner as indicating the imminent sinking of the city "because of the unstable foundations of the lagoon region, which caused continual movement" (Plant 2002, 237). In fact, the fall of the bell tower was related to a wall having been cut back and the collapse of a shaft, rather than the instability of the region's foundations (Pertot 2004, 97-9): yet fears that the bell tower's instability was symptomatic of decay in many Venetian buildings were chronicled by Massimiliano Ongaro in 1912 (Pertot 2004, 104). The tumbling of the Campanile bell tower captured international attention and stimulated nightmare visions that Venice was heading into a watery grave. "I saw the billows roll across the smooth lagoon like a gigantic Eager. The Ducal Palace crumbled, and San Marco's domes went down. The Campanile rocked and shivered like a reed. All along the Grand Canal the palaces swayed helpless, tottering to their fall," imagined the English writer John Addington Symonds (cited in Pemble 1996, 114).

Addington Symonds, Ruskin, Byron, Mary Shelley, Rogers, Wagner, Botta and Tom Moore assisted the creation of myths about Venice's expectant sinking. Pemble (1996) observes that nineteenth century intellectuals who became fixated with Venice sinking to its death, "Venetophiles", needed to construct this illusion because it was not impending reality: "An idea became an obsession and, Faust-like, the Venetophiles forfeited their souls by desiring eternity in a transient enchantment. They wanted a Venice that was perpetually dying, forever sinking to its grave; and since this was an impossibility, they connived at the manufacture of a fake and consoled themselves with illusion" (Pemble 1996, 143-4). The responses to the fall of the Venetian Republic and the toppling of the bell tower demonstrate that debates about Venice sinking have often been governed more by cultural outlook than physical reality: such fears were the product of a growing mood of reticence regarding man's impact on nature. Typically, Byron identified the source of Venice's enduring beauty in nature rather than the decaying man-made city and culture:

> Her palaces are crumbling to the shore,
> And music meets not always now the ear:
> Those days are gone – but Beauty still is here.
> States fall, arts fade – but Nature doth not die.
>
> (Byron 2008, 149)

The celebration of natural over man-made beauty contributed towards the growing influence of conservationism during the nineteenth century, aided by cultural romantics.

Part Two

FROM CONSERVATIONISM TO ENVIRONMENTALISM

Chapter Three

The Triumph of Conservationism

> "In the early 1800s Venice had been generally regarded as an odd and rather depressing wreck which could qualify as beautiful only when seen at a distance or by moonlight. By the end of the century the most fastidious sensibility was not only able but eager to contemplate the detail of its ruin. A transfiguring myth had developed, rooted in esoteric cults of art and literature; and as those cults became obsolete, the metabolism occurred that converts yesterday's highbrow conceit into today's middlebrow cliché. The myth lived on, providing a language and an iconography for advertising, journalism, and mass entertainment."
>
> John Pemble (1996, 1), British historian.

At the beginning of the nineteenth century, Venice was a shabby relic. We have recorded how it ceased to be politically significant and its cultural importance became paramount. Cultural representations of the city tended to emphasize its decay, degradation and death: lamented at first, but later celebrated across Europe and North America. This chapter examines how the victory of conservationism over modernization was fundamental to this shift in perception.

Enthusiasm for modernization was marginalized in Venice during the nineteenth century, and the city failed to keep up with the rapid modernization of numerous other European cities. By the end of that century, conservationists had gained the upper hand in establishing that buildings should be conserved rather than transformed: a trend that grew in Italy and elsewhere during the twentieth century and further weakened Venetian modernizers. The contemporary controversy over the mobile dams inherited this historical confrontation between conservationists and modernizers. Although the dam project is often interpreted as a victory for modernization, we will see how it follows the desire to modernize the outskirts of Venice and its lagoon in order to conserve the city center.[1] The triumph of conservationism in the center of Venice has contributed to its mythical status as a place to visit to escape from modernity.

THE LEGACY OF THE VENETIAN REPUBLIC

To understand the victory of conservationism during the nineteenth and twentieth centuries, we need to appreciate the traditions acquired during the Venetian Republic. The ancient Venetians predominantly built and remade their city through demolition, rebuilding, and intervention, in order to develop and to curb nature's impact. These traditions are contrary to the principles of conservationism and were re-fashioned over the two centuries following 1797.

Chapter One described how Venice was formed through a battle with nature to fortify dwellings on the islands in the lagoon. Venetians reclaimed land and extended their homesteads into the lagoon with little regard for conserving the natural environment, especially during the thirteenth and early fourteenth centuries. Maintaining settlements in the lagoon was a constant challenge in the face of flooding, silting, and also fire: "New building or rebuilding was frequently due to fires, which were the more devastating because the medieval city was so largely built of wood. Only gradually were wooden bridges and muddy lanes replaced by stone bridges and paved alleys and squares" (Lane 1973, 439). A fire in 1514 destroyed virtually all the shops of the Rialto area. Another huge fire devastated the Venetian Republic's Arsenal in 1569 but, as Andrea Morosini noted in 1719, "in a few months [the Venetian community] had rebuilt the towers and the walls of the Arsenal and returned it to its former splendor" (cited in Crouzet-Pavan 2002, 190). Destruction and rebuilding were constant features of life in Venice until the sixteenth century.

Although rebuilding was often necessary due to fire or other environmental hazards, it is important to note that the ancient Venetians also decided to transform many key buildings that had not been destroyed. During the eleventh century, St Mark's Church was rebuilt on a grander scale and the original wooden dome was replaced by five great domes, despite requiring the destruction of San Teodoro Chapel. Similarly, the Doge's Palace, considered by Ruskin as the central building of the world, was radically rebuilt in the mid-fourteenth century, and no visible traces remained of the palace of the ninth century. Demolition and radical rebuilding often characterized Venetian restoration. Francesco Sansovino recorded that substantial restorations to the legendary church of San Giacomo di Rialto were conducted in 1071 (cited in Howard 2005, 15). Further restoration to this church in 1531 was insufficient and it was completely rebuilt following a Senate decree in 1601. Scamozzi's circular church for the nunnery of the Celestia near the Arsenal was demolished in 1605 and rebuilt more conventionally. If improvements required the demolition of buildings, the ancient Venetians usually went ahead with little concern for conservation or nostalgia.[2] When new buildings were desired, there was scant regard for negative environmental impacts. Conservation was simply not on the agenda during the fifteenth-century building boom, when 200 palaces shot up beside the Grand Canal.

The creation of Venice in harsh natural conditions, its difficult maintenance and rapid development meant that Venetians learned formidable engineering and architectural techniques. These techniques developed through the city's prominent role in the Renaissance and the Enlightenment, which assisted the education and attraction of some exceptional architects. Before the Renaissance, the role of the architect had not been fully established and the creation of many ancient buildings in Venice was connected to their owners rather than designers. The term *architectus* was not widely used in Venice until the 1470s (Howard 2005, 64). We do not even know who oversaw the original designs of the Doge's Palace and Ca' d'Oro Palace. During and after the Renaissance many well-known architects became associated with the Venetian Republic, even if they hailed from outside Venice. Palladio conducted most of his work in the Venetian-dominated mainland and is often regarded as having had the greatest impact on Western architecture. Other prominent architects connected with the Venetian Republic include Bartolomeo Bon, Mauro Codussi, Antonio Gambello, Fra Giocondo, Baldassare Longhena, Giorgio Massari, Antonio da Ponte, Antonio Rizzo, Domenico Rossi, Michele Sanmicheli, Jacopo Sansovino, Giuseppe Sardi, Giovanni Scalfarotto, Vincenzo Scamozzi, Antonio Scarpagnino, Giannantonio Selva, Simon Sorella, Giorgio Spavanto, Tommaso Temanza, Andrea Tirali, Antonio Visentini and Alessandro Vittoria; many of whom approached their work in a manner that went beyond what we now think of as architecture.

Venice helped to pioneer an interdisciplinary approach to engineering and architecture. An early example was Fra Giocondo, who came from Verona and submitted many project proposals in Venice. He became a well-respected engineer and architect in the late fifteenth and early sixteenth centuries. This interdisciplinary approach to engineering and architecture developed during the Renaissance. The historians Manfredo Tafuri and Ennio Concina have emphasized the engineer's contribution to humanist and political tendencies in Venice during the Renaissance, particularly in naval architecture (cited in Plant 2002, 148).

Following the Renaissance, individuals who were both architects and engineers flourished during the Enlightenment. Venetian Andrea Tirali was appointed as the engineer in charge of Venice's waterways and sea defenses in 1688. He won admiration for his architectural work on the facade of the Theatine Church of San Nicolò da Tolentino, erected between 1706 and 1714. Tommaso Temanza was an architect for the church of the Maddalena and a garden pavilion behind Palazzo Zenobio at the Carmini. Temanza also designed bridges and supervised hydraulic work on water regulation for the Brenta River and on the great sea defense walls. River intervention and the building of sea walls added to the Venetian Republic's reputation for engineering and architectural ingenuity. The diversions of the Piave, Sile and Brenta rivers away from flowing into the lagoon began in 1324 and developed in the 1500s (see Figure 1.1). These diversions were largely concluded by 1683, although the definitive diversion of the Brenta was not completed until 1860. The construction of the sea walls started in 1744 and took 38 years. All these projects were celebrated examples of how the ancient Venetians intervened aggressively to prevent nature from destroying their lagoon. Goethe (1962, 99) visited Venice's sea walls in 1786 and complemented the Venetians on the preservation of their lands amid the water. Even though the sea walls were largely destroyed by the 1966 November storm surges, they were renovated, to stand as a "tremendous bulwark" (Norwich 1983, 604).

Venice's leading role in the Renaissance and the Enlightenment helped to form the role of the architect, often combined with engineering expertise: skills that were highly regarded in a society that prioritized development. The political stability, military victories, confidence and human ingenuity of the Venetian Republic contributed to a belief that Venetians had achieved an exceptional level of development, at least until the seventeenth century. Crouzet-Pavan (2002) writes:

> Natural calamities were surmounted; oppositions and conflicts were erased. Always miraculously overcome, they hardly ruffled the surface of Venetian life. Thanks to the excellence of its government, Venice could enjoy concord and peace, flourishing solidarities, and full employment in a harmonious and active society where everyone labored in his proper place and for the common good. Venetian historians stated, and Venetians probably believed, that Venice had reached a privileged state of development. (Crouzet-Pavan 2002, 190-1)

For most of the 1,000 year history of the Venetian Republic, the overriding concern was development rather than conservation. During the 200 years that followed the fall of the Republic in 1797, the relationship between development and conservation was reversed. In the remainder of this chapter, we outline the factors contributing to the triumph of conservation over modernization in Venice, starting with the redefinition of restoration, which changed attitudes towards the demolition, reconstruction and conservation of buildings in the city.

THE REDEFINITION OF BUILDING RESTORATION

The Venetian Republic developed rapidly and its engineering and architectural traditions meant the city was particularly well-suited to embrace modernization. However, as we saw in the previous chapter, stagnation dominated the Republic by the eighteenth century. Napoleon looted much of Venice's fine art and treasures after conquering the Republic in 1797, and Venice passed to Austrian rule in 1798 after Napoleon traded the city. When Napoleon regained control of the city in 1806, he recognized the need for infrastructural development.

Napoleon's 1807 Grand Tour of Venice resulted in an ambitious plan for urban regeneration. The most noticeable changes were concentrated around St Mark's Square:

> Under Napoleonic rule, the church of San Geminiano was removed, and an arcade similar to that of the Procuratie Nuove extended around the west end. Behind them was built a ceremonial entrance and ballroom for the palace of the Bonapartist King of Italy (now the entrance to the Civico Museo Correr). The grain warehouses which had been on the south side of the Procuratie Nuove were removed to create there a royal garden with an outlook over the Bacino San Marco. It is still called the Giardinetto Reale. (Lane 1973, 442)

It is claimed that the seven-year French rule of Venice after 1797 under the leadership of Napoleon resulted in exceptional demolition. Forty palaces were demolished and more than 80 churches and convents razed to the ground (Lauritzen 1986, 46). The established Venetian architect Selva led the bold new plans for urban redevelopment under Napoleon. Selva had no qualms about obliterating the island of San Cristoforo for his design of a cemetery on the newly-reclaimed land connected to the island of San Michele. Criticism within the city grew after Selva's work demolished four convents for the construction of new gardens at Castello and a new street, Via Eugenia (now Via Garibaldi). It has even been suggested that reactions against Napoleonic demolitions represented the birth of conservationism in Venice: "The genesis of public collections and the preservation of patrimony for all was now at issue," writes Plant (2002, 62). The changes to St Mark's Square provoked widespread condemnation, especially the demolition of its San Geminiano Church to build a staircase to the new palace ballroom, which "became an international issue studied by the academies of Europe" (Plant 2002, 66). In 1810, the famous sculptor Canova personally challenged Napoleon himself over the changes to St Mark's Square (Plant 2002, 72).

Undoubtedly, Napoleon's changes to Venice had significant environmental consequences. On the one hand, he supported the repair of the sea walls, which had been built between 1744 and 1782, to fortify flood protection. On the other hand, he renewed the dredging of major canals to permit larger ships to sail between the Arsenal and the sea. As we observe in Chapter Seven, contemporary environmentalists have claimed that Napoleon's dredging increased flooding and waves that damaged buildings, even though channels were also dredged during the Venetian Republic, between the Malamocco entrance to St Mark's basin and the Arsenal (Lane 1973, 454). Many of the changes Napoleon brought to Venice improved aspects of the city's environment, including the new cemetery, introducing municipal lighting and the creation of gardens: the lack of greenery in the city had been commented on by the Enlightenment thinker Montesquieu (Plant 2002, 59). There was also some restoration of central buildings, including St Mark's Church and the Doge's Palace. On balance, French rule can be characterized as a mixture of conservation and demolition in the interests of modernization (Plant 2002, 64).

A combination of demolition, restoration and development was also evident in the three periods of Austrian rule of Venice: 1798-1806, 1814-1848 and 1849-1866. During the second period of Austrian rule demolition was extensive, although without vision comparable to the Napoleonic grand plans (Plant 2002, 85). The peaks of Austrian demolition were during the 1820s and the 1860s (Pemble 1996, 124-125). Howard (2005, 268) identifies one specific reason for demolition under Austrian administrations: taxation penalized property owners so heavily that many palaces and houses were demolished because their owners could not afford their upkeep. Despite these demolitions of private residences, Pemble (1996, 124-125) notes that the demolition of churches was often accompanied by reconstruction and development. The church of Santa Lucia and the Scuola dei Nobili were destroyed to make space for a modern railway station. Some deconsecrated churches were reopened, others were restored, and a new slaughterhouse was built at San Giobbe. Numerous modernization projects affected buildings within urban Venice. Napoleon's improvements of the city's streets and canals continued under the Austrians, as canals were dredged, bridges were restored or rebuilt and streets were pleasantly widened. "Little by little, the ancient configuration of the city was being transformed. The simplest, cheapest method of creating new streets was the filling in of canals, and it was under Austrian rule that most of the city's *rii terà* were created. These 'land canals' are among the most attractive of the streets of Venice, being unusually wide," writes Howard (2005, 271).

Italian administrators took control of Venice in 1866. They discovered that modernization required demolition, just as French and Austrian governors of the city had found. "Two straight stretches of roadway, each a full ten metres wide, meeting each other at an angle at the church of San Felice, were cut through the city by demolishing all the existing buildings on the route," notes Howard (2005, 274). Venetian-controlled authorities also approved demolition to favor modernization while the city was under Austrian governorship. One such demolition project received little local opposition and was only shelved due to lack of funds, which Lauritzen (1986) identifies as a source of hope for those challenging the mobile dams:

> [T]he aggressive, pompous taste of all nineteenth-century Europe...cannot be blamed solely on the Austrians. Indeed, Venice's own town council showed itself all too anxious to improve the city at the expense of its former beauty. In 1858 it passed a resolution, with only a few dissenting votes, to tear down all the old buildings on the Riva degli Schiavoni overlooking the San Marco Basin and build there a gigantic international exposition centre of the Crystal Palace sort, beloved of nineteenth-century Europe. But despite the town council's vote of approval, the project was never realized: it simply cost too much for a city of Venice's extremely limited financial resources. Today's opponents of the mammoth project to close off the lagoon with moveable barriers might take comfort from this little cautionary tale now a hundred years old. (Lauritzen 1986, 47)

Although the international exposition center faced limited opposition, conservationists' campaigns became more vocal as the nineteenth century progressed. Initially, their campaigns concentrated on challenging the combination of demolition and restoration evident during the French, Austrian and Italian administrations of the nineteenth century. Restorers worked especially hard to transform Venice's decaying buildings under the direction of Austrian and Italian administrators: between 1816 and 1840, there was significant expenditure to improve the city's infrastructure, including restoring dilapidated churches and palaces. St Mark's Church was extensively repaired between 1843 and 1865. From 1853, this work was supervised by the architect Giovanni Battista Meduna, who also went by the name Giambattista: he acquired a reputation as a modernizer by rebuilding (with his brother Tommaso Meduna) the Fenice Theater after fire in 1837 and altering the magnificent Ca' d'Oro Palace. Giovanni Battista Meduna restored St Mark's Church north facade from 1842, addressed its subsidence and had tackled the southern facade by 1875. Another major transformation in Venice during the nineteenth century was of the thirteenth century Fondaco dei Turchi Palace, overseen by Federico Berchet. "Berchet did not hesitate to demolish whatever time had altered and, where necessary, he saw no problem in forcing the material to comply with his requirements," writes Italian architect Gianfranco Pertot (2004, 53).

Such restorations were welcomed by some, including Britain's *Building News* in December 1877. This publication admired Italians' determination "to restore to their primitive grandeur the monuments of national history" (cited in Pemble 1996, 130). By contrast, many famous foreigners celebrated ruined buildings over those that had been restored. Diderot, Mme de Stael, Benjamin Constant, François-René de Chateaubriand and Victor Hugo all expressed a preference for ruined buildings over ones that were in pristine condition. Pemble (1996) records these preferences: "Constant attested to the expressive power of dereliction", "Hugo held that a building was only truly complete when time had worked its alchemy upon it," and "[i]n Venice in 1833 Chateaubriand rediscovered the special beauty of decay and decided that the modern touch was a blight" (Pemble 1996, 127). Pemble adds that restoration was antithetical to these sentiments, explaining how it typically transformed buildings:

> It meant something much more than straightforward stabilization and repair. It meant completing a concept. 'Restoration' demanded an expert understanding of structure and a perfect command of architectural idioms, and it could lead to a building's (*sic*) looking as it had never looked before. The principle of notional integrity allowed the preservation of high-quality additions and alterations; but in practice anachronisms were ruthlessly minimized. (Pemble 1996, 128)

The leading opponent of restoration was Ruskin. Between his first visit to Venice in 1835 and the second visit in 1845, iron balustrades were installed on bridges and numerous buildings were demolished and repaired. Ruskin believed he needed to record the stones of Venice before they were ruined by restoration: "It is *impossible*, as impossible as to raise the dead, to restore anything that has ever been great and beautiful in architecture... That spirit which is given only by the hand and eye of the workman can never be recalled... Do not let us talk then of restoration. The thing is a lie from beginning to end" (cited in Pemble 1996, 131. Emphasis in original). Ruskin reacted against restorations overseen by Giambattista Meduna on the Ca' d'Oro, Ca' Giovanelli and St Mark's Church. "Something has to be said of the character of such works, their tendency to *ripristinare* or clean up, to restore a building to a past it perhaps should have had but never had had, involving the interpolation of extraneous forms, materials, types and images from a notional past," writes Pertot (2004, 47), sharing Ruskin's objections to Meduna's attempts to restore these buildings in pristine (*ripristino*) condition.

Ruskin's views on restoration appealed to critics of modernization around Europe. He gained many adherents within the British ruling classes, who appeared to be losing their appetite for industrial expansion and development as the nineteenth century progressed. In France, Ruskin was well-known as an art critic from the 1860s; in the early twentieth century, he was remembered by Proust as the defender of ancient Venice. Ruskin also had a profound impact in Venice and provided intellectual backbone to a flourishing conservationist movement in the city. In the 1870s, Venetian artists and writers began discussion groups about Ruskin's work. Plant (2002, 131) credits Ruskin with persuading Venetians to be fearful of modernist innovation. One of Ruskin's pupils was Giacomo Boni, who learned English to read Ruskin's work and met the man in 1876. Boni was one of 50 signatories to the 1882 pamphlet *L'Avvenire dei monumenti a Venezia,* which criticized restoration. In 1888, Boni was appointed as the Italian government's chief agent for the care of ancient buildings as the inspector of monuments in the *Direzione Generale delle Antichità e Belle Arti.* Another important Italian, Count Alvise Zorzi, was a disciple and friend of Ruskin, and Zorzi met Ruskin during the mid-1870s. In 1877 Zorzi became famous by starting a campaign in Venice to save the church of San Moise from demolition. Although this campaign was unsuccessful, Zorzi went on to publish a pamphlet that was highly critical of the restoration of St Mark's Church under the direction of Giovanni Battista Meduna. Ruskin wrote an introduction to the pamphlet and contributed financially. "This pamphlet was in effect a conservationist manifesto. It denounced current methods of restoration as vandalism, deplored the ruthless modernization of Venice, and pleaded for the removal of the profit-motive from the upkeep of ancient buildings," writes Pemble (1996, 135).

Zorzi did recognize that structural work was necessary to prevent St Mark's Church from crumbling, especially on the southern side of the building. Pemble (1996) comments that "[e]ven conservationists agreed that the state of this part of the edifice was critical" (Pemble (1996, 146). Zorzi emphasized that the new should be made old, and argued that the masonry of St Mark's Church should have been darkened with soot instead of cleaned (Pemble 1996, 147). Zorzi was supported by Pietro Saccardo, a member of the governing organization of St Mark's Church, and a growing body of international opinion. In 1879, the Ministry of Public Instruction in Rome suspended work on the Church while an inquiry was conducted by the Commission for the Preservation of Ancient Monuments: "Within a year the Commission had formally recommended a policy of conservation, and the minister ordered that the west front be repaired rather than rebuilt" (Pemble 1996, 148). Saccardo replaced Giovanni Battista Meduna after his death in 1883 and oversaw conservation work on St Mark's Church. When the hoardings were taken down in 1886, Saccardo had left the Church looking more or less as it had done in 1865. Conservationism also prevailed for work done on the Doge's Palace between 1876 and 1889, when visual changes were minimized.

The shift in Venice from restoration as transformation to restoration as conservation was influenced by growing international intervention. Concerns were raised about the restoration of St Mark's Church in Germany, France and Britain. In Paris, the *Comité des amis de Saint Marc* was established.

The Society for the Protection of Ancient Buildings (SPAB) in Britain led a campaign to stop the transformation of the church. The SPAB campaign was slow in getting off the ground, largely due to the initial indifference of SPAB founder William Morris, despite his visit to Venice in the spring of 1878. At the end of 1879, two and a half years after the publication of Zorzi's pamphlet, Morris spurred the SPAB campaign into action, having realized that the restoration of the church was an opportunity to promote his organization. He commented to Robert Browning that the restorers of the church "have given us an opportunity of appealing to people who might not otherwise be easy to move" (cited in Pemble 1996, 149). The SPAB campaign got into full swing with letters to the press, public meetings, scientific investigations, reports and a memorial with 2,000 signatures, which it addressed to the Italian government.

The campaign against the restoration of St Mark's Church continued for almost 30 years. Growing international involvement during the 1870s helped to turn Venice into a city of world patrimony. *The Times* (London) announced in 1879 that St Mark's Church was "the pride and possession of the whole world" (cited in Pemble 1996, 145). When the campaign became more muted in the summer of 1880, SPAB set up an international St Mark's Committee. The secretary was the painter Henry Wallis, who worked closely with Charles Yriarte in Paris and Charles Eliot Norton in Massachusetts. They recruited a wide range of notable artists, architects, historians and aristocrats, including Saccardo and Zorzi. Yet Zorzi was reluctant to promote the campaign in the Italian press because he felt there was insufficient opinion in favor of conservation and against restoration: "I fear that I shall be unable to make converts. I have many enemies as a consequence of being the first to raise and plead the cause of St Mark's...As for the newspapers, we should hold off until the Italian members of our society are more numerous, especially since almost all the papers are currently of an opinion opposite to mine in the matter of St Mark's and other Venetian monuments" (cited in Pemble 1996, 150). There was more enthusiasm outside Italy. For instance, the author Ivan Sergeyevich Turgenev joined the campaign and asked how he could promote it in Russia. The manifesto of the St Mark's Committee included over 100 members from Britain, France, the USA, Germany, Austria, Switzerland, Holland, Belgium, Poland and Russia.

In Venice and many other parts of the world, conservation had become highly influential in architectural opinion by the end of the nineteenth century. Restoration had ceased to mean transformation and had been redefined as conservation. "By the end of the century, in Venice as elsewhere, conservation was the new orthodoxy. When people spoke of restoration they now meant making the new look old, not making the old look new," explains Pemble (1996, 153). The redefinition of restoration was illustrated by the changing attitude of Camillo Boito, who was a prominent Italian architect, engineer and restorer. When describing Boito's early work in Venice, Pemble (1996, 129) remarks that "in the first half of his career he perpetrated restorations that showed all the rashness and brashness" of northern Italians. Boito's growing doubts about restoration were evident in a code of practice for restorers that he drew up in 1883: "It was full of compromise and pragmatism, but it endorsed the Ruskinian prescription of clearly dated and undisguised repair in lieu of architectural reproduction" (Pemble 1996, 136). This code of practice was adopted by the fourth Congress of Italian Architects in Rome in 1883.

The demolition and restoration of Venetian buildings during French, Austrian and Italian administrations of the city provided a focus for the launch of conservationism in the city. The redefinition of building restoration as conservation instead of transformation was a considerable breakthrough for conservationists, who extended their campaigns beyond building restoration to challenging other modernization projects.

BRIDGE BATTLES

In the early nineteenth century many Venetians were demanding modernization to tackle economic stagnation, which was largely blamed on Austrian rule. Pressure grew to build a bridge between Venice and the Italian mainland (*terraferma*), which would benefit the city by connecting with developing parts of Europe. Plant (2002) explains how debate about this bridge expressed desires for and opposition against modernization from the 1830s:

> The bridge project displayed the city's zeal to strike out across the lagoon and unite the city with the Terraferma, but this decade and the next also saw the inception of the 'Venice-as-modern' crisis that was to accelerate at the close of the nineteenth century, and frustrate the twentieth. Venetians pressed for modernisation against the strictures of Habsburg rule. On the one hand there was nostalgia, but on the other a new generation was responding to modern advances and increasingly noticed Austrian attempts to frustrate or delay. (Plant 2002, 117)

For those who dreamed of a return to the Venetian Republic, the bridge would confirm the death of the myth of Venice's virginal isolation from the mainland. When the bridge was eventually completed, it was the longest in the world and was widely acclaimed as a triumph for modernization. "The bridge was built, too slowly for the Venetians, between 1841 and 1846, and for some years afterwards was described by contemporary guidebooks as a miracle of modern engineering," notes Plant (2002, 110). The bridge and the establishment of railway services to key European cities created momentum behind the modernization of Venice, by connecting Venice physically and mentally to a modernizing Europe. Pemble (1996) describes how these developments brought about a shift in outlook:

> [R]epopulation of the Venice of the imagination reflected a change in actual circumstances. In the time of Byron, Shelley, Cooper, and Sand, Venice was neglected and physically isolated. It was accessible only by sea; and since its political independence and most of its commerce had been extinguished in 1797, the captains and the kings had departed. But then in January 1846 a viaduct across the lagoon was opened and the city was linked by rail to Vicenza. By 1857 the line was complete to Milan, and Venice was brought within the reach of the railway-travelling public of Britain, France, and Germany. (Pemble 1996, 15)

Venice's railway station provoked further controversy. In 1860, various historic buildings on the Grand Canal were demolished to build the station, including Palladio's church of Santa Lucia, which became the station's name. The bridge debate raged into the twentieth century because a road was added in the 1930s to accommodate vehicles. Plant (2002) outlines how conflicts over building the road emerged in the late nineteenth century:

> The land link was to be controversial, not only in its first stage as a railway, but later in the century, when plans were mooted for a carriageway – and then an automobile road – to the mainland. The controversy was to divide Venice into advocates of the bridge, the *pontisti*, and critics, the *anti-pontisti.* That debate crystallised the need to conserve Venice and its picturesque qualities by keeping the marks of modernisation at bay. (Plant 2002, 109)

The tension between conservation and modernization extended to conflicts over other bridges in Venice. Seventeen iron bridges were erected between 1850 and 1870, mostly replacing wooden or stone bridges. Before 1854, the Rialto Bridge was the only pedestrian crossing anywhere over the Grand Canal: in the 1850s, two new bridges over the Grand Canal at the Accademia and the Ponte degli Scalzi by the railway station were built. Despite being built during Austrian administration, the iron Accademia Bridge was designed by Englishman Alfred Neville and completed in 1854. This iron bridge became a victim of conservationist pressure and it was replaced with a wooden structure. "Neville's bridge had a relatively brief existence: it was demolished in 1932 as it was increasingly felt that its design was out of sympathy with the environment," explains Plant (2002, 147). The demolition was a shock to modernizers, who could not imagine the iron bridge as visually abhorrent: "Neville's Accademia bridge, one of the most anathematized features of Venice at the time of its demolition in 1933, was praised by the *Illustrated London News* as a 'handsome structure' with 'elegance of form'. It was inconceivable to the early Victorians that their civil engineering could be aesthetically offensive" writes Pemble (1996, 125). The Ponte degli Scalzi, an iron structure modeled on the Accademia Bridge in 1858, was replaced by a stone bridge in 1933-4.

Conservationism acted as a brake on modernization projects needed in the center of the city. By 1984, the wooden Accademia Bridge was declared unsafe and replaced by another temporary wooden structure in 1985-6. No one seemed to have the courage to come up with a modern replacement, as Lauritzen (1986) notes:

> The bridge was designed in 1933 as a temporary structure made of wood to replace the Austrians' iron bridge. In the fifty years since, no project for a permanent bridge, like the stone one at the railway station, has ever been adopted and the condition of the wooden bridge, which has come to seem less and less temporary to the Venetians, has deteriorated to such an extent that in 1984, it was decided to dismantle it altogether. When the operation began, there was actually no approved plan for the bridge's reconstruction. (Lauritzen 1986, 70)

A modern replacement for the wooden Accademia Bridge was still under discussion at the time of writing, although conservationist objections remain. We will return to this in Chapter Nine. Before we can understand the current bridge battles, however, we need to contemplate several stages in the evolution of conservationism in Venice.

PRESSURES TO MODERNIZE

Despite conservationist achievements with replacing iron bridges and redefining building restoration, considerable pressures to modernize Venice were exerted during the nineteenth and twentieth centuries. Napoleon's modernization projects have been noted, and the modern development that continued under Austrian rule included the construction of a breakwater system to combat tides between 1838 and 1857, and gas street lighting. The Austrians also reduced health risks from some of the more stagnant canals by filling them in. These modern improvements met with some criticism: for example, Ruskin derided Venice's railway bridge and gas lamps in a letter to his father (Norwich 2003, 101), but many Venetians continued to desire modernization and complained that Austrian rule was limiting it.

Modernist zeal concentrated opposition on Austria and became associated with emerging Italian patriotism. The Ninth Italian Scientific Congress was held in Venice during 1847 as a rebuke to Austrian restrictions on such events. In 1848, revolution in Vienna and the founding of the second French republic encouraged the Venetian revolution of 1848-9 led by Daniele Manin. Although the Venetian revolution captured the spirit of challenging Austria for better modernization, Manin was chided for his reliance on symbols associated with Venice's ancient elite, including the crest of St Mark (Plant 2002, 140). Venice returned to Austrian control when the revolution was defeated, and most modernization schemes between 1848 and 1866 remained on paper (Plant 2002, 146). By the end of Austrian administration in 1866, Venice had experienced some modernization compared with the start of the century. It is worth quoting Howard (2005) at length to appreciate how important modernization was for the city's survival:

> By the time that Venice was liberated from Austrian domination in 1866, the face of the city had changed considerably. The period is often remembered for the tragic demolitions of notable historic buildings, but as we have seen there were also many constructive changes. After all, Venice could not have preserved her medieval way of life unchanged in the context of modern Europe. Techniques of building and modes of transport, which had been little altered for a millennium or more, had to be modernized if the city were to survive at all. (Howard 2005, 272-3)

Modernization efforts were stepped up after Venice became part of Italy in 1866. "[T]he policies for modernizing the city, initiated by Napoleon and kept in motion by the Austrians, were given new life by the greater commitment of the new Italian administration," notes Howard (2005, 273). One priority for the new local council was to develop the streets and squares, as Pemble (1996) points out:

> After the incorporation of Venice into the Kingdom of Italy the main concern of the Consiglio Communale (*sic*) was to modernize the city by opening up its dense medieval texture. The first of several new avenues, the Strada Nuova, linking the Campo Santi Apostoli with the Campo Santa Fosca, was begun in 1868; and in 1871 an area was cleared at San Paternian for a new *campo* with a monument to the patriot Daniele Manin. (Pemble 1996, 125)

Another priority was the modernization of transportation infrastructure. This was necessary if Venice was to be revived through international trade, which was growing during the second half of the nineteenth century. The Austrians had considered the need for new port facilities to develop Venice and under Italian administration in 1869, work began on a new maritime station in the south-west corner of Venice (see Figure 1.2), opening in 1880 with quays, warehouses, offices, a customs house and rail access. "[T]he Stazione Marittima became the focus of new industrial development and led to a dramatic revival in overseas trade," notes Howard (2005, 274). Connections were created with new shipping routes and the opening of the Mont Cenis tunnel between Italy and France in 1871. Pemble (1996) writes:

> After the building of Continental railways, and the opening of the Suez Canal in 1869, the inland sea again became a great thoroughfare. It carried the traffic between Britain and its empire in North Africa and the East; and Venice, specially favoured by the opening of the Mont Cenis tunnel, found itself in the mainstream of modern communications. In 1872 it captured the India Mail, a major prize in the transit business. The Peninsular & Oriental Company contracted with the Italian government to provide a regular mail service between Venice and Alexandria, and this put Venice on the map as the main port of embarkation for British personnel travelling to India and Egypt. (Pemble 1996, 17)

Venice's port facilities and shipping developed further during the 1880s and 1890s. This required new jetties, dredging channels and creating new ones, especially between 1882 and 1892 at the Lido entrance to the lagoon. By 1890, cargo ships could sail to the new harbor at the maritime station. "As modern industry developed in its hinterland, Venice became again a great port, second only to Genoa among the ports of modern Italy," notes Lane (1973, 454). In 1895, the P. & O. Company signed a convention with the Italian government to establish a regular British cargo and passenger service between Venice and Port Said, linking Venice directly with Egypt, Aden, India, Ceylon, China, Japan, Australia and New Zealand. The maritime port was expanded in 1904. By 1913 it was handling two and a half million metric tons (2.75577 short US tons) of traffic compared with one million metric tons (1.10231 short US tons) in 1886.

The modernization of the port, railway and communications stimulated Venetian industry as well as trade. During the 1880s, shipbuilding was revived in the Arsenal and the glass and fabric industries prospered. Tobacco and beer enterprises started up and large mills were built near the city center. Next to the newly constructed maritime station at Santa Marta, a cotton mill was operating by 1882; on the Giudecca Island, the Giovanni Stucky flour mill began production in 1884. These were not the first mills in Venice: the Austrians had converted the deconsecrated church of San Girolamo into a mill with the old bell tower functioning as the factory chimney. The wave of modernization in Venice from the 1870s revived conservationist objections. The British were particularly influential supporters of Venetian conservationists, as Pemble (1996) explains:

> In the mid-1870s the British intellectual and artistic élite suddenly decided that, as the heirs of Byron, Turner, and Ruskin, they had a special mission to save Venice from the ravages of restoration and modernization. From this

> time, measures proposed for the sake of public health, efficient transport, and economic development were regularly reported in the British press and almost always condemned on aesthetic or moral grounds. Venetian conservationists were encouraged and patronized, and the Venetian authorities became a target of constant criticism. (Pemble 1996, 144)

One focus for conservationist criticism was the building of a factory to make railway carriages on the island of Sant'Elena. Venice City Council gave permission to the *Società Veneta d'Imprese e Costruzioni Pubbliche* to develop the factory and faced the wrath of British writers and newspapers. Robert Browning referred to it as "a deed of unutterable barbarism" and the *Builder* asked in 1889 "whether there was no site but Sant'Elena to found a carriage factory on" (cited in Pemble 1996, 144). Another development that provoked indignation was the start of steam transport in Venice. *Vaporetti* (steam boats) operated across the lagoon from 1872 and in 1880 services began on the Grand Canal. "It does seem lamentable," wrote the *Builder*, "that we cannot preserve some of the unique character of such a place from being destroyed by the all-levelling spirit of modern trade and its servant steam" (cited in Pemble 1996, 145). The introduction of gasoline powered motor boats around 1909 prompted fears that modernization would crush the gondolier: as the *New York Times* commented in 1909: "Above all things, the tourist requires rapid transportation in 'seeing Venice.' This the motor boat provides him. [The gondolier] may still remain for some time a picturesque personage until, like the horse cab in land cities, he gives way to the Frankenstein monster, which modern invention and the demands of modern utility have brought into being" (cited in Davis and Marvin 2004, 142).

Disapproval of modernization also came from Italians. Domestic conservationist concern was illustrated by the national restoration legislation proposed by Camillo Boito and adopted in 1883. Plant (2002, 193) credits Boito with instigating a new environmental conservationism in Venice that categorically rejected new building. By the late 1880s, Boito and international conservationist pressures helped shift the balance against modernization: "The city and its lagoon had become a battlefield where past and present were ranged as enemies, mutually destructive. With shifting degrees of intensity, the fight would continue through the next century; but the modern cause was largely lost" (Plant 2002, 194). The growing impact of conservationist opposition to new building in Venice was demonstrated by a campaign supported by Boito. In 1886, Venice Mayor Serego degli Alighieri presented an ambitious plan for slum clearance and redevelopment to the *Consiglio Comunale* (City Council). A campaign of opposition to the plan was organized by Pompeo Gherardo Molmenti, an important Venetian historian and public figure. An official commission of inquiry was established under Boito and the plan was severely diminished. "This drastically curtailed Alighieri's scheme. In the event only sixteen out of his forty-one proposals were implemented. From now on the Consiglio Communale (*sic*) paid as much attention to conservation as to modernization," writes Pemble (1996, 178-9). For instance, in 1890, the Consiglio Comunale bought the gondola boat yard at San Trovaso to stop speculators from developing the site. The influence of Molmenti was evident again in 1898 when he ran a successful campaign against proposals for a second bridge from the mainland to Venice.

In 1902 polemics intensified between conservationists and modernizers after the collapse of the bell tower on St Mark's Square. As discussed in Chapter Two, the destruction of the bell tower became an international concern; so did its replacement. Conservationists argued that the bell tower should be rebuilt exactly as it was previously, whereas modernizers saw an opportunity for improvement. "In any previous period," wrote the *Builder*, "if an important building fell down, the immediate desire would have been not to 'restore' it, but to erect something better and finer in its place" (cited in Pemble 1996, 158). Conservationists claimed that the bell tower should be rebuilt according to the principle of *'com'era, dov'era'* ('as it was, where it was'). Meeting on the same day as the bell tower collapsed, "the Venetian city council resolved to rebuild it 'where it was as it was'," notes Lane (1973, 444), although some internal differences with the original were introduced. The application of this principle to the rebuilding of the bell tower signaled the ascendancy of conservationism and confirmed the redefinition of restoration as conservation rather than transformation.

'As it was, where it was' became the guiding principle for building in Venice's historical center from then on. It was largely applied for the eventual restoration of the Fenice Theater in 2003, which burned down in 1996 (Pertot 2004, 201). "In the twenty-first century the newly rebuilt theatre, La Fenice, has been criticised by some Venetians as a contrived pastiche of the previous building destroyed by fire," observes Ackroyd (2010, 147). Since the early twentieth century, 'as it was, where it was' has prevented the innovative modernization of Venice's buildings. Of course, buildings were never reconstructed precisely: "Nothing can be rebuilt 'as it was'; the very fact of rebuilding precludes that possibility" (Ackroyd 2010, 282). Nonetheless, the principle became a conservationist tool against change. "*[C]om'era e dov'era w*ould become both Venice's defining credo and its eventual prison, setting it on the road to becoming the museum city it is today," remark scholars Robert Davis and Garry Marvin (2004, 218).

POLITICIZING MODERNIZATION

In the early twentieth century, modernizers launched political movements against conservationism. The first was started by writers, artists and architects who called themselves the Futurists, and whose main propagandist was Filippo Tommaso Marinetti, an Italian poet and novelist. Although the Futurist movement began in the arts, it also promoted an explicit political agenda that challenged conservationism. *The Founding and First Manifesto of Futurism* was reproduced on the front page of the Parisian newspaper *Le Figaro* in 1909, and in 1910, the Futurists launched a manifesto titled *Contro Venezia Passatista* (Against Venice of the Past). Marinetti opposed the replication of the bell tower and was supported by the painters Carlo Carrà, Umberto Boccioni and the composer Luigi Russolo. Marinetti and his associates scattered the manifesto *Contro Venezia Passatista* across St Mark's Square. They failed to prevent the completion of the replica bell tower in 1912. Yet the Futurists also campaigned against the general conservation of Venice as a ruin, calling for the replacement of decaying buildings with an industrial and developed city. "We want to prepare for the birth of an industrial and military Venice, with power to dominate the Adriatic, the great Italian lake," the Futurists declared (cited in Pemble 1996, 159), reviving memories of the Venetian Republic's celebration of domination over the sea. Marinetti went so far as to welcome the destruction of ancient Venice by flooding, addressing a full audience at Venice's Fenice Theater thus: "Oh how we shall dance! Oh how we shall cheer the lagoon, to incite it to destruction! And what a dance of triumph we'll perform around the illustrious ruin!" (Cited in Pemble 1996, 160, footnote).

The Futurists were strongest in Italy, although parallel movements were formed in Britain, France and Russia. Marinetti attacked "this deplorable Ruskin" in a speech at London's Lyceum Club, "with his hatred of machinery, of steam, and of electricity, this fanatic for antique simplicity is like a fully grown man who wants to go on sleeping in his cradle" (cited in Pemble 1996, 160). The Futurists considered the fascination with Venice by Ruskin and the British as stifling the city:

> The British fixation with Italy's past was attributed to his mummifying influence. And if Ruskin was the supreme pontif of *il passatismo*, Venice, the city with which he was most closely associated, was its *cloaca maxima*. It had become the temple of the cult that the Futurists most abhorred – 'l'adorazione della morte.' This was anathema, because it signified morbid enslavement to intellect... So the Futurists abolished death by decreeing the end of hypertrophied intellect and prophesying man the intuiting mechanism – a new Prometheus. (Pemble 1996, 160)

Despite taking their campaign outside Italy, the impact of the Futurists beyond the Italian peninsular was limited and artistic splits emerged. Moreover, the humanistic appeal of their message was compromised by its nationalistic and militaristic tones in the run-up to the First World War. Marinetti's chauvinism was upstaged by Benito Mussolini, who rejected socialism in 1914, and by Gabriele D'Annunzio, who returned to Italy from Paris in 1915. D'Annunzio and Mussolini launched the fascist movement, which also politicized modernization in Italy and Venice.

D'Annunzio sought to repress the masses using the politics of fascism, fearing the threat they posed to elite culture. The Italian Communist Party (PCI), which was inspired by the 1917 Russian Revolution, also embraced modernization. Unlike the fascists, the PCI theoretically championed modernization for the masses; but during the 1930s fascism prevailed over communism in Italy. In *Il Piacere* (1888), D'Annunzio disparaged of "the grey modern flood of democracy" that was "submerging so many lovely and rare things" (cited in Pemble 1996, 169). Like the intellectuals Nietzsche, Matthew Arnold and Ernest Renan, D'Annunzio worried that high culture would be crushed between the masses and an increasingly powerful, yet philistine, industrial bourgeoisie. For many elite intellectuals, a retreat into defending high culture seemed preferable to modern democratic politics and Venice was the perfect refuge.

In the early twentieth century, "[s]een from the shadows of the modern world, Venice acquired a new and enviable prestige" (Pemble 1996, 169). Not only did Venice represent high culture; its working class appreciated culture. In the central chapter of the second volume of *The Stones of Venice*, Ruskin described a utopia with laborers made happy by thought: for Ruskin, "Venice had possessed the ideal proletariat" (Pemble 1996, 169). A working class that included glassmakers, gondoliers, boat makers, fishermen and artisans was tolerable for cultural elitists, and the Venetian working class was often idealized by artists, as in the painting of Venetian fishermen by Ludwig Dill (1848-1940). "The threat to Venice did not come from 'the people'; nor was concern for the city limited to an élite of birth and intellect. The threat to Venice – the human threat that is – came mainly from a cadre of architects, engineers, town-planners, and sanitary scientists who were motivated by professional ambition rather than by political ideology," notes Pemble (1996, 174-5).

As Italy industrialized, Venice was perceived to be threatened by modernization. "She is the Andromeda of Europe," declared Francis Marion Crawford in 1905, "chained fast to her island and trembling in fear of the monster modern progress, whose terrible roar is heard already from the near mainland of Italy, across the protecting water" (cited in Pemble 1996, 145). This pressure from the mainland to modernize was held at the borders of Venice's historical center.

MARGINALIZING MODERNIZATION

Modernization in Venice was marginalized from the city's center by administrators and business people who decided to develop the outskirts, especially Lido Island and Marghera's industrial area. The first bathing facilities opened on the Lido in 1857, and the 1886-91 plan for Venice's revival provided substantial funds for developing Lido Island with a road network. The luxurious Hotel des Bains and Hotel Excelsior Palace were completed in 1900 and 1906. With additional new hotels, Venice's Lido became one of the most fashionable European seaside resorts, with golf, tennis and riding facilities established by 1920. Lido airport opened in 1926 and the grand casino started in 1938. Tourism was boosted by the road link between Venice and the mainland and the first multi-storey car park at Piazzale Roma (see Figure 1.2), both completed in 1933. Tourism and tourist services also increased in Venice's historical center. But the building of huge tourist hotels and facilities were concentrated on the Lido Littoral Island.

Modern industry was diverted away from Venice's historical center. This was achieved through the creation of the first Marghera industrial zone on the mainland bordering Venice in 1903 (see Figure 1.2). Un-reclaimed marshland on the western side of the lagoon was bought by a consortium of Venetian investors led by Count Giuseppe Volpi, who aimed to restore Venice's past glory as a maritime leader. Instead of creating a conventional port to support distant industries, the plan was that materials would be directly loaded between ships, factories and refineries, resulting in considerable savings on transportation costs. The industrial zone was completed in 1932 with the port operational from 1917. "Volpi's conception proved so profitable that within the first year of the industrial port's operation, its users found that they had cut their costs by fully ninety per cent," records Lauritzen (1986, 54). The industrialists in the consortium benefited from Marghera's business: so did Venice and its people. "Without the major source of income created by the industrialization of the lagoon, Venice would probably have become a ghost town, unable to support even a fraction of the present population," notes Howard (2005, 280).

Despite these benefits, the Marghera industrial area became another point of conflict between modernizers and conservationists during the 1920s and 1930s. Its existence intensified the debate about adding a road alongside the railway bridge between Venice and the mainland. Count Volpi and his partners sought to make commuting between the industrial zone and the city easier for workers by providing a bus service along the road. The completion of the road link in 1933 was invaluable for working-class Venetians getting to new jobs, but it alarmed conservationists. Venice became increasingly divided by a conserved ancient city center and its developing outskirts. "The gradual abandonment of large-scale industry in Venice and Murano was accompanied by the development of Marghera on the nearby mainland as a major industrial centre, specializing in oil refining and the manufacture of chemicals," notes Howard (2005, 280). The Santa Marta cotton mill and the Stucky flour mill proved to be the last factories in the city center. Similarly, "[t]he foundry and factory on Sant'Elena were replaced with working-class housing, and the old city, now designated 'il centro storico' ('the historic centre') was left in comparative peace – a counterfeit but venerated relic, enclosed on one horizon by a garish suburban seaside, and on the other by a grey industrial megalopolis," writes Pemble (1996, 179).

Diverting development to the outskirts of Venice marginalized modernization in the city center. In addition, the energies of modernizers were increasingly detached from city development during the twentieth century as they were drawn towards the arts and design. An international modern art exhibition was established in the city's public gardens (see Figure 1.2) after several artists, including John Singer Sargent, displayed their work at a National Exhibition of modern painting in 1887. It became the *Biennale d'Arte* in April 1895 when over 200,000 visitors attended. Numerous European countries opened permanent pavilions in the gardens for exhibitions between 1907 and 1914. Many modern design projects were never fulfilled but Venice's modern art and design exhibition became an outlet for the modernist imagination: "Modern traits were possible in the design of buildings that by and large stayed on paper. To accommodate the modern, Venice had the Esposizione d'Arte, an initiative that became known as the Biennale. For the duration of the events Venice appeared to embrace contemporary art" (Plant 2002, 234). Venice also engaged with modern culture through the Cinema Biennale film festival, established on the Lido in 1932, with its now famous Golden Lion award. Similarly, music and theater were added to the Biennale in the 1930s.

In combination with the growing importance of Venice's educational institutions, the Biennale helped central Venice emerge as a site of culture and education, rather than industry. Venice's University at Ca' Foscari was founded in 1829 and grew during the nineteenth and twentieth centuries. The *Accademia di Belle Arti* and various technical craft schools assisted the promotion of Venice as a place of learning. For instance, the *Accademia di Belle Arti* founded a special course of architecture in 1923, which led to the establishment of a school of architecture in 1926, becoming the internationally-renowned Venice University Institute of Architecture (IUAV). The development of modern education, design, art, tourism and the industrial zone affected the expectations of Venetians. A more positive outlook emerged in the early twentieth century, which was reflected in modern architecture:

"These very fundamental changes in the means of livelihood of the inhabitants of Venice were the background against which the architecture of the period evolved. Gradually, the nostalgic, retrospective mood faded, and a more adventurous, forward-looking approach to design set in" (Howard 2005, 282).

A futuristic approach had also been adopted by many supporters of industrial modernization in Venice during the period before the First World War, including Volpi, Piero Foscari, D'Annunzio and Vittorio Cini. As well as being a partner in the Marghera industrial complex, Cini played a leading role with shipping companies based in Venice, the CIGA luxury hotel chain with its headquarters in the city, and companies to supply northern Italy with electricity. Although Cini inspired industrial modernization, he supported the restoration of Venice's heritage. The most significant example was the rejuvenation of the San Giorgio Maggiore Island, its Palladian church, and its monastic complex, after a long period of deterioration.

Despite the balance struck between industrialization and restoration by Venetian industrialists like Cini, they became increasingly unpopular as the twentieth century progressed. The influence of conservationism was one reason for this. Another was that, following the First World War, Venetian industrialists' modernization proposals were compromised by association with fascism (Plant 2002, 281). Marghera's industrial area was strongly identified with local and national fascist politics (Plant 2002, 277). Mussolini ratified contracts for the port's construction in 1924. Volpi became Mussolini's Minister of Finance in 1925. In 1933, Mussolini decided to hold his first meeting with Adolf Hitler in Venice. Volpi had become President of the reconstituted Biennale in 1932, and was famously photographed with Mussolini and Hitler at Venice's Biennale exhibition in the public gardens. The photograph acted as a reminder of the relationship between fascism, modernity, art and architecture. "The arrival of the modern movement in architecture, in Venice as elsewhere in Italy, was linked with the growth of Fascism, which found a potent means of expression in the public buildings of the 1930s," writes Howard (2005, 283). Fascist architecture was evident in the styles of the casino and cinema on the Lido, but the road bridge between Venice and the mainland was the most graphic display of modern development during Mussolini's rule.

Venice's fascist past was partly cleansed during the Second World War by the city's anti-fascist movement, especially 45 days of famous Partisan resistance in 1943. After the War, some modern artists in Venice sought to rectify the pre-war association of modernist art with fascism. In 1945, artists formed the political art movement *Fronte Nuovo delle Arti*, launching a manifesto in 1946. The American Peggy Guggenheim was one of the founders of this movement, along with the famous *avant garde* artist Emilio Vedova. The movement helped to politicize the Biennale. The 1948 Biennale had an anti-fascist theme; in 1968, Vedova withdrew his work from the Biennale and international students protested in Venice against its "authoritarian, racist" character (Plant 2002, 326). The legacy of fascism continued to haunt modernist art in Venice. Nonetheless, modernists who had been critical of fascism were highly regarded during the post-war period. Venetian Carlo Scarpa had disparaged nationalist architecture in a public letter in 1931 (Plant 2002, 307), and became one of Venice's leading modernist architects, although with "too few" buildings (Plant 2002, 349). Scarpa represented the increasing tendency for modernist architects in Venice to concentrate on interiors, thereby avoiding conflicts with conservationist restrictions on exteriors. Scarpa's interior restorations, his designs of his own house and an Olivetti shop on St Mark's Square won admiration, as did his glassmaking. But the limited application of Scarpa's ideas to the exteriors of Venice's buildings illustrated the continued weakness of modernization in the post-war period, when it had been compromised by its association with fascism and restricted by conservationism.

The dominance of conservationism over modernization in the center of the city after the Second World War was demonstrated by the failure to build three significant projects in the 1950s and 1960s. The ambitious student accommodation, hospital and conference center designed by Frank Lloyd Wright, Le Corbusier and Louis Kahn were all victims of conservationist pressures. Conservationism enforced rigid planning limitations: "[P]lanning restrictions prevented the erection of what might well have been the finest example of twentieth-century architecture in the city. In 1953 America's greatest architect, Frank Lloyd Wright, was commissioned to build the so-called Masieri Memorial House, a centre for foreign architectural students on the Grand Canal" (Howard 2005, 286). A compromise dwelling was constructed that preserved the facade of the existing building on the site. It partly collapsed.

Despite the overall dominance of conservationism in Venice after the Second World War, there were rare examples of modern building within the city. In 1954, a new railway station was completed in a modern style that made no concessions to nearby historical buildings. In the same year, the redevelopment of the Rio Nuovo area created space to build a modern office block for the electricity company *Società Adriatica di Elettricità*. Likewise, a new headquarters for the bank *Cassa di Risparmio di Venezia* was designed in 1964 and invaded the historical center at Campo Manin. More recently, a few modern hotels have crept into the peripheral areas of Venice's center, including the Hilton Mulino Stucky and Cipriani's Hotel on the Giudecca. But these were atypical examples of modernization in the center. Defenders of modernization were isolated and subject to hostility in the city by the time the Venice problem emerged after the 1966 high floods. A notable example was Wladimiro Dorigo (1973), author of an historical study on Venice and an active politician from the late 1950s into the 1970s. Dorigo defended the bridge across the lagoon and Marghera's industries, and was sued for defamation by conservationists in Italia Nostra (Dorigo 1973, 55). Nevertheless, modernization proceeded on the city's outskirts with the expansion of the Marghera industrial area during the 1960s. "In the 1960s, a Second Industrial Zone spread southward from Marghera over what until 1953 had been tideland. By 1970, the industries at Marghera – chemicals, petroleum, plastics, etc. – provided nearly 40,000 jobs," notes Lane (1973, 454).

The expansion of the Marghera industrial area occurred as Italy experienced an economic boom in the 1960s and 1970s. Job growth at Marghera helped stimulate the construction of apartment blocks near Venice in Mestre, which usually included modern comforts, such as central heating and plumbing. For many working-class Venetians, Mestre's modern apartments were cheaper and more attractive than ancient Venetian houses that were limited by coal- or wood-burning kitchen stoves. In Mestre, modern bathrooms connected to sewage systems were a considerable improvement on much Venetian housing. According to Lauritzen (1986), even in 1986, "forty-five per cent of houses in Venice today are without a bathroom and their inhabitants have no option but to resort to the public baths. The toilet, behind a screen or curtain in the kitchen, drains through a pipe directly into the canal outside. Fully sixty-five per cent of Venetian drains empty into the city's canals" (Lauritzen 1986, 63).

The lack of modernization in Venetian houses was exaggerated by the city's conservationist laws, some of which had been introduced by Italian fascists interested in historical preservation:

> Although many Venetians were eventually driven away from the city by the inexorable deterioration of their houses, an increasing determination on the part of the authorities to preserve the monuments of the historic centre was also threatening to make Venice uninhabitable. Some of the earliest, serious preservation laws had been enacted under the Fascist regime in an attempt to capture the reflected glory of an otherwise vanished past. (Lauritzen 1986, 66)

According to the 1938 Instructions for the Restoration of Monuments, "[i]n the restoration of monuments and works of art, any completion or *ripristino* or any addition of elements not strictly necessary for the stability, conservation and consolidation of the work is to be ruled out absolutely" (cited in Pertot 2004, 128). A law concerning streets, the water network and sanitary conditions was passed in 1939 and continued to govern detailed regulations through the 1970s (Pertot 2004, 175-6). Law number 1089 passed in 1939 guaranteed protection to buildings of recognized artistic or historic value, even though it could be extended to any building (Pertot 2004, 148). For the 1984 Special Law for Venice, a list of monuments subject to the protection of Law 1089 was required. A government official for Venice (*Soprintendente)*, Margherita Asso, listed all Venetian buildings (Pertot 2004, 185).

Conservationist laws meant that, even in the 1980s, over 95 percent of Venice's buildings were listed historical monuments (Lauritzen 1986, 66). This made it extremely difficult for most householders to modernize their dwellings, especially as laws prohibited alterations to the exterior or the interior of historic monuments. Illegal alterations became common, and many Venetians were driven out of the city: "The difficulties of modernizing, or even improving, a Venetian house are legion and for the poorer working-class Venetian in the 'fifties and 'sixties, they proved overwhelming. So with the offer of an attractive alternative in Mestre's modern housing developments, Venetians began to desert the historic centre," writes Lauritzen (1986, 67). The movement of working-class Venetians from inadequate and protected housing in Venice to modern apartments in Mestre illustrated the marginalization of modernization from the city in the 1960s and 1970s.

During the 1980s and 1990s, there were attempts to rectify the lack of adequate modern housing in Venice. These decades witnessed significant construction, although governed by strict sustainable criteria. In 1979, the use of reinforced concrete was banned in Venice, even though exceptions have since been made. Furthermore, the post-modern character of building was emphasized (Plant 2002, 402-5). Disused industrial structures were converted for accommodation, such as the former factories of Mulino Stucky and the Dreher beer company. Between 1980 and 1986, houses were built to replace former cement works at the western end of the Giudecca, behind the Mulino Stucky. Several housing estates were constructed, mostly without violating Venetian traditions and not using modernist designs that might have clashed with ancient architecture. A waterfront housing complex was built on the northern fringes of the city at San Gerolamo between 1987 and 1990, with traditional Venetian architecture in mind: "Here the references to the local vernacular are blatant: we see inset windows framed by Istrian stone with dark-green shutters and even traditional chimneys, which no longer serve their function as cinder traps since the use of solid fuel is forbidden," explains Howard (2005, 297). A similarly sensitive waterfront housing complex was erected between 1982 and 1989 on the eastern edge of the Sacca Fisola. Although a less traditional housing complex called the Area Saffa was built between 1981 and 1994 in Cannaregio, modernist housing was isolated to the outer islands in the lagoon. For example, between 1979 and 1986, modern dwellings were constructed on the island of Mazzorbo.

This emphasis on marginalizing modern development to the city's outskirts has also been relevant for the mobile dams, which are located at the inlets between the littoral islands that separate the lagoon and the sea. The construction of the dams has therefore been concentrated on these outer islands, including the Lido, which has been a site of modern development since the late nineteenth century. In addition, the building of modern dams on the borders of the lagoon has been justified by its supporters as a way of conserving Venice's ancient center. Andrea Rinaldo, Professor of Hydraulic Constructions at Padua University, points out that the dams will assist "the conservation of the city" (Rinaldo 2001, 80). This is similar to the original case put forward in favor of creating the Marghera industrial area. Although Marghera was an emblem of modernization, it gained some acceptance in the early 1900s by emphasizing the preservation of the city's historical center. "In the 1920s Porto Marghera was established as the new industrial and economic hope of the old city; in the 1970s it was castigated as the cause of pollution and destructive high tides – *acque alte* – in the historic city. Yet in the early twentieth century the scheme was won with the argument that historic Venice would be protected by the removal of industry and the establishment of a separate port," records Plant (2002, 234). But conservationists did not accept development on Venice's borders to help preserve the city. They successfully challenged the expansion of Marghera with a Third Industrial Zone, which was legislated for in 1965 and canceled by the Special Law of 1973. Even modernization on the margins of Venice appears to have been too much.

RECENT RESTORATION

Although the triumph of conservationism in Venice led to serious limitations on development, recent restorations have often not respected conservationist criteria. Restorations of buildings, as well as artwork and monuments, proliferated after international funds poured into Venice in response to appeals following the 1966 high floods. More than fifty organizations joined the Association of Private Committees (APC) to sponsor restorations in the aftermath of 1966. Now the APC includes 24 organizations from 10 countries.

American financial assistance for restoration in Venice has been particularly beneficial and generous. The American fundraising committee the International Fund for Monuments (IFM) helped restore the Scuola Grande di San Giovanni Evangelista after it was identified as in need of attention by the Superintendent of Monuments in 1969. The IFM Los Angeles chapter restored Venice's former cathedral, the church of San Pietro. Work on the restoration of the huge Franciscan church of the Frari was supported by six groups, mainly from the USA. American not-for-profit organizations have helped restore important artwork: many ceiling frescoes have been beautifully restored, such as the Tiepolo ceiling in the church of the Pietà, which was financed by the American Kress Foundation. Since the IFM became Save Venice Inc. in 1971, it has financed many restorations, including Donatello's statue of St John the Baptist. It was estimated that Save Venice funded 60 to 70 percent of US$1 million (€763,125)[3] annual APC collective outlay in the years leading up to 2004 (Davis and Marvin 2004, 221 and 225). Most significantly, Save Venice restored the church of Santa Maria dei Miracoli between 1987 and 1997 at a cost of almost US$4 million (€3,052,457). Substantial American funding has been assisted by the United States government granting an income tax write-off for tax-paying citizens who contribute to American charities. Numerous countries have made significant contributions to restoration in Venice.

Many restoration projects have enjoyed combined international support. The abbey of San Gregorio was restored and opened in 1968 as a laboratory to restore artwork. The American Committee to Rescue Italian Art (CRIA) gave photographic and X-ray equipment for the laboratory. Lighting was provided by German donors. The British Italian Art and Archives Rescue Fund (and its successor, The Venice in Peril Fund) raised large amounts of money for the laboratory and helped arrange for the National Gallery of London to create a scientific department. Contributions for the laboratory were also forthcoming from Italia Nostra and the *Nederlands Comtié Geeisterde Kunstaden Italie.* The San Gregorio laboratory has restored a vast number of paintings, as has the *Accademia delle Belle Arti*, including Tintoretto's 'Crucifixion'.

In addition to international projects, there are many examples of stand-alone restorations supported by organizations from individual countries. The baroque interior of San Stae Church was restored between 1977 and 1978 by the Swiss *Pro Venezia Foundation,* which went on to spend eight years on the church of Santa Maria Del Giglio. The *Comité Francais pour la Sauvegarde de Venise* restored the church of Santa Maria della Salute, notably rescuing the 'falling angels' on one facade. There are also many instances of restoration work supported by Italian organizations, such as work on the Gothic Santo Stefano Church by the *Comitato Italiano per Venezia*. Many critical restorations and regular monument maintenance have been funded through Venice's special laws, regional government, Venice City Council, the state lottery and semi-public Italian bodies.

Although much of this restoration work in Venice has improved ancient buildings, there has been a lot of poor restoration work. In particular, external restoration work has often been based on misconceptions about how air pollution affects building restoration. During the 1960s and 1970s, pollution became a highly politicized issue and restoration work was often driven by political concerns about pollution rather than scientific understanding of restoration techniques. Reports and newspaper articles repeatedly identified the oil refinery at Marghera as causing air pollution that damaged buildings and statues. Similarly, in an interview with the author, environmentalist Cesare Scarpa linked the oil industry with the deterioration of Venetian stonework, speaking about the Marghera industry of "petro-chemicals that in many previous years have destroyed a lot of work, with the emission of colossal fumes in the air that have transformed a lot of stone decorations in the Venetian palaces into chalk" (Scarpa 2005). In fact, Venice's stonework has suffered relatively little pollution from Marghera, with sea winds blowing most fumes towards the mainland and away from Venice. But this has not prevented the identification of Marghera with pollution in the city:

> The petro-chemical refineries were singled out as the villains, despite the fact that Marghera lay a full three and a half miles from the historic centre across open water. In most cities, this three-mile gap would be filled with the pollution from automobile traffic or the oil-fired central heating of houses built on the outskirts. The fact that Venice's prevailing winds, like those of all port cities, are on-shore winds which help keep Marghera's pollutants at bay, was never mentioned in the reports. (Lauritzen 1986, 105)

Sixty percent of air pollution in Venice was previously caused by central heating systems using oil (UNESCO, cited in Lauritzen 1986, 105), until they were supplied by methane gas. This change was made after the city applied clean air restrictions, following the 1966 Clean Air national legislation. Subsequently, "Venice could boast the cleanest air of any city in Europe, though of course this was never mentioned in the foreign press" (Lauritzen 1986, 111). As part of its campaign to save Venice, UNESCO highlighted pollution in a manner that had an impact on restoration.

In the Introduction, we explained how UNESCO led the politicization of the Venice problem. UNESCO's campaign was instigated with a major report on the city's problems, the 1969 *Rapporto su Venezia*, the cover of which sported an illustration of a blackened marble head, which appeared to have been damaged by pollution. The UNESCO report "did feed the fires of the mounting anxiety over living – and breathing – conditions in Venice with charts, graphs, statistics and photographs of deteriorated marble statuary. It neglected to point out, however, that all the calculations were based on the deterioration of the saccharoidal type of Greek or Carrara marble and applied only to a minute percentage of the city's stonework," writes Lauritzen (1986,

105). Long before concerns about pollution, the ancient Venetians understood that corrosive salt sea air and a damp climate meant these types of marble were unsuitable for their statues. So, for most of their statues, the ancient Venetians went to the quarries of Istria across the Adriatic Sea to get white marble-like limestone, which is highly resistant. When parts of the statues built with Istrian stone blackened, mistaken foreign pressure was applied to clean them up:

> The areas of Istrian stone statuary that are not washed white by rainwater will turn black: this is part of the stone's organic nature and not a matter of surface dirt. These black areas also attract a higher concentration of corrosive elements, but the organic blackening serves as a protective skin on the stone. As soon as the Italian authorities realized the implications of this, they became justifiably reluctant to see Istrian stone washed clean until such time as the restorers could also guarantee the stone's future protection. The foreign press constantly represented the caution of the Italian authorities as the negative attitude of obstructionist bureaucrats. (Lauritzen 1986, 108)

These misconceptions and tensions between Italian and foreign opinion were evident for numerous restorations of Istrian stone that had blackened over time. For instance, "[t]his cautious attitude prevailed when one of the many American committees applied to clean the Istrian stone facade of the church of Santa Maria del Giglio" (Lauritzen 1986, 109). Cleaning techniques have advanced and foreign organizations have financed successful restorations: the restoration of the small pavilion known as Loggetta di Sansovino, situated on St Mark's Square, was effectively cleaned up with the support of The Venice in Peril Fund. This organization has made many positive contributions to restoration work in the city. One commendable example is the restoration of the Madonna dell'Orto Church, completed by the British Italian Art and Archives Rescue Fund. Pertot (2004) is positive about recent restorations conducted by private committees like The Venice in Peril Fund, although he maintains that most building works in the city are blind to conservational issues (Pertot 2004, 218-9). "For a long time conservation was merely proposed, urged, but not applied; now it has begun to be applied systematically, though still within a limited sphere," concludes Pertot with regard to conservation in architecture (Pertot 2004, 198).

Other reviews of recent restoration work in Venice have found conservationism pervasive, while advising that it be further strengthened. This was the recommendation of artistic scholar Wolfgang Wolters (2010) in the introduction to a book about restoration in Venice. Wolters referred to research by Francesco Trovò (2010) of IUAV revealing that the debate about transformation and conservation continued to influence restoration projects between

1984 and 2001 in the city. Data analyzed between 2000 and 2008 indicates that there have been wide-ranging changes to the use of Venetian properties (Gibin and Tonin 2009, 62-66). The growth in the number of properties in Venice during this period has been predominantly facilitated by internally splitting up existing units into smaller ones, rather than by new building. Wolters (2010) writes that doors, windows and architectural elements on facades remain immune from substantial changes, and criticizes the antiquated legal definition of historical monuments: "The concept of 'monument' used by the legislator is based on criteria and inventories that are now out of date and do not reflect the reality of an urban fabric such as that of Venice" (Wolters 2010). Wolters welcomes the application of conservation criteria for restorations of non-'monumental' houses and is disappointed by the failure to define all of Venice as a protected historical zone: "The proposal made some time ago that the whole of Venice should be statutorily designated an area of special architectural and historic interest and that its status should be used to draw up a framework of intelligent regulations and structures to guide all action to restore the habitat unfortunately fell on deaf ears" (Wolters 2010).

In theory, conservationism dominates legal definitions of restoration in Venice. These definitions build on the nineteenth century redefinition of restoration as conservation, not transformation. In practice, recent restoration projects in the city center have adopted both transformative and conservationist approaches, but the city center has remained largely free of modern constructions. This triumph of conservationism in Venice has mostly been due to local circumstances, although national trends have also impacted upon the city.

THE GROWTH OF CONSERVATIONISM IN ITALY

National preservationist associations and legislation helped shape the evolution of conservationism in Venice. We have observed how many conservationist pressures on Venice came from foreign sources. Likewise, Italian conservationist organizations and laws were conditioned by outside influences, especially in the late nineteenth and early twentieth centuries. The initial environmental associations established in Italy concentrated on educating the public about the country's natural beauty and introducing conservation policies. Touring Club Italiano was one of the first organizations, started in 1894. It remains popular today, although largely for recreational purposes. The origins of the Touring Club Italiano and the mountaineering society formed in 1863, Club Alpino Italiano, reflected the expansion of leisure pursuits. The British scholars Philip Lowe and Jane Goyder (1983) describe how the timing and characteristics of the first Italian environmental groups closely resembled those in Britain, Germany and the USA. The inclusion of English words 'Touring' and 'Club' in the first environmental associations betrays the roles of British climbers in early Italian mountaineering.

Similar associations were founded towards the end of the nineteenth century and in the early twentieth century. The Association for Mountains and Forests was formed in 1898, the National Association for Picturesque Landscapes and Monuments of Italy in 1913 and the National League for the Protection of Natural Monuments in 1914. As their names suggest, these organizations campaigned to conserve natural and man-made features of the environment. Pressure from these associations was partly the reason for the 1939 government act on conservation, which protected key architectural and natural monuments: according to the Italian scholar Raimondo Strassoldo (1993), Italy's act was based on English and German examples. Moreover, the idea that the state should protect the landscape, enshrined in the 1947 Italian Constitution, was borrowed from the Weimar Constitution of 1919. These foreign influences over Italian conservationism were consequences of the weak and unstable character of the Italian elite.

The Italian elite experienced considerable instability after the nation's formation in 1861, fascism, and the reconstitution of the Italian Republic following national collapse during World War Two. This affected the coherence of conservationist groups, which were largely made up of members of the elite. For instance, the Italian sociologist Mario Diani (1995, 20) notes the elitist composition of the National League for the Protection of Natural Monuments. When conservationist associations first emerged in Italy, the Italian elite was still in the initial stages of developing a coherent identity and was vulnerable to foreign influences; by the early twentieth century, the formation of new conservationist organizations suggested a more confident assertion of Italian national identity.

It is noticeable that many of the names of the early twentieth century Italian conservationist groups contained the word 'national' and they were completely Italian (the above names are translations). This indicates a self-conscious attempt within the Italian elite to distance itself from foreign guidance. The rise of fascism, with Mussolini becoming Prime Minister in 1922, led to a more aggressive form of Italian nationalism (Duggan 1994, 233-9). Following the defeat of fascism, the reconstitution of the Italian elite after the Second World War included reasserting Italian national identity. The name of one of the most important environmental organizations in Italy and Venice reflects this trend: *Italia Nostra*, translated as Our Italy, was formed in 1955. Its name projected a sense of independence for the new post-World War Two Italian Republic following subordination to Germany and then American-led liberation.

Strassoldo (1993) refers to Italia Nostra as "the first really indigenous, important *Italian* conservationist *social* movement" (Strassoldo 1993, 458).

Despite efforts to establish an Italian identity within conservationist organizations, foreign connections continued to affect Italian conservationism in the post-Second World War period. During the mid-1950s, an Italian branch of the International Union for the Conservation of Nature and Natural Resources (IUCN) was founded at the highest cultural-scientific level (the National Research Council). This acted as a channel for conservationist ideas from Anglo-American culture into the Italian elite (Strassoldo 1993). Foreign input into Italian conservationism became more significant with the opening of the Italian chapter of the WWF in 1966. The Italian National Bird Association (LIPU) was created in 1964 and, like the WWF, was predominantly induced by the English community in Florence (Strassoldo 1993).

The growth of conservationism in Italy is usually explained as a reaction to industrialization and modernization. "At the basis of this view lay notions of defence and preservation. Environmental action was mainly read as a reaction to the innumerable threats which industrialisation, urbanisation and human misconduct posed to places of natural beauty," writes Diani (1995, 22). Following economic improvement during the 1950s and 1960s in Italy, the Italian scholar Antonio Cederna (1975) characterized environmental action up until the late 1960s as tiny cliques of intellectuals concerned with the degradation of nature. "The birth of *Italia Nostra* in 1955 was the first public reaction to the destruction wreaked by the economic development in the aftermath of World War II," declare the scholars Angela Liberatore and Rudolf Lewanski (1990, 14). Italia Nostra's most prominent campaigns between 1955 and 1975 concentrated on cities famous for their art and architecture, like Rome, Florence and, notably, Venice. In Rome during the 1950s intellectuals who helped form Italia Nostra campaigned against what they believed was the "second sack of Rome" by construction companies rebuilding substantial parts of the city's center (Diani 1995, 21). This conservationist opposition to development often came into conflict with popular enthusiasm for modernization: "The large majority of the Italian people and politicians considered quick economic growth as the only way to overcome the historical backwardness of the country" (Diani 1995, 20-1). As in Venice, conservationism increasingly acted as a brake on modernization.

The rise of conservationism in Italy during the late nineteenth and twentieth centuries undoubtedly had an impact in Venice. The creation of national conservationist organizations and legislation influenced the evolution of environmentalism within the city. This was best illustrated by the 1959 opening of the Venice chapter of Italia Nostra, which became the most prominent green group and led the initial campaign against the mobile dams. In addition, Venice had to abide by national legislation such as the 1939 act on conservation. The tension between conservationism and development was also fraught in Rome, where emblems of modernization like the car and subway trains invaded the Roman city center and, by 2010, Rome's first skyscraper, the 120-meter (373.7-foot) high Eurosky Tower, was under construction. However, Venice was subject to stronger and earlier foreign conservationist pressure than was evident nationally: it became more of a focus for conservationism than other Italian cities, and represents a clearer expression of the triumph of conservationism.

FROM DEVELOPMENT TO CONSERVATION

Before the fall of the Venetian Republic, Venetians typically built and remade their city through demolition, rebuilding and intervention to curb nature's impact and to develop. The ancient Venetians believed they had achieved a privileged state of development and transformation was considered more important than conservation. Although there was scant development during the eighteenth-century decay of the Venetian Republic, modernization was a priority when France, Austria and Italy controlled Venice in the nineteenth century. Some notable advances included street lighting, the cemetery island, bridges, new streets, steam boats, the maritime station, port development, the railway and factories. Yet periods of economic stagnation and the increasing influence of conservationists placed limits on modernization.

Conservationist campaigns grew in strength during the nineteenth century, bolstered by foreign support. Conservationists succeeded in redefining restoration so it no longer meant making old buildings and monuments look new. By the end of the nineteenth century, restoration was synonymous with conserving the past. The rebuilding of the collapsed bell tower on St Mark's Square according to the principle of 'as it was, where it was' confirmed the redefinition of restoration as conservation, rather than transformation, and this has subsequently become the guiding principle for restoration in Venice.

Modernization in the period before the First World War was less rapid than it had been in the early and mid-nineteenth century, as Pemble (1996) records: "Venice altered during the thirty years before the First World War – but not beyond recognition. The rate of change was probably slower than it had been under the Austrians, and was certainly slower than it had been under the French. Venice underwent nothing comparable to the *sventramento* (evisceration) of Rome and Milan" (Pemble 1996, 178). So modernizers escalated pressure for development, and political movements were launched. Modernization gained momentum as industrialists and fascist politicians helped establish the Marghera industrial area and the road bridge. The Italian Communist Party also welcomed modernization, especially at Marghera where it led to new opportunities for labor unions and workers' employment. But modernization in Venice was largely compromised by its strong relationship with fascism, and central Venice specialized in culture, tourist services and education rather than industry.

As the twentieth century progressed, modernization was marginalized to the Lido, Marghera and the mainland. Development on the outskirts was justified according to the need to conserve Venice's historical center: and of course, preserving and restoring ancient buildings, monuments and artwork are vital endeavors. Unfortunately, conservationists have also prevented the construction of modern housing for working-class families, a new hospital and student accommodation in the city; and conservationist legislation, with over 95 percent of Venice's buildings listed as historical monuments, has become a burden on the city.

Although the rise of conservationism in Venice and Italy has typically been explained as a straightforward reaction to local industrialization, we have observed how foreign influences assisted the flourishing of conservationist consciousness. This trend occurred earlier and was more advanced in Venice than in other Italian cities. Venice achieved a mythical status as a retreat from modernity in the minds of foreigners, especially for nineteenth century Anglo-Saxons who felt estranged from modern change and sought refuge in culture. "Generations of Britons and Americans hungry for culture made the city their Jerusalem," writes Pemble (1996, 10): compared with other leading Italian cities, "Venice was better able than they to match the idea of a precious residue, refined by and set apart from the turbid flow of history" (Pemble 1996, 10).

During the nineteenth and most of the twentieth centuries, conservationism focused on the negative aesthetic impact of development, and too little consideration was given to improving the lives of people and their societies. "From a conservationist perspective, the relationship between mankind and natural resources/historical tradition was mostly framed in aesthetic and ethical terms, with little attention to its social and political implications," explains Diani (1995, 22). This changed in the 1960s, when conservationism was politicized and 'green' political parties were formed in many European countries. Conservationism was transformed into environmentalism, which involved the application of green thinking to an array of social and political issues.

NOTES

1. Gibin and Tonin (2009, 49) define the city center as the zones within Venice 1: San Marco, Castello, Sant'Elena, Cannaregio and Venice 2: Dorsoduro, Santa Croce, San Polo, Giudecca, Sacca Fisola. The central areas can be viewed in Figure 1.2.
2. This discussion is expanded in Chapter Seven.
3. All currency conversions were correct on 21 December 2010.

Chapter Four

Transforming Conservationism into Environmentalism

> "The drift from old-style conservationism reflects the development of scientific understanding of ecology at least as much as contact, collaboration, or competition with newer, more radical EMOs[1]...In 1980, the WWF expanded its agenda to embrace development issues and first introduced the term *sustainable development*...As a direct result of the UN Conference on Environment and Development (UNCED) process and the 1992 Rio Earth Summit, the WWF widened its ambit to work with other NGOs to form a common agenda on development and environment."
>
> Christopher Rootes (2005, 30. Emphasis in original), Reader in political sociology and environmental politics, University of Kent, UK.

The changes experienced by the WWF illustrate the broader transformation of conservationism into environmentalism. Although Rootes's analysis focuses on the British environmental movement, he notes that the WWF is a transnational organization (Rootes 2005, 39). The adjustments made at an international level by the WWF affected its practice in countries where the organization is active. In the previous chapter, we noted that the WWF was one of the first transnational conservationist organizations to become established in Italy; here, we discuss how the WWF, along with other conservationist associations, modified its campaigns in Italy to embrace a broader environmental agenda.

After the radical uprisings of the late 1960s and early 1970s, conservationist issues were redefined due to the impact of new radical environmentalists and the international emergence of the sustainable development agenda. Distinctions between man and nature eroded as a wider range of issues were incorporated into a holistic definition of environmentalism. This process is crucial to the evolution of environmental campaigns in Venice.

THE BIRTH OF RADICAL ENVIRONMENTALISM

During the late 1960s and early 1970s, radical environmentalists emerged from the civil rights movements in the USA and from social movement uprisings in Europe. By the 1980s, these actors had significantly altered green campaigning in Italy and elsewhere. In the USA, the publication of Rachel Carson's 1966 book *Silent Spring*, and the writings of reputable scientists such as Barry Commoner (1971) and Paul Ehrlich (1971[1968]) drew attention to environmental problems. Anna Bramwell (1989), the author of several books on environmental history, documents how Berkeley students, who were disappointed by the failure of the 1968 student movement, found environmental issues attractive (Bramwell 1989, 225-6). Students, feminists and other activists were captivated by books written by the radicals Murray Bookchin (1963) and Ivan Illich (1971), as well as Charles Reich's *The Greening of America* (1971). Environmental utopia novels also had a meaningful impression on American activists, including *Ecotopia* (1975) by Ernest Callenbach and Edward Abby's *The Monkey-Wrench Gang* (1975).

Radical social movements had major repercussions for many European societies after 1968. Formidable Italian social movements lasted well into the 1970s and involved the working class, students, and women. Environmental issues were taken up by new left groups and labor unions with a perspective that was very different from the conservationist approach that we analyzed in the previous chapter. Antimo Farro (1990), an Italian scholar of social movements, describes how protests developed against industrial pollution in some working class areas in northern Italy and became part of the political agenda for workers' and students' movements. Shifting from left-wing concerns with factory conditions to environmental community issues was not a huge jump for radical activists. "One by-product of the 1960s' leftist student and worker movements was criticism of the working conditions in factories and their effects on workers' health. The movements' focus quickly shifted to the poor living conditions caused by industrial pollution in the areas surrounding the factories and, subsequently, to urban areas in general," explain Liberatore and Lewanski (1990, 14). It was through the issues of health in the workplace and urban housing that environmentalism began to establish a foothold within the workers' movement. Yet environmentalism was limited in its penetration of the dominant political cultures in Italy at this stage: "The two political cultures dominant in the country, Catholic and Marxist, had almost no place for naturalist musings and showed little inclination to succumb to the environmental motif. What came closest was the concern for the unhealthiness of the work environment, or the lack of adequate green space in working class residential districts," writes Strassoldo (1993, 459).

Over time, new subcultures evolved to supplement the well-founded 'red' (Marxist) and 'white' (Catholic/Christian democratic) political cultures that dominated during the 1950s and 1960s (Poggi 1968). Diani (1995, 24) suggests that environmental consciousness spread through new "countercultures" of neo-religious groups, youth subcultures, anarchist currents, and alternative food, medicine and yoga lifestyles, which developed the earlier romanticism of conservationism. "[T]he 'counter-culture' of the period was unmistakably imbued with romantic motives (naturalism, ruralism, orientalism)," notes Strassoldo (1993, 459), referring to the 1960s. Conservationist romanticism helped stimulate environmental consciousness in Italian subcultural groups. But these groups were insufficiently developed in Italy during the 1970s to break the hold of the dominant red and white political cultures. Diani (1995, 24) notes that new Italian subcultures were never consistent enough to build up an "alternative" sector, as in Germany. Environmental consciousness and protests were a minor part of the alternative movement of the late 1960s and early 1970s. Sidney Tarrow (1989), Professor of Sociology at Cornell University, conducted a detailed survey of 5,000 Italian protest events identified by newspapers between 1966 and 1973: only 50 dealt with environmental issues (Stefanizzi 1987).

Farro's (1990) analysis confirms Diani's (1995) assessment that environmental campaigns during the 1970s indicated that a movement had not yet been formed. The absence of an assertive environmental movement was illustrated by the way that green policies were largely absent from the agendas of the major political parties in the 1960s and 1970s. The Italian elite was too preoccupied with problems such as terrorism, which threatened its immediate existence, to take environmental issues seriously. The emergence of radical environmentalists during the late 1960s and 1970s altered the landscape of environmental campaigning, as radical environmentalists highlighted green concerns about health in the workplace and urban housing and Italian conservationists made connections with radical environmentalists through these new issues. A significant issue was that of nuclear power: in their campaigns against nuclear power plants in 1974, conservationist associations Italia Nostra and the WWF formed links with the radical groups *Democrazia Proletaria* (Democratic Proletarian) and *Avanguardia Operaia* (Workers' Vanguard). These links altered the conservationist campaigning tradition of lobbying politicians for change. Conservationists began to experiment with mobilizing the public, "thus detaching at least partially conservation from classical styles of pressure politics and rendering it more similar to radical examples of collective action" (Diani 1995, 3). Radicals brought changes to environmental campaigning; however, overall cooperation between conservationists and radical environmentalists during the 1970s was minimal. Diani (1995, 103) notes that radical political ecology groups were more likely to work with non-environmental organizations than conservationists during the 1970s. One of the major reasons for the lack of cooperation between conservationists and radical environmentalists was divergence in their framing of environmental issues.

THE RADICAL AND CONSERVATIONIST FRAMING OF ENVIRONMENTALISM

In the 1970s, Italian radical environmentalists and conservationists framed environmental problems very differently. Conservationists in Italia Nostra, the WWF and other associations highlighted threats to historical buildings, artistic treasures, animals and places of natural beauty. The radical interpretation of environmental problems paid little attention to these issues, focusing on urban, industrial and social environmental questions. Radical environmentalists and conservationists also differed in their framing of problems. Conservationists tended to stress the defense of nature as an individual ethical question rather than emphasizing its relationship to wider society (Diani 1988, 47). By contrast, in the radical framing of the environment, "the main aim shifts from defence to the transformation of the structural element responsible for the environmental decay" (Diani 1988, 58). Industry in general and capitalism in particular were identified as causing environmental problems.

Presenting environmental issues using anti-capitalist vocabulary created a partial alignment between radical environmentalism and the Marxist master frame[2], which, alongside Catholicism/Christian democracy, was the dominant political frame of reference in Italy during the 1970s. For a specific interpretive frame presented by claims-makers to have a broader political impact on society, there needs to be an alignment with the dominant interpretive frame of the era; and the Marxist master frame had the greatest influence on how counterculture political groups framed issues in 1970s Italy (Della Porta and Diani 1999, 77-80). Thus for radical environmentalists, "[t]he capitalist mode of production was held responsible for the unrestrained waste of natural resources" (Diani 1995, 25). The differences between the radical and conservationist environmental critiques led to divergent strategies: conservationists concentrated on education and the rationalization of individual and institutional behavior, while radical environmentalists organized political protests for a drastic transformation in the distribution of social power. In contrast to conservationists, the aim of radical environmental action was to create "a decentralised, self-governed society, where small social and economic production units could be run directly by citizens and workers. To this purpose, a much greater emphasis was placed on explicit political action, both conventional and unconventional, including confrontational tactics" (Diani 1995, 26).

By the mid-1970s, radical environmentalism had developed into a significant political force and came into conflict with conservationism. This was illustrated by a campaign in July 1976 that began after dioxin gas escaped from a chemical factory producing pesticides and herbicides in the town of Seveso, 32 kilometers (20 miles) from Milan. Approximately 600 people in Seveso and nearby small towns and villages were evacuated. About 2,000 people were treated for dioxin poisoning and many people complained of feeling ill. Although no human deaths were directly attributed to the accident, many farm animals and pets died or were slaughtered. Radical environmentalists, mainly from Milan, played an active role alongside several new left groups in setting up a local committee aiming to exercise some control over the local authorities: this committee, as the Italian scholar Giovanni Lodi (1988) describes, proved unable to cope with the moderate approach of conventional ecological associations.

The tensions between conservationists and radical environmentalists were due to drawing upon very different political traditions. The conservationist perspective was based on traditions of protecting buildings, monuments, animals and landscapes, as discussed in Chapter Three. The radical environmental approach, by contrast, was heavily conditioned by the political conflicts of the late 1960s and 1970s. While the Marxist master frame continued to dominate countercultural political organizations, it was extremely difficult for conservationists to create shared understandings with radicals of environmental problems. Moreover, while radical environmentalism drew upon the Marxist master frame, it was based on substantial criticisms of the Marxist traditions that had developed in Italy in the early twentieth century. The radical critique of the environment castigated leftist groups that were in favor of industrial growth for the benefit of the working class: "criticism was . . . directed at the pro-industrial view adopted by traditional workers' movements, who shared with their capitalist opponents, an emphasis on the purely quantitative concept of development," explains Diani (1995, 25). The culmination of the radical environmentalists' criticisms of Marxism was the formation of Legambiente in 1980, founded as a branch of the large leftist cultural and recreational association ARCI, which had many links with the Italian Communist Party (PCI). Legambiente took up positions on several issues that conflicted with the PCI, especially on nuclear energy and restrictions on hunting. In 1986, Legambiente detached from ARCI to become a formally independent organization, although many links remained.

Numerous environmental campaigns were initiated by radical environmentalists who distanced themselves from Marxism. The Radical Party (RP), the non-Marxist heir of the 1960s social movements, became the first political party to take green issues seriously. The RP leader, Marco Pannella, stressed that animals and nature were in need of liberation. In 1972, the RP created environmentalist flanking organizations, including the Italian chapter of Friends of the Earth (FoE), the League against Hunting (LIC), and the League against Vivisection (LAV). The link between the RP and FoE was later suppressed. Diani's (1995) research found it probable that the RP helped steer conservation groups in more radical and political directions (Diani 1995, 25).

Gradually, the framing of environmentalism by those in radical and conservation groups began to converge. Both conservationists and radical environmentalists had experienced considerable difficulties in aligning their framing of the environment with the dominant master frames, yet their ideas were insufficiently developed to become alternative political frames of reference. Similarities between the conservationist and radical framing of the environment largely grew due to changes that took place within conservationism. Environmental mobilizations after the 1960s underwent a general shift from rural and natural ecology towards urban issues (Diani 1995, 112), which stimulated greater conservationist interest in the urban questions that concerned radicals. In addition, conservation groups began to get involved with more explicit political issues and those that were considered left-wing (Diani 1995, 25). The best example of this convergence was the WWF's adoption of the anti-nuclear issue in 1974. Nuclear power was also a prime target for left-wing organizations, as it fitted their anti-military agendas (Strassoldo 1993), and through the anti-nuclear cause radical and conservationist environmentalists came together in an unprecedented manner.

THE CONVERGENCE OF RADICAL AND CONSERVATIONIST ENVIRONMENTALISM

Following the 1973 international 'oil shock', Italy needed to secure new energy sources due to limited natural resources and an over dependency on oil imports. The 1975 national energy plan envisaged expanding the number of nuclear power plants from one functioning plant at Caorso to 20. In the first phase of the anti-nuclear campaign, radical environmentalists and conservationists created some shared assumptions about the consequences of nuclear expansion, especially the growth of social control, the centralization of power, and the possible militarization of surrounding areas (Diani 1995, 31). However by 1981, this first phase of anti-nuclear protests died down, when the Constitutional Court ruled that a referendum proposed by Friends of the Earth three years previously to prevent new nuclear facilities was incompatible with national law and international agreements.

An upturn in anti-nuclear protests occurred due to two developments. Following a government decision to host cruise missiles in 1983, there was a shift in focus to the military rather than the civilian use of nuclear power. Public opinion also swung against nuclear power after the Chernobyl nuclear accident in Ukraine on 26 April 1986. During November 1987, three referenda against nuclear power were held, and the anti-nuclear campaign won all of them. Existing nuclear plants were made dormant, the construction of new ones stopped, and the whole nuclear program was put on hold. This second phase of the anti-nuclear campaign brought significant changes to how Italian environmentalism was framed.

As the Marxist master frame was waning, radical environmentalists in Legambiente were beginning to frame the nuclear issue without anti-capitalist rhetoric. This narrowed the differences between the radical and conservationist framing of the nuclear issue (Diani 1995, 38). For conservationist organizations, the anti-nuclear movement provided an opportunity to expand the domain of issues that they campaigned about beyond their traditional focus on protecting landscape, buildings and wildlife. Domain expansion can prove to be fundamental for the development of a social movement organization: "Social movements cannot afford to succeed and then relax; once members believe a cause is won, they may drift away. Domain expansion and elaboration let movement leaders argue that the battle continues, that work remains to be done, and that the cause needs continued support," explains Best (1999, 169). For domain expansion to be successful, it is usually necessary for a movement to convince new audiences that issues that concern them are related to the movement's existing interests. Snow and colleagues (1986) refer to this process as "frame extension," whereby a movement attempts to "enlarge its adherent pool by portraying its objectives or activities as attending to or being congruent with the values or interests of potential adherents" (Snow et al. 1986, 472). New audiences will not be convinced when an organization is seen to be 'jumping on the bandwagon' by adding a popular social problem to its campaigns if the issue appears unrelated to the organization's core values.

In the case of the WWF, frame extension was effective. The theme of protecting nature proved flexible, in extending from wildlife to territorial planning and protection from nuclear plants. Nationally, Italia Nostra found it more difficult to expand its domain of campaigns from its conservationist roots. Even though Italia Nostra grew through its involvement in the anti-nuclear campaign, its rate of growth was more moderate than that of WWF Italy. Diani (1995, 43) records that Italia Nostra did not expand its domain definition of environmental problems during the 1980s beyond the nuclear issue and its traditional focus on conserving artistic and architectural treasures, and instead of mass pressure activities, it concentrated on education and lobbying. However, the anti-nuclear campaign did mean that Italia Nostra began to address issues outside the traditional conservationist domain.

Domain expansion by conservationist associations and the toning down of anti-capitalist rhetoric by radicals assisted greater cooperation between environmentalists. Yet the convergence of radical environmentalism and conservationism through the anti-nuclear campaign was also facilitated by creating a broadly-defined frame of reference. Radicals and environmentalists were brought together under what Lodi (1984, 138-150) describes as a frame of peace and natural equilibrium. Equilibrium became a key concept in Italian environmentalism that was especially influential in Venice, as we explore in other chapters. The differences between radical and conservationist environmentalists were not completely dissolved, but the anti-nuclear campaign nonetheless signaled the arrival of environmentalism at the center of Italian political debate. Under this broadly framed campaign, radical environmentalists and conservationists united to create an environmental social movement that set the national political agenda for the first time.

THE ENVIRONMENT MOVES TO THE CENTER OF ITALIAN POLITICS

The anti-nuclear movement of the 1980s established environmentalism on the national political agenda. This was assisted by the emergence of various environmental issues that we consider in subsequent chapters, including pollution and global warming. Wider frame realignments in Italian politics also created a space for the environmental movement to have an impact. As Della Porta and Diani explain, "A *realignment frame* combines the perception of good opportunities for independent action with the decreasing capacity of traditional alignments to support collective identities and structure political action" (Della Porta and Diani 1999, 81. Emphasis in original).

Realignments in Italian politics occurred after conflicts from the 1970s failed to be resolved. The combination of an elite that was not strong enough to defeat the workers' movement, and the inability of that movement to mount a decisive challenge to the elite, resulted in a protracted stalemate. This was illustrated by the 'historic compromise' that took place between 1976 and 1979, when the PCI agreed to back the Christian Democrat-led government.

The 1980s witnessed a gradual erosion of support for the political parties associated with the ideologies of Christian democracy and Marxism: "The percentage of the electorate voting for the two main parties declined from more than 66 percent in 1975 to around 45 percent in 1990," notes Patrick McCarthy (1997), Professor of European Studies at the John Hopkins University Bologna Center (McCarthy 1997, 132). The collapse of communism in Eastern Europe in 1989 revealed the ideological bankruptcy of political parties connected with Marxist traditions. Christian Democrats also suffered, having relied heavily on the negative threat of Marxism, rather than on a positive political program, for their public support. In the early 1990s, these weaknesses led to the implosion of Italian political parties directly linked to the ideologies of Christian democracy and Marxism.

Environmental groups were able to take advantage of these frame realignments when they united as a social movement through the anti-nuclear issue. The existence of such political opportunities is an important component in what Tarrow (1994) identifies as a "cycle of protest", through which "[p]olitical opportunities are both seized and expanded by social movements, turned into collective action and sustained by mobilizing structures and cultural frames" (Tarrow 1994, 7). The most important changes in opportunity structure identified by Tarrow were the opening up of access to participation, shifts in ruling alignments, the availability of influential allies and cleavages within and among elites (Tarrow 1994, 86). As Tarrow outlines in an earlier work (Tarrow 1989), all these elements helped shape the opportunities for the environmental movement during the 1980s. The establishment of the Greens as a new player in Italian politics illustrated both the existence of new political alignments and the way in which the environmental movement grasped this opportunity. The Green Lists, the Green candidates put up for election, were formed by activists in provincial and local council environmental campaigns. These activists were mainly drawn from the WWF, FoE, and Legambiente, combined with some smaller associations. For the European elections in 1989, Green factions reached 6.2 percent of the Italian national vote: following which, in 1991, the groups united to form one party.

Despite the expanding political influence of the green movement during the 1980s, the movement's social base remained limited. "[O]ne may safely conclude that a distinctive feature of the ecology movement in Italy in the 1980s was to be (a) its lack of capacity to provide a broad potential constituency with effective incentives for action, and (b) its failure to define and articulate issues sharply enough to set a major political challenge," writes Diani (1995, 34). These weaknesses partly explain why the Green share of the vote fell to 2.7 percent in the 1994 general elections and to 2.5 percent in the 1996 general elections. The green movement also suffered from a general climate of political passivity in Italy during the 1990s, as compared to the 1980s (Biorcio 1992). This passivity particularly affected political parties that were deeply rooted in social movements. Diani (1995) notes that members of environmental organizations between 1985 and 1990 became increasingly involved in non-political activities, such as cultural and leisure associations, sports clubs, charities and voluntary groups (Diani 1995, 165). After the end of the anti-nuclear campaign, there was less emphasis on collective action and people began to relate to environmental associations more passively and individualistically (Diani 1995, 76), for example by focusing on personal change in the form of vegetarianism and green shopping. WWF and Greenpeace, the organizations that were growing most, had a strong emphasis on animal protection, to which people often related in a highly personal manner, and even the radical environmentalists in Legambiente found a new interest in animal rights (Diani 1995, 185). During the 1990s, the ideological and political origins of radical environmental organizations became less influential in the issues upon which they concentrated, and conservationist associations became more focused on institutional change than mobilizing the public (Lewanski 1997, 147).

However, despite greater passivity and declining electoral support for the Green Party, the environmental movement penetrated Italian political institutions as never before. The Green Party was propelled into national government for the first time in 1996. Although the Greens' share of the national vote fell, their inclusion in Romano Prodi's Olive Tree coalition secured their participation in Prodi's government between 1996 and 1998. The Green Party gained the Environment Ministry in the Prodi government and maintained control of this Ministry in governments until 2001. This enabled the Green Party to block Venice's mobile dam project between 1996 and 2001 and impact upon its evolution in subsequent years. The inclusion of the Green Party in government elevated the debate about Venice and its environment in national politics. By blocking the mobile dam project for five years, the Green Party, as well as environmental organizations working closely with it, demonstrated how environmentalism had become central to Italian political life. It was also an indication that the Green Party and environmental organizations had managed to seize the opportunities presented by political realignments and participate in institutional decision-making. This partial institutionalization of the environmental movement was assisted by the adoption of sustainable development policies by many environmental organizations, governments and international bodies from the late 1980s.

SUSTAINABLE DEVELOPMENT

Conservationism was changed by the sustainable development agenda in three ways. The adoption of sustainable policies by many conservationist organizations and institutions brought greater cooperation between them. Conservationist associations developed a more political orientation, partly due to their greater involvement with political institutions. Finally, conservationists expanded the domain of their campaigns through addressing broader social and economic questions raised by sustainability.

i) Partial Institutionalization

Environmental organizations, including those that have campaigned against the mobile dams in Venice, assisted in the formation of the sustainability agenda. Referring to the WWF and Greenpeace, environmental social movement scholar Andrew Jamison (1998) remarks:

> It is largely their conceptions of global environmental problems that have influenced the programmes of the United Nations agencies and the World Bank, and it is their conceptions, as well, that have helped set environmental policy agendas in many developing countries. In this latter respect, the active collaboration between the WWF and the IUCN[3], which led, in 1980, to the formulation of the World Conservation Strategy, with its articulation of the idea of 'sustainable development', has been an important force. (Jamison 1998, 232)

As the scholar Wolfgang Sachs (1999, 77-78) explains, sustainable development differs widely in terms of the interpretation of its content and strategies. The best-known definition originated with Gro Brundtland, who wrote the 1987 report *Our Common Future* for the World Commission on Environment and Development: "Sustainable development is development that meets the needs of the present without compromising the ability of future generations to meet their own needs" (Brundtland 1987, 43). The Brundtland Report became a point of reference for institutions and organizations regarding sustainable development policies. Yet in *The International Handbook of Environmental Sociology* (1997), British scholars Michael Redclift and Graham Woodgate question the cultural assumptions made in the Brundtland Report about definitions of "needs" and "development" (Redclift and Woodgate 1997, 56). Assumptions in the Brundtland Report definition, plus ambiguity about what exactly needs to be sustained, led to contradictory interpretations of sustainable development by people from different backgrounds (Redclift and Woodgate 1997, 56). This is despite the fact that the Brundtland Report was drawn up by a combination of representatives from governments and non-governmental organizations.

Notably, the ambiguous definition of sustainability has assisted its flexible deployment in claims for and against the mobile dams in Venice. Venetian environmentalists have rejected the dams for lacking sustainability, while Stavros Dimas, the former European Environment Commissioner, in 2009 exerted pressure for the dams to be built and operated according to sustainable criteria (*Il Gazzettino di Venezia,* 15 April 2009).

The agenda of sustainable development has helped to bring institutions and environmental activists together. Sachs (1999) describes how state representatives have tended to develop institutional forms of sustainable development management and accounting, while representatives of civic culture have sought to mobilize knowledge of sustainable development in policy-making. Where the sustainable development agenda of state officials and activists in environmental organizations have met, this has provided a powerful force towards the institutionalization of environmentalism. In Italy, extensive environmental legislation was enacted by regional and national administrations in the 1980s (Liberatore and Lewanski 1990; Strassoldo 1993). Lecturer Martin Rhodes (1995, 168) concludes that environmentalism had a wide-ranging political impact in Italy in this decade. Then the weak legitimacy of Italian state institutions in the early 1990s made environmentalism appealing as a means for the elite to forge some new points of public contact (Diani 1995, 28; Melucci 1996, 164). The European Community exerted further pressure on the Italian state to incorporate environmental policies. The increasing institutionalization of environmentalism in the 1990s fulfilled a prophecy by the late sociologist Alberto Melucci (1996, 282) that environmentalism could act as a resource in attempts to modernize Italian institutions.

Internationally, the institutional adoption of environmentalism since the late 1990s warrants its establishment as a master frame in its own right (Eder 1998, 206-7), comparable with master frames that were influential in the past, as "capitalism", "communism", "national identity", and "human rights". The sociologist Maarten Hajer (1998) convincingly demonstrates that international institutions have developed an integrated approach to nature and society, noting the World Bank's "ecomodernist stand" (Hajer 1998, 252). Adopting environmentalism as a guide to policy has provided elites with direction and greater legitimacy (Eder 1998, 205). Environmental sociologist Bronislaw Szerszynski (1998, 105) explains how environmentalism has been transformed from a countercultural to a mainstream discourse. This transition indicates that the environmental movement has played a role similar to many past social movements in reconstituting socially-organized knowledge. For example, many nineteenth century political and cultural currents, including romanticism, utopianism and socialism, began as opposition movements, and later contributed to the reconstitution of knowledge within dominant cultural outlooks and leading institutions (Jamison 2001, 53).

The establishment of environmentalism as a master frame and the partial institutionalization of environmental protest raise the question of whether environmental collective action continues to have meaning. The Italian sociologists Diani and Francesca Forno (2003) write that Italian environmental actions in 1988-97 only partially evoked the image of a movement engaging in sustained, large-scale collective challenges to authorities: rather, they were promoted by a set of public interest groups, mostly institutionalized in their forms of action (Diani and Forno 2003, 164). Researcher Paul Statham (1995) characterizes the environmental movement as a "cultural pressure group," which has faced multiple competitors (including state institutions) (cited in Eder 1998, 204). Sociology Professor Klaus Eder (1998) concludes that since the 1980s "environmentalism has also been appropriated by the movement's opponents" (Eder 1998, 203).

ii) Politicization

The greater involvement of environmental organizations with political institutions and governments has resulted in strategies that have aimed at political policy-making. Environmental organizations' campaigns have certainly been more directly political than conservationist protection projects. Moreover, the sustainable development agenda has been instrumental in the politicization of environmental organizations. "The use by FoE and the WWF of the concept of sustainable development to promote a comprehensive reformist agenda is clearly a more overtly political project than the simple collection of environmental issues with which they began," writes Rootes (2005, 32).

We have seen how the social uprisings of the 1960s and 1970s encouraged conservationists to pursue campaigns that were more political than beforehand. The subsequent emergence of the sustainable development agenda accelerated the politicization of environmentalism. By widening the environmental agenda through sustainable development, environmental organizations like FoE became more attentive to international political concerns. As Rootes (2005) notes, "Since the mid-1980s, FoE has broadened its portfolio to include deforestation and mainstream political issues such as economy and health, and has become increasingly involved in campaigns to promote human rights and economic development in the global South" (Rootes 2005, 32).

iii) Domain Expansion

In 1980, the WWF expanded its domain of campaigning issues to include those connected with the theme of sustainable development. Rootes (2005,

30) points out that this organization's European name change in 1986, from World Wildlife Fund to the World Wide Fund for Nature, reflected the wider scope of its activities, especially those connected to sustainable development. The WWF has refined the theme of sustainable development while retaining an interest in more traditional conservationist issues. Such domain expansion arose from the recognition in environmental movement organizations that campaign issues could no longer be limited to traditional conservationist concerns (Rootes 2005, 32).

To appreciate how Italian conservationist organizations achieved domain expansion beyond their traditional concerns, it is useful to review the issues on which they were campaigning. Research by Della Porta and Massimilliano Andretta (2000) reveals that the largest national environmental organizations were active on a wide range of claims during the 1990s, the decade when sustainable policies became more prevalent. Della Porta and Andretta received completed questionnaires from 18 national Italian environmental groups about the types of claims they were concentrating on. Six categories of claims were identified and the first four were more frequently identified than the last two:

- Industrial pollution/urban ecology
- Nature protection
- Alternative production and technology
- Personal change/lifestyle
- Weapons and military
- Animal rights

Many groups evidently addressed a wide range of claims. Thirteen organizations worked in more than three of the above categories of claims, and four organizations campaigned in five or six categories. Compared with the conservationist issues focused on before the 1970s, these data suggest that "domain expansion" (Best 1999, 169) or "frame extension" (Snow et al. 1986, 472) into a broader range of environmental concerns had developed by the 1990s. Campaigns were no longer limited to conservationist protection: conservationism had been transformed into environmentalism, with the assistance of the orthodoxy of sustainable development.

Over the past 30 years, environmental politics has changed in fundamental ways. As Jamison (2001, 94) writes, the 1987 Brundtland Report involved environmental activists in redefining the meaning of ecology. Since then, activists and governmental officials have tended to move away from emphasizing the protection of nature towards making the whole of society more environmentally-conscious through the sustainable development agenda. Let us now consider how this affected Venetian environmental campaigns.

CHANGES TO CONSERVATIONISM IN VENICE

Sustainable development and radical environmentalism fundamentally changed international and Italian green organizations, and had profound repercussions for Venetian conservationism. The unique geography of a city built in a lagoon and its historical heritage made Venice particularly suitable for the application of sustainability. Yet it was the city's 'red' subculture that meant radical environmentalists made a considerable impression on conservationists.

As discussed in this book's Introduction, the conservationist organization Italia Nostra was highly active in the formation of the Venice problem after the high floods of 1966. Italia Nostra's conservationist campaigns to protect Venice's buildings and monuments and to reduce pollution continued through the 1970s. As these campaigns evolved, they were shaped by radical tendencies in Venice. The city developed a deep-rooted red subculture, allowing radical environmentalism to have a formidable impact. Sociologists have usually located the red subculture in the center of Italy and the Catholic/Christian democratic white subculture in the North East, where the state has been relatively weak (Trigila 1986, 48); but the city of Venice in the North East has been more influenced by red subculture than the white subculture prevalent in the rest of the Veneto region (Putnam 1993, 7). An indicator of the preference for white or red subculture was the relative local electoral support for the Christian Democratic Party (DC) compared with the PCI or the Italian Socialist Party (PSI) (Caciagli 1988). From the Second World War until the implosion of the post-war political parties in 1992-94, Venice was dominated by the parties of the left, especially the PCI and the PSI. In the 1970s, the PCI consistently polled a quarter of all votes in Venice and even the gondoliers were militant labor unionists (Fay and Knightley 1976, 76). The DC and its successors, including Forza Italia and People of Freedom (PDL), have been relatively weak in the city.

The radical left climate that flourished in Venice during the 1960s and 1970s was heavily conditioned by the student movement. In an interview with the author, Giorgio Sarto, former Senator in the Italian Senate for the Green Party and former member of Venice Safeguarding Commission, described how the student movement born in the 1960s had a momentous impact at the Ca' Foscari University and IUAV in Venice (Sarto 2005). Radical students and academics formed close relationships with workers, who were employed in Venice's industrial area and service sector, as well as in the bordering mainland city of Mestre. Radical environmentalist campaigns in Venice concentrated on the issues of health and pollution related to Marghera's industrial zones (Keahey 2002, 127-8), plus working-class housing (Dorigo 1973, 95).

The red subculture in Venice had a discernible effect on the evolution of conservationist campaigns. Yet it would be wrong to characterize this effect as radicalizing conservationism. Rather, as the national green organizations experienced, the convergence of radical and conservationist frames led to a new environmental synthesis. Radical input expanded the range of claims addressed by environmental activists. Cristiano Gasparetto, a member of Italia Nostra's Venice board of directors, personified this trend, bringing his extensive experience in the PCI into the organization to increase its consideration of social issues: "I think that in the past Italia Nostra was very cautious with cultural and social problems. It used to be too light on social problems. Due to my personal history, I hope I have brought greater sensitivity towards social problems inside Italia Nostra" (Gasparetto 2005a). Environmental claims-making became more focused on launching campaigns directed at political institutions, one example being Italia Nostra's 1998 appeal to the EC against the CVN's monopoly of the mobile dam construction. Environmentalists in Venice consistently lodged appeals at the local TAR tribunal and petitioned Venice City Council against the mobile dams.

Although Venice's red subculture meant that environmental campaigns experienced domain expansion and politicization due to radical influences, engagement with sustainability also affected campaigns. Venetian environmentalists with conservationist backgrounds explicitly engaged with the discourse of sustainable development in their claims about the mobile dams. For instance, Giannandrea Mencini of VAS told the author that the costs of building and maintaining the dams would be almost unsustainable for the city (Mencini 2005a). Giorgio Sarto linked sustainability to incorporating justice and equality into planning about the dam project: "They should take into consideration sustainability in the environment and justice and equal relationships" (Sarto 2005). This comment illustrates how environmentalists' claims made in the campaign against the mobile dams incorporated concerns that went far beyond the conservationist protection of nature. Other important players in the debate over the mobile dams have also used the framework of sustainable development: for example, Alfonso Pecoraro Scanio, the former Environment Minister and green movement leader, responded to a defense of the dam project by stating: "it contradicts sustainable development" (*Corriere della Sera* 2006).

Framing claims about the mobile dams in terms of sustainability builds on the debate about the wider future of Venice that has been under way since the high floods of November 1966 (see UNESCO 1969; Dorigo 1972; Costa 1993). In a contribution to this debate, Ignazio Musu (2001, 1), who was on the Commission of International Experts that reviewed the Environmental Impact Study of the mobile dams during the 1990s, commented that an awareness of sustainable development in discussions about Venice's future has emerged in recent years. One research project on sustainability claims that Venice's unusual history of man maintaining a city in a lagoon makes it a special case study in sustainable development (Bon et al. 2001, 28). Likewise, Rinaldo (2001) comments:

> The Venice issue is indeed becoming a case study of paramount importance, not only for the intrinsic importance of the historic, artistic and architectural value of the city which *per se* commands global attention, but also because it is a paradigm of a new complexity in the interactions among economics, society and the environment. It is such complexity that motivates the need for defining and pursuing environmentally sustainable development. (Rinaldo 2001, 61)

Environmentalists, politicians, academics and others debating Venice's problems have assisted with the construction of Venice as an international symbol of sustainability. Chapters Five and Six will discuss how sustainability and radicalism have shaped the campaign and claims against the mobile dams. To appreciate the evolution of these campaigns and claims, first we need to draw some conclusions about the transformation of conservationism into environmentalism.

Modern environmentalism has eroded the conservationist dichotomy between nature and society. The sociologists Scott Lash, Bronislaw Szerszynski and Brian Wynne (1998) argue that the ecological movement of the late 1960s questioned the modernist distinction between man and nature that had been predominant since the Enlightenment (Lash, Szerszynski and Wynne 1998, 11). Early conservationists perceived their role as one of protecting nature; after the 1960s, as Jamison notes: "Nature was no longer to be conceived in oppositional contrast to society, as a refuge from society as had been characteristic of conservationism; with the concept of the environment and the related concept of the biosphere, the focus was shifted to the interactions between nature and society, their mutual dependencies and interrelationships." (Jamison 1998, 229)

The domain expansion that took place from the 1970s was an important innovation because it assisted the emergence of a holistic definition of environmentalism (Jamison 1998, 229), a perspective based on viewing man and nature as one. Since the 1980s, the widespread application of sustainability has led to the further incorporation of social, political and economic issues into environmental agendas; and by the turn of the millennium, a holistic environmental culture had become pervasive in many Western countries.

NOTES

1. 'EMOs' abbreviates 'environmental movement organizations'.
2. A 'master frame' has been defined as a package of political vocabulary that gives meaning to specific events, issues or claims throughout a society, as the social movement scholars David Snow and Robert Benford (1992) explain: "[M]aster frames can be construed as functioning in a manner analogous to linguistic codes in that they provide a grammar that punctuates and syntactically connects patterns or happenings in the world" (Snow and Benford 1992, 138).
3. The IUCN is the International Union for the Conservation of Nature, previously the International Union for the Protection of Nature, which was partly established by Julian Huxley (the first director of UNESCO).

Part Three

CLAIMS-MAKING ABOUT VENICE'S MOBILE DAMS

Chapter Five

Environmentalists Challenge Venice's Mobile Dams

"The clash started half way through the 90s . . . We started our campaigning against Project MOSE in 1996." Michele Boato, Director of the Eco Institute, 16 December 2005.

While environmentalists have been skeptical about the dam project since it was first proposed, they did not concentrate on campaigning about its environmental impact until the 1990s, as Michele Boato indicates above. This chapter analyzes environmental campaigning in Venice since the high floods in 1966. These floods became the catalyst for debating the Venice problem, and a popular mythology has emerged that situates the late 1960s to the early 1970s as the period when environmentalists started challenging the dam project.[1] Here, we question the idea that environmental activists have been protesting against the dams since they were formally proposed in 1970, and argue that the campaign against the dams was constructed over five distinct phases.

The changing character of the campaign against the mobile dams was conditioned by the triumph of conservationism in the city and its subsequent transformation into environmentalism. In the 1970s, the campaign concentrated on conservation against the threats posed by pollution, shipping and the Marghera industries; during the 1980s, it incorporated the wider themes of sustainability. The partial institutionalization of the national environmental movement also impacted upon environmental organizations in Venice. We end this chapter by looking at the creation of an institutional assembly against the mobile dams in 2005, and exploring how this was based on the experience of previous environmental campaigns in the city.

FIVE PHASES OF ENVIRONMENTAL CAMPAIGNING IN VENICE

Phase One (1966-1984): Safeguarding Venice

In 1966, the only significant environmental organization in Venice was Italia Nostra. The Venice chapter of Italia Nostra was formed in 1959 and its initial work focused on restoration projects: for example, in 1965 the organization funded the restoration of a row of houses in the Dorsoduro quarter (Lauritzen 1986, 33). This reflected Italia Nostra's concentration on urban architecture and aesthetics in the early years after the national organization's formation in 1955 (Strassoldo 1993). Following the 1966 high floods, the priorities of activists in Italia Nostra started to move away from restoration and heritage towards wider political questions of protecting the environment. Maurizio Zanetto joined the Venice chapter of Italia Nostra following the 1966 floods, and became its president. He explains:

> We cannot be limited to the renovating of monuments. It is work that should be done...it would be a disaster if there were no private committees that spend a lot of money. Now for us the biggest problem is a political problem. I don't know if the battle is lost, but it is now that we will decide the future of Venice and whether some of the laws will be accepted. We cannot decide this with a single renovation. (Zanetto 2005)

After 1966, prioritizing restoration work seemed to make little sense unless the bigger questions of protecting historical sites from flooding and pollution could be addressed:

> The disaster of 1966 made it clear to everyone interested in saving Venice that the lagoon was at the heart of the matter. Every problem the city faced – flooding, subsidence, brick and stone decay, pollution – either began with the lagoon or was somehow involved with it. Individuals and voluntary groups were able to help with the restoration but, admirable though this work might be, it would turn out to be a waste of time and money unless something was done about the lagoon. (Fay and Knightley 1976, 156)

The first phase of the campaign was broadly defined in terms of protecting Venice's lagoon. A lively campaign including demonstrations on boats and legal appeals was fought between 1969 and 1973, as outlined in an illuminating book by Venetian historian Giannandrea Mencini (2005b). The principal issues highlighted were pollution, damage from ships navigating the lagoon, and the Marghera industrial area. Marghera's industries were linked both to pollution and to dredging specific channels in the lagoon. Channels are necessary for large boats to cross the shallow lagoon: the Bocca di Lido-S. Nicolò channel was deepened by dredging from 3-4 meters to 12 meters in 1966, and the depth of the Bocca di Malamocco channel was increased from 4-5 meters to 15 meters during the 1960s. Currents led to further deepening, to an estimated depth of between 16 and 20 meters in places (Perlasca 2001, 61). Dredging the Malamocco channel meant that the majority of larger ships bound for Marghera entered the lagoon at the Malamocco inlet rather than the Lido inlet, thus taking these vessels away from Venice's central islands. Yet concerns were raised about the threat of potential pollution and damage from oil tankers on their way to the refineries on the mainland side of the lagoon, especially if one of these had an accident. Large ships also generate waves that batter against buildings and monuments, exacerbating erosion. The waves caused by these vessels and other boats became a key issue highlighted by Italia Nostra.

Another prominent claim emphasized by Italia Nostra during the early 1970s was the contribution of Marghera's industries to the sinking problem. In the aftermath of the 1966 floods, Italia Nostra stated that the Marghera industrial area was the main cause of the 1966 flood (Plant 2002, 358). Substantial evidence supporting this claim has been provided by Ghetti's (1988) research, as noted in this book's Introduction. Ghetti established that the principal reason for subsidence in Venice during the twentieth century was the extraction of groundwater from under the lagoon, especially between 1930 and 1970. In 1971, when the subsidence problem had been understood, the Italian government swiftly ordered the closure of a large number of groundwater wells. Marghera's industries were subsequently supplied from surface watercourses, with the result that connecting these industries to the sinking problem was largely neutralized as a campaigning issue after 1971.

Conservationists also challenged plans for the expansion of the Marghera industrial area. A proposed third industrial zone would have required filling parts of the lagoon to provide additional land where new industrial units could be built. Radicals and activists in labor unions saw advantages in terms of new job opportunities for workers. Italia Nostra campaigned against the third industrial zone, reversing its previous position on this issue: "Italia Nostra that had in previous years endorsed the go-ahead for a second and third industrial zone at Marghera, and for the 'tanker canal' (even criticizing the delays in the work on it)" (Pertot 2004, 152). The 1973 Special Law for Venice prevented the development of the third industrial zone. This was regarded as a victory by conservationists; however, the law did not require Marghera's port to be closed, which was considered a setback. The failure to close Marghera's port and ban ships from the lagoon was partly due to a lack of popular support, despite the campaign's inclusion of people from a wide spectrum of political backgrounds. Cristiano Gasparetto of Italia Nostra explained, in an interview with the author: "Immediately after the great flood of the 1960s, Italia Nostra built a front, which was called 'the front to defend the lagoon', which put together some different political opinions. It was formed with men like Indro Montanelli, who was a great journalist on the right. But together with some people from the left, a front was built that could allow moving against the political lobbies" (Gasparetto 2005a).

Several environmentalists interviewed by the author described this front as marginal to the interests of the majority of Venetians. Michele Boato, who joined the campaign later, said: "Only Italia Nostra supported the campaign by Montanelli against the oil channels. We were not around then. So we were outside of the campaign. But the sensation was that the population of the city was disinterested and it was only the fishermen that really saw the sea coming into the lagoon with the oil" (Boato, M. 2005). The 1970s campaign to ban oil tankers failed, but by 2001 oil tankers without protective double hulls had been prohibited from traveling through the Venetian lagoon.

In a broader sense, the 1970s campaigns to protect Venice and the lagoon sowed the seeds of the campaign against the mobile dams, as we discuss later in this chapter. Here it is important to record that activists of the 1970s were relatively uninterested in the dam project, compared to the issues of pollution, Marghera's industries, and damage caused by shipping. This observation questions the idea that environmentalists have been battling against the mobile dams since they were proposed.

As discussed in the Introduction, the 'competition of ideas' for proposals to protect against high flooding was initiated in 1970 and the first mobile dam feasibility study was produced in 1971. Yet the project was regarded as little more than a vague idea by environmental campaigners. "This project goes back to 35 years ago. Then it was not a project that they were going to execute. It was just a drawing on a piece of paper," commented the experienced activist Fabio Cavolo (2005). Campaigners were aware of the plans: the bulletin of Italia Nostra interventions in Venice (2007: 1957-1999) archives record entries in 1970 considering the closure of the three port inlets with mobile dams. "Since the beginning of the 1970s, when the project of closing the dams started, we have always been quite dubious about this, the Project MOSE," claimed Italia Nostra's Venice President Alvise Benedetti (2005).

Plans to build the dam system were first given legal recommendation in the 1973 Special Law. Meanwhile, Italia Nostra chose to concentrate on pollution and the more general inadequacies of safeguarding measures related to the 1973 Special Law. Safeguarding became a defining concept in the response of Italia Nostra to the 1966 floods. "When the association was created in the mid-50s, in 1966 there was the huge flood and we didn't know what to do," said Benedetti (2005): "The theme of safeguarding . . . was a theme that was acutely felt by Italia Nostra." 'Safeguarding' has proved to be a durable concept and has become a defining feature of the Venice problem, even though its meaning has changed over time. The laws that preceded the high floods of 1966 were not framed in terms of safeguarding, but the term featured frequently in the laws for Venice enacted by the Italian state after the 1966 floods in 1973, 1984, 1991, 1992 and 1995 (Vianello 2004).

In the first phase of the campaign in Venice, the safeguarding concept was developed according to the early conservationist framing of the environment, in terms of defending Venice's historical heritage. It was defined by the 1973 Law as "safeguarding of the environment (landscape, historical, archaeological, artistic features)" (Fletcher and Spencer 2005, 671. Appendix 2). In the 1970s, there was less concern about defending nature in terms of vegetation, birds and wildlife habitats, and activists used the 1973 Special Law as a point of reference in their campaigns (for example by Boato, S. 2005). After the 1970s, the definition of safeguarding expanded from pollution to a variety of questions about protecting the lagoon's complex ecosystem. "The environmental problem is becoming increasingly perceived as one which goes beyond the reduction in pollution; it also involves the maintenance and preservation of a whole set of natural values which find their unique expression in the lagoon ecosystem," writes Musu (2001, 2). The progression during the first phase of campaigning was away from restoration projects towards pollution and the wider political issues connected with safeguarding the lagoon. Although the dam project was largely ignored, there were indications that Italia Nostra was broadening its campaigning beyond its traditional domain of conservationist issues and addressing more political questions. These trends advanced during the second phase of the campaign.

Phase Two (1984-1994): Seeking Re-equilibrium

A new stage in the Venetian environmental movement began in the early 1980s. In an interview with the author Stefano Boato, who has been described as a pivotal figure in the campaign against the mobile dams, explained how activists called for measures to achieve general re-equilibrium as a response to firmer official proposals for the mobile dams. 'Re-equilibrium' was to be achieved by applying the 1973 Special Law recommendations to improve the balance between the city, the lagoon, and the sea. According to Boato, re-equilibrium was de-prioritized by the Venice Water Authority and the CVN (New Venice Consortium) after 1984, when agreements were approved to study the construction of an immovable dam near the Lido lagoon inlet:

> The battle started in 1984/5. Then there was the big Project that aimed to build fixed dams. Now they are only mobile. They wanted to start the Project more in terms of dams than re-equilibrium. There was a period of this general vision that was clear for years. Italia Nostra did a long campaign that made it clear that the re-equilibrium of the lagoon was more important. That became law and already, the state had said that. It was in the Law of 1973. (Boato, S. 2005)

The 1973 Law called for the protection of hydraulic and hydro-geological equilibrium. Measures for re-equilibrium in the 1973 Law included building up coastal defenses and improving the watercourses feeding the lagoon. During the 1980s, environmental campaigners began to demand the implementation of re-equilibrium measures because they felt the 1973 Law was not being put into practice. "There was a request to apply the laws that were not applied. In 1973, there was a law about re-equilibrium and this law was not applied. The first point of the Italian state law was not applied and has still not been applied. It was to re-equilibrate the environmental disaster," remarked Boato (Boato, S. 2005). Although environmentalists used the 1973 Special Law as a point of reference in their campaign for re-equilibrium measures, the Law's emphasis on conserving Venice's heritage appeared outdated by the 1980s. Sociologist Luigi Scano (2004) writes, in his review of safeguarding laws for Venice: a "[f]ew years after they were promulgated, the 'special law' 171/1973 and the related DPR 791/1973 (about the conservative restoration of lagoon historical centres) already show their limits, their weaknesses, the cultural obsolescence of their inspiration. Their influence upon the upgrade of the urban historical texture has been scarce, almost void relatively (*sic*) to the purpose." (Scano 2004)

Given the limited character of safeguarding through the 1973 Special Law, environmentalists advocated a wider definition of safeguarding than was applied during the 1970s, emphasizing that the implementation of safeguarding should go beyond protecting buildings against pollution to include the re-equilibrium of the whole lagoon. "I think for the safeguarding of the city there have been some aspects that have been central, like trying to take away pollution and to re-equilibrate the morphology of the environment in the lagoon. These are the big environmental problems," said Orazio Alberti, member of VAS (*Verdi Ambiente e Società*) and former Green Party Councilor (Alberti 2005). The 1970s campaign priority of pollution was not forgotten: indeed, a rapid increase of algae in the Venetian lagoon during the 1980s exaggerated pollution. But pollution, like other environmental concerns, came to be treated as part of a complex eco-system.

The 1980s campaign for re-equilibrium stressed the complexity of interrelated issues, including Marghera's industries and high floods. In an interview with the author Cesare Scarpa, former director of Legambiente in Venice and active figure in the No MOSE permanent assembly, responded to a question about the definition of Venice's environmental problems by describing their complex relationship: "This is a difficult question because to protect the lagoon environment is going into some different and complex things. On one side if we have the need to defend the lagoon of Venice with the city in the center from high waters, we have the need to defend the city of Venice from another dangerous presence: that is petro-chemicals . . . There are a lot of aspects; there are hydraulic, morphological aspects" (Scarpa 2005).

Complexity became the dominant theme in safeguarding Venice during the mid-1980s. Although it was a new theme in the Venetian environmentalists' campaign, it was unoriginal. Complexity, with its stress on the interactions between natural and human factors, is rooted in the theory of Gaia, coined by James Lovelock in 1972 (Tickell 2006, xii). Venetian environmentalists developed the theme of complexity beyond the 1970s conservationist emphasis on safeguarding from specific threats, with the result that complex safeguarding brought wider issues regarding the city and the lagoon's ecosystem into focus. Rather than isolating environmental problems, their interrelationship was paramount. This is why the general re-equilibrium of the city, the lagoon and the sea were considered more important than singular interventions, including the mobile dams. Beppe Caccia, former Green Party Venice Councilor, explains that safeguarding Venice is complex because it is also a social question and cannot be solved with hydraulic intervention:

> We say that the Greens historically have been against, not only this single project, but against the idea that it can be a big project which can help confront the problems of safeguarding. Sometimes, we speak of complexity to avoid the problems. Here, instead, we speak of complexity to get into the real problems of safeguarding. We need to take into consideration this possibility when we think about the problems of safeguarding as a system of interventions. It is a system of interventions that cannot be considered only as hydraulic systems. It is a system that deals with the physical safeguarding of the city and also the social safeguarding of the city. (Caccia 2005)

Expanding environmental problems from the protection of buildings to social questions is a local example of the shift from conservationism to environmentalism analyzed in the previous chapter. The sustainable development agenda was having an impact in Venice by increasing the range of environmental concerns, and through the discussion of complexity, the Venice problem could be extended beyond the issues of flooding and sinking. As has been noted, there are multiple causes of high floods in Venice, but the local problem was largely identified with the storm surges from the sea in November 1966. Edoardo Salzano, Italia Nostra member, relates the multiple causes of the 1966 flood to the inappropriate concentration on blocking the sea with dams:

> Another single thing is that during the great flood of 1966, a very strong contribution came from the rivers. We forget about it. We think that everything depended on the sea water. All the work that should have been done on the rivers has not been done. So if the contribution of the rivers is significant, Project MOSE would really be an obstacle. So there are a lot of reasons to worry about the damage that could be created by Project MOSE. (Salzano 2005)

The singular and simplistic solution of the MOSE dams was contrasted with the complexity of the Venice problem. "I think that MOSE is too much of an easy answer to a complex problem. For this reason, it is not able to solve the problems of Venice," stated Marco Favaro, WWF member and Green Party provincial Councilor (Favaro 2005). Venetian environmentalists believe that Project MOSE is an unsuitably simple solution, given the complexity of Venice's problems. Stressing complexity allows them instead to argue for general re-equilibrium, and to challenge the typical definition of the Venice problem and its solution. "[I]n national and international public opinion, the idea is that the 'Venice problem' can be immediately identified with high water and, through a simple equation, that the solution to the problem of high water can be *sic in simplicity* the MOSE," writes Vianello (2004, 159. Emphasis in original). The underlining of Venice's complex problems during the 1980s helped to redefine safeguarding. The safeguarding measures in the Special Law of 1992 accentuated defending nature rather than Venice's historical heritage, and specified protecting the natural characteristics of the lagoon, including restoration of lagoon morphology, opening up closed lagoon areas to tidal expansion, and halting the lagoon's deterioration.

Although the environmental campaign of the 1980s was dominated by the discussion of general re-equilibrium, challenging the mobile dams became more important after 1988. In that year, the CVN revealed its prototype for the mobile gates project, the MOSE model. The Italian news agency *ANSA* (1997) reports people flocking to watch as MOSE was placed at the Arsenal near Venice's Lido inlet, where it was put into operation for research:

> This was an enormous steel caisson measuring 20 × 17.5 metres mounted on a hull 32 × 25 metres on which four piers rose to a height of 20 metres to support a crane. The crane was used for the replacement of defective parts. The hull also housed the control centre and personnel . . . when it came time for Moses to get his feet wet, there was considerable anxiety for a time because a huge air bubble trapped in the gate caused the caisson to swerve. Then everything worked according to plan. Approaching noon the yellow gate had been raised and allowed to float for a quarter hour before sinking it beneath

> the surface, where it disappeared like a sounding whale. This event, historic in some ways, rekindled debates and polemics – and not only among the environmentalists. (*ANSA* 1997)

The placement of the MOSE model became an historic event due to the way it was framed, which was influenced by environmentalists. But others shaped this event too. *Mosè* is Italian for Moses, and the way the above report links the dam to the biblical parting of the Red Sea illustrates the role played by the media in the interpretation of the MOSE model. In addition, there was a prevailing sense that the event was poorly handled in public relations terms by the CVN. "They did an experimental model at the Arsenal. Everybody knows about it. It is what was sought for Project MOSE. It was really bad publicity for the CVN. They finished it 15 years late. People said 'what is that thing here in Venice?'" said Maurizio Zanetto (2005). The response by environmental campaigners and the media to the huge physical structure of the MOSE model added to the negative public reaction. Representatives of the CVN expressed frustration that Venetians had failed to understand that the MOSE model included above-water structures, but that the final gates would be below water (Keahey 2002, 164).

At this point, concerns about the impact of Project MOSE on the lagoon environment became more vocal. In an interview in 2010, Alvise Benedetti of Italia Nostra described how the battle against MOSE began in 1989 with appeals to the National Council of Public Works and the local TAR tribunal (*Il Gazzettino di Venezia*, 10 January 2010). In 1990, the Public Works Ministry declared that the MOSE Project should be delayed until the lagoon could be environmentally cleaned up. Hence, environmentalists' pressure on institutions to consider the wider issues of re-equilibrium had some success. As doubts were raised by institutions about the environmental impact of the dam project, campaigners called on the government for an environmental impact assessment.

Phase Three (1994-2001): Pressure for an Environmental Impact Assessment (EIA [VIA])

In 1994, the mobile dam project advanced to its final design stage. By then, the Public Works Ministry was satisfied that the project should proceed and authorized detailed plans for construction, but Venice City Council began to exert pressure to assess its environmental impact. "A more precise battle started in 1994 to go and verify this Project. In 1994, there was another phase that, at the request of the Council, was to go and verify the environmental impact," explained Stefano Boato, who was a Green Party Councilor in Venice City Council during that period (Boato, S. 2005).

The prominent role of local city councils was facilitated by the 1992 Special Law that made it a requirement to obtain the opinions of the Veneto Regional Council and those of the Venice and Chioggia City Councils for safeguarding strategies and measures. At this stage, the role of environmental associations in challenging the MOSE dams was relatively limited. "Campaigning around Project MOSE at the beginning, I am talking about the end of the 1980s, the early 90s; it was mainly a problem for local councils or scientists. The debate was a very closed debate, also for the environmental associations," confirmed Alberti (2005). Pressure to assess the environmental impact of Project MOSE was led by Venice City Council, especially after 1993, when Cacciari was first elected as Venice Mayor. Venice City Council "proposed subjecting the project to an assessment of environmental impact" (*ANSA* 1997). The Comitatone agreed to this in 1995.

When the Environmental Impact Study (EIS) proceeded in 1996, it was considered a victory by Venetian environmentalists. "With that achievement, the maximum that we could do was to create a situation where all the institutions, not only the citizens, had to verify it in detail," said Stefano Boato (2005). This comment also illustrates how it was important for Venetian environmentalists that institutions verified the environmental impact of Project MOSE, confirming the highly institutional orientation of the campaign during the third phase. This is not surprising given that Venice City Council was leading the campaign for assessing environmental impact. Nonetheless environmentalists, as well as European bodies, had a role in putting pressure on the government to conduct an EIS, as Keahey (2002) records: "By the mid-1990s, pressure from the European Union and from environmentalists and gate opponents had been building against the Consorzio's gates plan and against its monopoly status. Yielding to this opposition, the national Italian government's interministerial committee for the safeguarding of Venice in 1995 ordered the Venice Magistrato alle Acque to conduct an environmental impact study." (Keahey 2002, 231)

There were both positive and negative interpretations of the EIS by different committees. The Green Party's control of the Environment Ministry led to the blocking of Project MOSE through the 1988 Ronchi-Melandri decree, which was based on a negative assessment of the EIS by a commission set up by the Environment Ministry. However in July 2000, the TAR tribunal canceled the Ronchi-Melandri decree. The failure of the EIA (VIA) process to stop the mobile dam project was especially disappointing for those who had become involved in the campaign through providing information to the Environment Ministry commission for the VIA. In an interview with the author, Federico Antinori of LIPU explained: "In 1998, I started because there was the VIA commission that examined the Project. All the associations decided to contribute and present some observations to this committee and we did the same" (Antinori 2005). Environmental associations identified their involvement with the Environment Ministry commission from 1996 as an important stage in the escalation of the campaign. "In 1996 there was acceleration when a big meeting was organized and the Minister of the Environment took part. Since then we really started a big campaign against Project MOSE to inform people about the damages of MOSE," described Paolo Perlasca of WWF (Perlasca 2005). Similarly, Michele Boato gave 1996 as the date when his Eco Institute began campaigning against Project MOSE (Boato, M. 2005). Yet as the Project moved to its definitive phase in the early 1990s, it was Cacciari's new Venice City Council administration that led the campaign for an EIS. It appeared from these efforts that Cacciari had adopted a highly critical attitude towards the Project; but a surprising development came in 2000, when Cacciari resigned as Mayor, decided to run for President of the Veneto Regional Council, and declared that the mobile gate project should proceed to its next stage. Keahey (2002) records this event:

> Late in 1999, Venice mayor Cacciari, who along with his city council had long shared the Environment Ministry's skepticism about the gates and had driven much of the demand for an environmental study, resigned two years early. He wanted to run in the April 2000 elections for the office of regional president. A few weeks later, on January 12, 2000, he dropped his bombshell in the middle of a city council meeting that erupted in pandemonium, declaring that it was time for the mobile-gates project to advance to the Executive Project phase. (Keahey 2002, 242)

Paolo Costa was Cacciari's successor as Mayor of Venice. Costa had formerly been the Pro-Rector of Venice University and had served on the College of Experts retained to review the EIS on the mobile dams. Costa had to be replaced on the College of Experts while he served as Minister for Public Works in the mid-1990s, but he continued to play an important role during the early stages of preparing the mobile dam project. Costa was staunchly in favor of the mobile dams during his administration in Venice City Council, and publicly criticized Cacciari for delaying the dams between 1994 and 2001 (*Il Gazzettino di Venezia*, 3 January 2010). According to Paolo Perlasca (2005), Costa's election radically changed the relationship of Venice City Council to the campaign against the dams, because Costa and the majority of the ruling political parties in the Council supported the project.

Over the course of the EIS, environmental organizations provided information and opinions about the dams and the lagoon. In this way, they became participants in the institutional judgements of the EIS, especially in Venice City Council and the commission established to assess the EIS by the Environment

Ministry. Environmental organizations also engaged a wider range of institutions as part of their campaign for an environmental impact assessment. They launched many appeals to the EC, TAR, and Judicial State Council, and legal appeals became a new feature of the campaign during the third phase. From 1994 until 2001, the campaign priority had been to pressurize institutions to conduct an EIA of the mobile dam project. The next phase of the campaign was dominated by responding to the project's approval and the initiation of building work.

Phase Four (2001-5): Opposing the Start of the Mobile Dams' Construction

When the dam project proceeded to its Executive Phase in December 2001, activists challenged the final institutional approval processes; and when construction work began in 2003, they highlighted the environmental damage this caused. The change of direction was prompted by the election of a government led by Prime Minister Silvio Berlusconi in 2001. Berlusconi decided to make several infrastructure projects high-profile initiatives, and a new infrastructure law, referred to as the 'Strategic Objectives' or 'Objective Law', was drawn up to approve and fund various projects. This law was applied to the mobile dam project in Venice, which had been put on ice by the previous government led by Prime Minister Amato. "At the end of 2001 and the beginning of 2002, the Berlusconi government with the help of the Mayor of Venice (who was previously the Minister of Public Works) went against the negative assessment of the Council committee that voted no and the Comitatone voted yes," Stefano Boato told the author in October 2005 (Boato, S. 2005).

This new phase of the campaign was signaled by the establishment of an umbrella committee in 2001 called 'Save Venice and the Lagoon'. Many organizations were involved, although environmental associations had leading roles. While challenging the mobile dams was a priority for this committee, its name reflected the fact that environmental groups retained a wider frame of reference related to Venice and the lagoon. Many of the initiatives launched by environmental organizations in this phase focused on challenging the Objective Law, rather than specifically opposing the dams. Environmentalists criticized the way the Objective Law changed how decisions were made about safeguarding Venice, and concerns were raised about how the Objective Law would affect the operation of the Comitatone, which governs decision-making about interventions in Venice and the lagoon. Beppe Caccia described how the 2001-5 Berlusconi government changed the *modus operandi* in the Comitatone, especially reducing the Venice Mayor's power:

> Until Berlusconi, no one would have ever dreamt of thinking in terms of a majority in the Comitatone. The praxis inside the Comitatone is that, certainly formally, every minister that goes to the Comitatone and the President of the Council has one vote, like the vote of the Mayors of Venice, of Chioggia, of Cavallino, etc. But in the normal life of the Comitatone until Berlusconi, the vote of the Mayor of Venice was the decisive element in the material constitution of the Comitatone. This logic was swept away by the Objective Law and the new laws introduced by the Berlusconi government and the internal activities of the Comitatone. (Caccia 2005)

Prime Minister Berlusconi believed that the Objective Law enabled him to speed up the start of the dams' construction. On 14 January 2003, he stated: "[W]ithout the objective law we would be giving the okay to the works two and a half years late, in the fall of 2005, while today, we can give the okay to the realization of the work and I will have the honor of participating, at the beginning of February, in Venice, in an inauguration ceremony" (cited in Giannini, *Il Gazzettino di Venezia*, 15 January 2003). In the same newspaper article, Green Party Member of Parliament Luana Zanella pointed out that Berlusconi was giving the go-ahead to Project MOSE's inauguration before the Comitatone meeting scheduled to discuss this on 30 January. In fact, the approval of the Comitatone was delayed, as was the inauguration. But the application of the Objective Law to Project MOSE marked a decisive change in how the Comitatone operated. "The Objective Law and the way it was used to approve MOSE in 2003 declared the 'political death' of the Comitatone," observes Vianello (2004, 101).

The legal implications of the Objective Law were questioned by environmental associations, to undermine its implementation for the approval of the dams. For example, the WWF wrote to the then Italian President, Carlo Azeglio Ciampi, asking him not to approve the Objective Law because it "not only contradicts European laws, but has a central impact that empties significant recent reforms of the Constitution on the federal form of the State" (cited in *La Padania* 2001). In 2002, Italia Nostra and the WWF launched an appeal to the Venice TAR tribunal against the inclusion of Project MOSE in the Interministerial Committee for Economic Programs (CIPE) funding of projects based on the Objective Law. Legambiente appealed to the European Commission against the Objective Law in January 2003.

While the campaign between 2001 and 2003 focused on challenging the final approval stages of the dam project, activists also began to highlight the environmental damage caused by the start of the building work. Appeals to various institutions were launched, developing the negative judgments about the environmental impact of the dams that had been promoted by the Green Party and environmental organizations during the 1980s. In February 2003 and January 2004, appeals were placed at the TAR tribunal by the associations in the Save Venice and the Lagoon committee, led by the WWF and Italia Nostra. The appeals claimed that the Executive Phase of Project MOSE did not include a legitimate EIA. In January 2004, the Green Party launched an appeal to the EC against the approval of Project MOSE without an EIS. In April 2004, the Save Venice and the Lagoon committee appealed to the EC and TAR against the positive EIS assessment of Project MOSE by the Veneto Regional Council. However, the campaign emphasis on the EIA issue was affected by the May 2004 TAR court's rejection of all eight appeals against Project MOSE, which had been placed by environmental associations, Venice City Council, and Venice Provincial Council. These judgements were appealed and, in December 2004, rejected by the Judicial State Council.

As institutional appeals faltered, activists turned to organizing more protests. In September 2004, environmental associations, citizens' committees and some political parties cooperated in a high profile 'No MOSE' protest on 30 boats along Venice's Grand Canal. The citizens' committee 'The Damages of MOSE' set up its first official meeting at Punta Sabbioni in December 2004, calling for the MOSE building work to stop. The committee held a sit-in of approximately 100 people there in February 2005. The more prominent role of citizens' committees and their instigation of protests in this period suggested frustration with the highly institutionalized campaigns by environmental associations. Moreover, these campaigns had failed to prevent the start of the dams' construction. This could have led environmentalists to question the institutional orientation of their campaigns. Instead, Stefano Boato explained that the limited power of environmentalists during this fourth phase of the campaign was due to the dominance of political forces in favor of the Project in the key decision-making institutions (Boato, S. 2005). Despite facing formidable opposition, many activists felt that new opportunities to exert pressure within institutions were presented by the Venice City Council election in 2005. This led to a new phase in the campaign.

Phase Five (2005-2010): The No MOSE Campaign

In June 2005, activists decided to focus explicitly on campaigning against the mobile dams when the Save Venice and the Lagoon committee was wound up and the permanent 'No MOSE' forum[2] initiated. "Probably now with the new forum a new phase of the campaign is starting. The former phase was with the committee 'Save Venice and the Lagoon'," Marco Favaro told the author (Favaro 2005). According to Beppe Caccia (2005), progress with the construction of the dams stimulated the new level of mobilization, while Orazio Alberti suggested that activists decided to concentrate explicitly on Project

MOSE because they would lose the opportunity to stop its construction unless this was made a priority before the 2005 national election:

> Either we stop Project MOSE now or we will not stop it. The work has now started. Soon there will be the political election and if we do not block Project MOSE between now and the election, that is it. That will be the end of it. We are very late. The work has started and blocking Project MOSE now is very difficult. Going beyond now, it will become really impossible to carry on fighting against MOSE in the next years. We are like the Japanese in the jungle after the war that continued to shoot. This is why we are concentrating on MOSE. It is not that we have forgotten the other things. This is the time to fight against Project MOSE. If we lose this time, we will not have another opportunity to fight against Project MOSE. (Alberti 2005)

Constructing social problems at the right time is a vital part of effective claims-making. There was general agreement among the environmentalists interviewed that the establishment of a permanent assembly and the specific campaign focus against Project MOSE was long overdue. Fabrizio Reberschegg, Vice-President of the Green Party in the Venice municipality, explained the lack of urgency about concentrating on Project MOSE as typical of Italian lethargy (Reberschegg 2005). Federico Antinori (2005) blamed it on the deliberate concealment of information about how the Project would evolve and be built.

Whatever the reasons for not focusing attention on challenging the dams until 2005, the decision to launch the No MOSE assembly undoubtedly defined a new phase in the campaign. This new phase was also initiated by political realignments in the aftermath of the Venice City Council elections in April 2005. The Green Party did not gain sufficient votes to be included in the new administration. Marco Favaro predicted shortly after the election that the Green Party, in opposition, would be able to work more closely with the associations challenging Project MOSE (Favaro 2005). The Refounded Communist Party (RC) also received a low proportion of votes and sought to rebuild its political legitimacy through greater involvement in the environmental citizens' committees, especially those based on the Lido Littoral Island. "Given that the government has changed in Venice and the Refounded Communists are in opposition, the committees of the Lido are more active. You can see the work being done and the Refounded Communists must build legitimacy as a political battle," observed Stefano Boato (2005).

The RC was also highly active alongside anarchist squatters and the 'No Global Disobedients', who became prominent in the No MOSE permanent forum and dominated the campaign protest events in Venice during 2006. Stefano Boato believed the replacement of the Save Venice and the Lagoon committee with the permanent No MOSE forum facilitated a shift away from the domination of the campaign by the environmental associations towards control by the RC:

> These associations have coordinated in different ways. Before they had some coordination, then they had the committee Save Venice and the Lagoon. And now it is a question of hegemony. Before everything was coordinated by the environmental associations Italia Nostra and WWF. Now everything is coordinated by people from the Refounded Communists. It is important that all these forces are there for different motivations. (Boato, S. 2005)

Although environmental associations became increasingly marginalized in the protests against the dams, they worked more closely with Venice City Council to challenge the dams. The April 2005 Venice City Council elections led to Cacciari's re-election as Mayor of Venice, replacing Costa, a supporter of the mobile dams. Cacciari had been inconsistent in challenging Project MOSE, and while he did not promise to stop the construction of the mobile dams, his re-election created an opportunity for environmentalists to rebuild institutional links with Venice City Council. "I must say that Cacciari on Project MOSE is doing what he can because now it is very difficult to go back. He understands we can talk about it and there can be some initiatives that can help. We are trying to put together the bridge that was destroyed while Costa was in the Council," commented Alvise Benedetti (2005).

Campaigners against Project MOSE were particularly encouraged when Cacciari launched a commission to review the Project and included members of environmental associations. In May 2005, Cacciari put together a group of experts to evaluate alternatives to Project MOSE for closing the three inlets to the lagoon. Michele Boato was positive about the inclusion of a senior member of his Eco Institute in this commission, although he was skeptical about the commission's objectives (Boato, M. 2005). Orazio Alberti (2005) spoke in favorable terms about meeting the person in charge of the technical review initiated by Mayor Cacciari. According to Cesare Scarpa (2005), the involvement of the No MOSE permanent forum in the City Council review strengthened links between the administration and campaigners.

The re-election of Cacciari as the Mayor of Venice and the creation of the No MOSE permanent forum combined to bring environmentalists into a closer relationship with the Venice City Council. Sometimes campaigners were directly involved with the council, especially regarding the investigations of alternatives to Project MOSE. Many of the protest activities were part of a strategy to exert pressure on the City Council: for example, the No MOSE forum launched a petition against the mobile dams in August 2005. Italia Nostra's leading representative in the No MOSE permanent forum,

Cristiano Gasparetto, explained that the forum's petition against MOSE was partly to put additional pressure on the Mayor (Gasparetto 2005a). In May 2006, 300 people in boats protested against Project MOSE outside a Venice City Council meeting before it voted on a document calling for a verification and review of the Project; when the vote was postponed, the protests were reorganized to take place during the rescheduled meeting. Many such protests appeared to be challenging Venice City Council, and aiming to exert pressure on Council policy-making. During the fifth phase of the campaign, the No MOSE protests outside the Venice City Council acquired a highly institutional character.

This follows a pattern of environmental organizations concentrating on lobbying institutions, which was identified on a national level in the 1980s by Diani (1995, 189) and Rhodes (1995, 168). From 1988 to 1997, Italian environmental protest was mostly institutionalized (Diani and Forno 2003, 164). Activists in the campaign against the mobile dams developed an orientation towards putting pressure on city councils: "We will carry on pushing the local administration because they should take their responsibilities. There should be evaluations and they should take responsibility for what they are going to decide," stated Fabio Cavolo, Director of the environmental citizens' committee *Associazione Rocchetta e Dintorni* (Cavolo 2005). Research by Della Porta and Andretta (2000) into 18 national environmental organizations in Italy during the 1990s found that protests were often not designed to be disruptive or to mobilize the public, but were directed at policy-makers. However, the relationship between activists and policy-makers is often tense, and this was the case in Venice with the City Council. "Let's say that the relationship between popular committees and the Council is not always easy. Either a committee is useful for what a local council administration is doing and they are helped and satisfied with the requests that they present. Or they are seen as groups of people who are a hassle that time after time say something which has not got any scientific basis," said Cesare Scarpa (2005). While many of the contributions by environmentalists cited here express a positive attitude towards Cacciari's Venice City Council administrations, Maurizio Zanetto suggested that the Council had a tendency to take advantage of environmental associations:

> The Council, in particular, has always had an extreme and deplorable ambiguity. So I did not want to cooperate. It would be useful to take part in some consulting processes and relations. But I think, it is my own personal thought in my name (not Italia Nostra), I think that they always try to involve the association to clean themselves in order to involve the association to justify what they are doing. After they have heard the environmentalists they can do what they want. But they can say, yes, the environmentalists were there during the decision-making process.

> But our ideas are not taken into consideration because they took their decisions elsewhere. So I do not like being involved in this. This used to be my position. Maybe it is a mistake. But I do not think this has changed. (Zanetto 2005)

These comments also indicate an appreciation by Venice City Council that discussing matters with environmental associations was perceived to add to the legitimacy of decisions. The weak legitimacy of Italian state institutions in the early 1990s (Diani 1995, 28) made involving environmental organizations appealing for political elites. This might involve those organizations being manipulated and not gaining real influence, as Maurizio Zanetto feared; and Edoardo Salzano (2005) considered the opportunities presented by Cacciari's second administration to be limited. Yet when Marina Corrier, an activist in the No MOSE permanent assembly, was asked about the significance of Cacciari's re-election, she felt that it opened new possibilities for environmentalists to work with state institutions:

> I think that since the previous Mayor went away who blocked any dynamic of discussion in the city, the environmental associations, for example the permanent assembly against chemical risks, they started to understand it was possible to do something against Project MOSE. Beforehand the situation between the state, Region and the City Council was blocked. They were all in favor and had all the mass communications in their power. When the Council changed with a Mayor that during the electoral campaign said that he was against Project MOSE and wanted to block this work, even the environmental associations that thought it was difficult to have a movement, represented themselves. (Corrier 2005)

Cacciari's re-election encouraged the permanent assembly focused on chemical risks at Marghera to initiate a similar assembly against Project MOSE. The idea of a permanent assembly against Project MOSE was first reported in early June 2005 at an environment day organized by the Gabriele Bortolozzo Association to highlight concerns about Marghera's industries (Testa 2005), and was created on 22 June 2005. The creation of the permanent No MOSE assembly and the partial institutionalization of the campaign through the assembly thus followed the model established by other local campaigns.

HOW THE CAMPAIGN AGAINST THE DAMS BUILT ON OTHER CAMPAIGNS

There were two campaigns involving environmentalists in Venice that played a role in constructing the campaign against the mobile dams. The first was the campaign about Marghera; the second was the campaign against the EXPO exhibition in Venice. Through both of these campaigns, fundamental themes were developed and organizational lessons were learned for the campaign against Project MOSE.

The Marghera Campaign

A range of environmental problems connected to Marghera's industries were identified by conservationists after 1966, and we have seen how important themes emerged from these problems that developed during the campaign against the mobile dams. One prominent theme was that of complexity, which, according to Marco Favaro, was also identified by Green Party activists as a theme in the campaign against the Marghera industrial area:

> The Green Party started in Venice around the mid-1980s. The principal campaign focused on Marghera. So the chemical pollution produced by these industries. From this campaign a new vision of the complexity of the environmental problems of the lagoon was built. It was not enough to focus on Marghera Port, because the pollution depended also on the farm activities in the dry land and the impact of some kinds of fishing activities and the general problems of the physical aspect of the lagoon. All these complexities have to be considered linked, not separated things because this would be a reductionist approach which was the same approach that produced these problems. (Favaro 2005)

Although complexity has been an important theme in both the campaigns against Marghera's industries and the dams, risk has been more conspicuous. In the 1980s and 1990s, concerns about exposure to risks were highlighted by claims-makers who opposed the Marghera industrial zones. New risks from pollution were debated after five tonnes of light fuel spilled into the lagoon on 29 November 1995 and drifted for four days. Beppe Caccia described how risk subsequently became the dominant theme in safeguarding Venice from the Marghera industrial area:

> Two important questions are the battle for the transformation of the chemical heavy industry at Marghera Port and the physical and social safeguarding of the city of water. These are the two greatest battles where we, as Greens, have sought a role to promote initiatives at the base of the social movements for these great environmental questions. Now there is a permanent assembly against chemical risks at the Marghera Port since November 2003. It is a popular body, a body created from below, created after a big accident at the petrochemical plant of Marghera that risked producing huge effects with a chain reaction and a domino effect. Everybody knows about these risks. (Caccia 2005)

Indeed, environmentalists usually refer to the Marghera assembly as the assembly against chemical risks, as Marina Corrier's comments reveal above. During the 1980s and 1990s, environmentalists also made claims about the risks of working in Marghera for workers' health, as leading Green Party member Gianfranco Bettin (1997) outlines. According to Strassoldo (1993), it was through the issue of health in the workplace that environmental and left-wing concerns began to find common ground. Diani (1995) records this trend in Marghera: "Within factories, workers demanded wider control over the production cycle rather than mere wage rises, paying special attention to health and safety issues at work. Mobilisations run against the Montedison chemical plants in Marghera (Venice) provide a good example of both tendencies" (Diani 1995, 24). Marghera proved to be a very high-profile workers' health campaign. The companies that operated the chemical plant at Marghera were accused of being responsible for the cancer deaths of 150 workers after 1973, and 600 cases of people with work-related illnesses (Jim Morris 1998). Research by former worker Gabriele Bortolozzo and environmental campaign groups, especially Greenpeace, led to 28 former Marghera managers being charged with manslaughter and polluting the lagoon with dioxin. After a trial lasting three years, they were acquitted in 2001. Campaigns against Marghera's industrial area by Greenpeace, Italia Nostra, the WWF and Legambiente continue at the time of writing. Greenpeace has become especially professional in gaining media attention about Marghera's industries, most notably when its famous ship, the SV Rainbow Warrior, pulled into the Venetian lagoon, enabling television cameras to film four Greenpeace protesters leaving the ship to 'occupy' a Marghera plant (Greenpeace 1999). "Greenpeace denounced the death of the lagoon's ecosystem, the death of workers of the Petrochemical (plant) and the risk of the inhabitants of the zone," note Della Porta and Andretta (2000, 24).

Although Greenpeace has played a well-publicized role in campaigns about Marghera, the organization has not been involved in the campaign about the mobile dams. Fabrizio Reberschegg told the author that Greenpeace made a decision not to participate in this campaign because the organization did not consider it an international issue (Reberschegg 2005). However, Fabio Cavolo (2005) mentioned that many of the other organizations involved in the campaigns regarding risks at Marghera did help establish the permanent assembly against Project MOSE. When the No MOSE assembly was inaugurated, the newspaper *Il Gazzettino di Venezia* reported; "it already has a model: the permanent assembly against chemical risk at Marghera" (Testa 2005). A Green Party member quoted in this article pointed out that the initiation of a permanent assembly against MOSE was also modeled on the permanent assembly formed for the EXPO campaign.

The EXPO Campaign

Venice was among three cities considered as a possibility to host EXPO 2000, a four-month international exposition of modern achievements. In 1990, Italy's Foreign Minister Gianni De Michelis, a Venetian, led the efforts to host the exposition and was backed by 40 leading Italian companies. By June 1990, Italian Prime Minister Giulio Andreotti announced that Venice would no longer be considered, following a European Parliament vote in May 1990 that called on Italy to withdraw its candidacy. The EC Environment Commissioner, Italian Carlo Ripa di Meana, had demanded environmental studies for the Venice EXPO proposal and influenced this vote. Ripa di Meana is the former national President of Italia Nostra. Alongside other environmental groups, Italia Nostra put pressure on the Italian government regarding its decision about the EXPO. The opposition to holding EXPO 2000 in Venice was also supported by private preservationist organizations working through UNESCO (including Save Venice, The Venice in Peril Fund and the World Monuments Fund). The principal objection was that the influx of tourists would place an unbearable strain on a city already overstretched by tourism.

The EXPO campaign was important for the campaign against the mobile dams regarding the organizational lessons that were learned. Activists appreciated that the role of Venice City Council was decisive and were encouraged by winning a campaign. "We succeeded once, in the early Nineties, against the EXPO we succeeded. Even the experience of the first Council led by Massimo Cacciari was created with the fight against EXPO. We succeeded in building an alternative idea of the cultural development of the city in contrast to the idea of the big EXPO that would have brought 20 million tourists to Venice," commented Beppe Caccia (2005). Activists understood that cultivating international support was crucial in stopping the EXPO in Venice, and the campaign against the MOSE dams has been highly active internationally. "So the battle against the EXPO was won in the city by eradicating the opposition in favor of the EXPO...We will need to produce the same effect on the question of Project MOSE. We will not win only in Venice. We will not win only on a national or international level. There is a linking of levels," said Caccia (2005).

Though activists were concentrating on building the campaign against the EXPO in the early 1990s, they were simultaneously developing concerns about the mobile dams. The dam project became associated with danger: a theme that became more prominent as fears about global warming grew, as we will explore in Chapter Seven. Cesare Scarpa emphasized that during the EXPO campaign there was awareness that Project MOSE presented dangers, which increased when MOSE's construction began:

> We started an important committee with five people at the start that reached its apex in a demonstration on the Grand Canal with 500 boats and the Mayor and there was a group of Venetians that went to Paris to block EXPO in Venice. We are talking about 1990/91. Already, there was a lot of criticism against Project MOSE. For 15 years that information in diverse ways developed and this has carried on. In a way it has become more organized, stronger and obviously there is new knowledge of the dangers since the start of the works began two or three years ago. (Scarpa 2005)

Despite some popular mobilization against the EXPO, the contributions above indicate that this campaign had a highly institutional character. Venetian activists against the EXPO concentrated on appealing to the EC and the Italian government. Venice City Council also played a leading role in blocking the EXPO, alongside environmental and preservationist organizations. The establishment of a permanent assembly against the EXPO was reproduced for the Marghera permanent assembly, and both functioned as models for the No MOSE permanent assembly. The creation of the No MOSE permanent assembly aided institutional recognition of the campaign and involvement of environmental activists in City Council decision-making. Nevertheless, the City Council in 2006 failed to gain acceptance from the national government of alternative systems for blocking the inlets to the lagoon, and the campaign against the mobile dams subsequently dwindled as their construction progressed. The No MOSE assembly organized protests in May, September and October 2007, but from late 2007, many of the No MOSE assembly demonstrations became integrated into other protest campaigns. In mid-December 2007 and September 2008, the No MOSE assembly participated in demonstrations against the expansion of the 'Dal Molin' American air base in the city of Vicenza, which is 129 kilometers (80 miles) from Venice. No MOSE protestors joined a No Global Day in May 2008 and a march by 2,000 people in Venice in December 2008. Approximately 100 No MOSE activists joined other demonstrators in Venice on 28 May 2009, yet the campaign lost steam as the construction of the mobile dams advanced.

This does not imply that the No MOSE campaign is dead. The No MOSE assembly organized a series of protests on the Venetian island of San Servolo in April 2010 (Brunetti 2010). When the author interviewed Flavio Cogo for a second time in March 2010, he confirmed that the protests against the dams would continue, even though their construction was 60 percent completed (Cogo 2010). If the dams are finished, their inauguration is likely to be met with further demonstrations. However, since early 2009 there has been no real expectation that environmentalists could stop the dam project in a similar manner to EXPO 2000. "We lost the battle against MOSE," remarked former Venice Mayor Cacciari in an interview during January 2009 (Dina 2009a).

CONCEPTS AND THEMES IN THE CAMPAIGN AGAINST THE DAMS

As this chapter has demonstrated, the idea that environmentalists started challenging the dam project after it was proposed in 1970 is more myth than reality. This idea has become part of the mythology of the Venice problem that emerged after the 1966 high floods (James 2002). The reality is that environmental campaigners in Venice since 1966 have changed focus in different periods, and did not dedicate their attention to the dams until the 1990s.

The evolution of Venetian claims-making has been shaped by the trends examined in the preceding chapters. We have observed how the triumph of conservationism and its transformation into environmentalism helped to establish and change the direction of green campaigning in Venice. This change involved broadening environmental issues connected with sustainable development. Protecting Venice became a social as well as a natural question. The roles of environmental associations altered as they became partially institutionalized, which developed from highly institutionalized campaigns against the EXPO and Marghera's industries. The close relationship between environmental associations and institutions allowed other groups to dominate the protests against the dams. We saw how equilibrium has been a core concept in the campaign against the dams, particularly during the 1980s; now it can be appreciated that equilibrium in Venetian campaigning connected with the national environmental movement of the 1980s. The concept of equilibrium is especially formidable in Venice because it builds on the Venetian Republic's myth of constitutional harmony: political equilibrium during the Republic was destroyed by Napoleon, apparently alongside Venice's environmental equilibrium; the city subsequently became associated with environmental disequilibrium in cultural representations of the city. Safeguarding has also been identified as a key concept in Venetian environmental campaigning, and we have traced how this concept was redefined at different stages, especially due to the transformation of conservationism into environmentalism. Through analyzing each stage of the campaign against the mobile dams, several themes emerged, including complexity, risk and danger. The content of environmental claims about the dams forms the subject of the next chapter.

NOTES

1. See the quotation by James (2002) at the beginning of this book's Introduction.
2. The No MOSE permanent forum is also referred to as an "assembly."

Chapter Six

Environmental Claims about Venice's Mobile Dams

> "We have done some actions. We have tried to do some specials in our magazine *Green and Environment*. In that magazine there is a lot of debate about national heritage. We have tried to interview some experts on conservation and asked them about their own points of view on Project MOSE. Sometimes we try to interview some people who are for and some people who are against Project MOSE so we can underline the contradictions that come out on this theme. High water is obviously a problem in Venice. But it is not the only problem in Venice. We try to highlight what are the other problems in Venice and try to say how we can protect Venice with other works that are more gradual and irreversible and not fixed like MOSE is. Obviously, Venice is a city that has a lot of problems and spending all the money on MOSE is a risk."
>
> Giannandrea Mencini (2005a), VAS.

Environmental activists have worked hard to present problems with the MOSE dams to the public that extend beyond conservation. They have been acutely aware of the need to motivate audiences to reject the dams by linking them to problems of resources, risk and other factors. Activists have also been conscious of how important it is to offer alternative solutions to the flooding problem.

Our analysis of Venetian environmentalists' claims is divided into diagnostic, motivational and prognostic framing of the problem. In the diagnostic framing of the basic problems with the dams, environmentalists have highlighted the impact of building on the landscape, wildlife, habitats and vegetation in the lagoon, and the draining of Venice's resources. The resource claim has been framed to motivate audiences with the argument that the dams are distorting the economic and social life of the city. Emphasizing how the dams redistribute resources away from other projects in Venice has assisted the creation of highly symbolic victims and villains: environmentalists have been constructed as the protectors of Venice for its suffering citizens, while the CVN consortium overseeing the project has been depicted as defending businesses that have been developing an amusement park or museum city for tourist revenue. Abandoning the interests of Venetians in favor of tourists has been presented as unsustainable for the long-term life of the city. The expansion of environmental claims about the dams beyond conservationist concerns to wider social and economic questions has enabled activists to frame these claims using the internationally powerful topic of sustainable development. Connecting claims against the dams with sustainability has made them more appealing to public audiences, and linking claims to risks has provided environmentalists with a cultural resource that has helped them motivate audiences against the dams. We examine how activists have referred to international conceptions of risk, but have relied heavily on local definitions. Environmentalists' prognostic framing of solutions also is discussed in this chapter, including campaigners' proposed alternatives to the mobile dams for protecting Venice against high flooding.

Grouping activists' claims into diagnostic, motivational and prognostic categories reflects how claims-makers typically give meaning to social problems, as explained by Sociology Professor Donileen Loseke (2003):

> Claims-makers must give meaning to the facts by constructing a *social problem frame*, which has several components. The *diagnostic frame* constructs the meaning of the condition. (What type of a problem is it? Who or what causes the problem?) The *motivational frame* constructs the reasons why audience members should evaluate the condition as intolerable. (Why should audience members care?) The *prognostic frame* constructs the solutions to the problem. (What should be done?) (Loseke 2003, 59. Emphasis in original)

While categorizing the claims in this way assists understanding of them, the categories are not always clearly defined. The claims made by activists often move quickly from describing the basic conditions of the problem to identifying who is to blame and proposing alternative solutions. Nonetheless, dividing analysis of claims about the dams into diagnostic, motivational and prognostic framing contributes to an evaluation of the balance of myth and reality.

DIAGNOSTIC FRAMING

Since the placing of the MOSE dam model near the Arsenal in 1988, activists have raised conservationist concerns about the aesthetic impact on the landscape. As discussed in Chapter Five, this problem was highlighted when some Venetians did not understand that the final MOSE dams would be predominantly below water (Keahey 2002, 164). In more recent interviews for this book, environmentalists acknowledged that the mobile dams were designed to limit any negative aesthetic impact on the view, until they rise temporarily to block high tides. "Everything is under water to avoid a clash with the environmental and landscape aspects because one does not see it," confirmed Stefano Boato (Boato, S. 2005).

Although the mobile dams are placed under water, the creation of new structures will bring visible changes at the three inlets. The most significant structure to which environmentalists have objected is the creation of an artificial island to assist the operation of the mobile dams at the Lido inlet. "The Lido Outlet would be transformed with the creation of an artificial island, 9 hectares (22 acres) in size, for buildings and workshops, as well as a 20 metre (60 foot) high smokestack," declared the Italia Nostra Venice website (2007b). It is true that the Lido inlet, as well as the Malamocco and Chioggia inlets, have been physically changed. Diagrams and descriptions of the plans were reproduced in a 2009 book jointly edited by architects Andrea Groppello and Paola Virgioli from IUAV, which includes a useful description by Carlo Magnami of how the artificial island was designed with varying heights on different sides to see views of Venice and the lagoon:

> The morphological reasons for the appearance and shape of the new island are rather found in the form of the bottom and the flow of the currents and channels. On the one side it looks to history, to the linear nature of the coastline and the gradual reduction of the front of Sant Erasmo island as waterfront, on the other to recent formations, to the beaches, to the shallows where the sea floor in some places emerges, and perhaps may be said to be emerging, to establish new relations. So the reference heights of the island in section are different. Toward the sea it has the necessary height to be protected from the waves, but toward the lagoon the banks are lower and take on a curved shape, allowing easy mooring and creating a new beach. It is thus almost part of the archipelago in formation and the existing one. The plant sections are protected by differences of height and most of the land is a park, with dunes crossed by footpaths leading to the sea. (Magnami 2009, 125)

Although the artificial island does have one tower, it has been designed with aesthetically pleasing views, parks and dunes for the public to enjoy. It is notable that while Italia Nostra criticized the impact of the 20-meter (65.6-foot) "smokestack" tower on the view, they have not made complaints about lighthouses of similar heights that are scattered near the inlets to the Venetian lagoon - such as the Rochetta Lighthouse near the Malamocco inlet, which is 24 meters (78.74 feet) high.

Conservationist concerns have also been raised about potential pollution from the functioning of the dams, especially from zinc released into the lagoon. "When the dams are completed, anodes to protect their huge metal gates will release about 10 tonnes of zinc into the Lagoon each year. The toxic metal could accumulate in the food chain," stated the Italia Nostra Venice chapter website (2007b). The Commission (*Collegio*) of International Experts (1998) analyzed this problem in detail when assessing the mobile dams' potential environmental impact, and established that constant monitoring should prevent damaging pollution levels:

> Another potential impact is that of the release of chlorine and zinc from the gates to sea water. The amounts of chlorine to be released with the cooling water are small and unlikely to have any effect. Large quantities of zinc, in absolute terms, will be dissolved from sacrificial anodes located for cathodic protection inside and outside the gate element. It is estimated that up to 12 tons could be thus released yearly over the three inlets. This would occur continuously for the anodes on the outside of the gate elements and in bursts when the gate elements are voided of their water when being raised. The zinc concentration inside the elements will depend on the frequency of closures. It may allegedly reach 25 milligrams/liter in a total water volume of 53000 to 71000 m3. Modelling the dispersion estimates a zinc concentration of 25 micrograms/liter at 1.2 km from the gates, whereas background concentration is now 1 to 1.5 micrograms/liter. Zinc, an essential element, has low toxicity: the EU limit value for fish is set at 200 micrograms/liter. The impact may thus be considered as small but it should be recommended, if the project is implemented, to monitor zinc in the mussels farms in view of the strong concentration power of mollusks for this element. Impurities (cadmium, etc.) in the zinc used for the anodes would seem to be too small to be a cause of concern. (Commission of International Experts 1998, 33)

The CVN's website has also responded to concerns about pollutants from the dams. Based on the experience of similar systems, constant monitoring was assured: "Throughout the world, a great number of structures with metal components have been in operation in a marine environment for decades and their long-term efficiency is guaranteed both by the quality of the materials used and by appropriate and constant maintenance" (Consorzio Venezia Nuova 2009d).

As the construction of the dams has progressed, claims about the negative impact on the landscape have increasingly focused on damage to areas where building work has been concentrated. "From the individual point of view, yours or mine, I think that Project MOSE means we have lost some very nice landscape. Now the works by the CVN at Pellestrina are horrible. They have transformed the internal beach and the ocean," remarked Edoardo Salzano in an interview with the author (Salzano 2005). Many other conservationist claims against the mobile dams have highlighted their negative impact during construction, especially upon wildlife, habitats and vegetation. "Before the dams actually start operation, Venetians will have to undergo at least eight years of construction with high environmental impacts (for example, the sediment released into the Lagoon's waters could devastate shellfish harvested)," speculated the Venice Italia Nostra website (2007b). Environmentalists' claims about the dam building sites are supported by the designation of surrounding areas as Sites of Community Interest (abbreviated to SCI in English and SIC in Italian) by the EC. Since the 1992 Rio Convention on Biodiversity, SCI directives about birds and habitats have been incorporated into various EC laws that should be adhered to by member states. "MOSE will produce many direct and indirect impacts, direct impacts especially during construction in the areas around the three inlets of the lagoon and paradoxically important naturalistic venues are concentrated in these areas – in fact they are SIC (European Community protected)," predicted Marco Favaro (2005). "The coastline would be devastated at Ca' Roman, whose beaches are protected as a natural area under EU law," noted Italia Nostra's Venice chapter website (2007b).

Claims about damage to these SCI-protected areas in the early stages of building the dams were included in an appeal launched by Italy's Green Party to the EC on 5 March 2004. In December 2005, the EC opened an investigation into whether Italy was guilty of breaking a 2000 EC nature directive by failing to set aside protected areas for birds that live around the lagoon. Reassurances from the Italian government laid this investigation to rest in April 2009. The EC decided to 'archive' the case, which led to the approval of €1.5 billion (US$1.96532bn)[1] from the European Bank of Investments (BEI) for the dams (*AGI,* 14 April 2009); but in April 2010, the BEI froze its investment in the mobile dams. Intensive lobbying by the WWF convinced the bank that further environmental assurances needed to be made before finalizing this funding, especially regarding the nature directive for birds (*La Nuova Venezia,* 7 April 2010). Challenging such conservationist claims became important for the CVN and Italian governments to secure funds for the dam project to progress.

In contrast, environmental campaigners linked the concentration of resources on the dams to reduced spending on maintaining the Venetian lagoon and supporting Venetians. "The only connection is that this Project will dredge money that probably could be spent in so many different ways. So you will build here something that probably for the next 10 or 15 years will dredge all the money that you could invest, I will repeat, in natural ways for the preservation for everybody," suggested Gustavo De Filippo (2005). In addition to building funds, the running and maintenance costs were identified as a drain on conserving Venice's heritage. "All the economic resources will end up in Project MOSE and there will be no resources left to renovate the monuments in Venice. The costs of maintenance of Project MOSE are really, really high," remarked Paolo Perlasca (2005).

In these claims about the basic resource problems with Project MOSE, the CVN consortium of companies overseeing the Project has been constructed by campaigners as the principal diagnostic frame of blame. In particular, activists have emphasized the illegitimacy of the control and financial gain by private companies involved in the CVN consortium. Through the CVN, companies have enjoyed a virtual monopoly over the design and building of the mobile dams. "So they created a pool of building companies with Fiat first in line, which in the past had a huge role in the Italian economy, because this Project is a great boost for the builders and the Italian economy. So the CVN is a representative for big Italian building companies and their big international interests. It is a way of making money circulate," claimed Fabrizio Reberschegg (2005). The accumulation of private profits at the expense of local communities is a common target, especially for radical environmentalists: as Bramwell (1994) notes, "Ecologists talk of the need to put community before profit" (Bramwell 1994, 178). The CVN's private profits are considered especially problematic due to the belief that the funds for the dams come out of central state provisions for the whole Venetian economy: "The big risk is that I think this money will be monopolized by the CVN which would create a monopoly on work in the city. Also it would aggravate the position of the city as a dependent, which is like an assisted and helped city. This is a city that gets a lot of capital investment from the state and is managed by a monopoly, the CVN," commented Orazio Alberti (2005).

In many ways, it is not surprising that the sole concession granted to the CVN was later regarded as an unacceptable monopoly. To understand this claim, it is useful to consider the historical circumstances in which the CVN was established. The CVN was created in a period when the Italian Socialist Party (PSI) dominated Italian politics, holding the offices of President between 1978 and 1985 and Prime Minister between 1983 and 1987. The PSI was characterized by extensive corruption, the awarding of contracts to favored associates, and the use of the state to benefit private interests. From the 1960s, the PSI discarded ideological attachments to socialism, which was replaced by a highly corrupt form of business-orientated politics by the end of the 1980s; "In the absence of any ethical basis to its politics, the PSI attracted and encouraged the ruthless, the ambitious and the unscrupulous. A new category of 'business politicians', able at using private resources to build personal support, took over the apparatus at local level," explains Stephen Gundle (1996, 87). The CVN officially came into being in 1984 when the

PSI's leader Bettino Craxi was Prime Minister. He predicted the mobile dam project would be completed in 1995 as "the first of a new generation of public works" that would place technology at the service of the environment (*ANSA* 1997). The PSI was the principal victim of the Tangentopoli corruption investigations in the early 1990s; Craxi was convicted for a corruption case from the 1980s and fled into exile in Africa (Gundle 1996, 85-6). Although the legitimacy of such monopolistic bodies as the CVN had been called into question by the 1990s, it was difficult to discard the consortium, as it had overseen the development of the dam project. However, environmentalists chided state institutions for failing to exercise control over the CVN.

Fabrizio Reberschegg (2005) argued the CVN is so financially powerful that it has been able to monopolize Project MOSE without state involvement. Several environmentalists interviewed by the author criticized the weakness of the state body the Venice Water Authority (MAV) in holding the CVN to account. MAV's weakness in its relationship with the CVN is perceived as the abdication of public interest by the state, and the domination of private economic interests. The conviction that the state is a force for good that should be mobilized to control private business interests is connected to the leftist backgrounds of radical activists at the forefront of campaigning against the CVN. There is a long tradition of advocating a stronger state among Italian left-wing activists, especially those involved with the Italian Communist Party (PCI). The PCI's former leader Enrico Berlinguer brought the party into an 'historic compromise' with the state during the 1970s, helping to "instil into the PCI the sense that Italian public institutions were precious and must be defended" (McCarthy 1997, 103). Chapter Four analyzed the extent to which the campaign in Venice was influenced by the city's radical subculture, and 11 environmentalists interviewed by the author had backgrounds in leftist organizations. Here, we can see how Venice's radical subculture and environmentalists' backgrounds conditioned claims about the CVN. For Marco Favaro, MAV has become the CVN's partner in crime by building Project MOSE as a simplistic solution for the narrowly defined problem of high water:

> The CVN and Water Authority were able to simplify the problems and all the situation of the complexity of the environmental problems of the lagoon could be reduced to a simple equation. Venice has the problem of high water. To save Venice we have to close high water. To close high water we have to close the three inlets. All the other considerations about transformation are considered philosophical questions by the people. The realistic approach, the very concrete approach of the Water Authority is winning at the moment. (Favaro 2005)

As outlined in the previous chapter, Venetian environmentalists have challenged the simplification of Venice's complex problems. The CVN's representatives contest this claim. Alberto Scotti (2005), a leading engineer for the CVN, describes the mobile dams as only one part of a complex combination of interventions by the CVN to protect Venice against flooding:

> The solution chosen to provide a full response to the complex problem of high waters involves a combined system of interventions including: temporary closure of all three lagoon inlets by means of a row of mobile gates; local measures to raise shores and banks, compatible with the architectural and socio-economic structure of the individual built-up areas; and wide area morphological measures compatible with the environment to protect against the most frequent flooding. (Scotti 2005, 252)

In addition, Scotti records the progress with the CVN's protection plan against medium flooding before 2005, which did not include the mobile dams (Scotti 2005, 253). Yet despite these extensive prior interventions, environmentalists have claimed that the CVN is building the mobile dams as its solution to a narrow definition of the flooding problem. Moreover, the above comments by activists reveal how they have constructed the CVN as the primary adversary. Constructing a villain to blame for a social problem is an intrinsic part of creating a diagnostic frame (Loseke 2003, 59), and although other adversaries, such as Prime Minister Berlusconi, have also been highlighted by environmental campaigners in relation to the mobile dams, the CVN has been a constant enemy. The CVN has been overseeing the mobile dam project since councilors and scientists started to lobby for an assessment of the project's environmental impact in the 1980s, and was one of the companies in the consortium organized to promote the EXPO for Venice in 2000: the focus of another environmentalist campaign (Vianello 2004, 164). Chapter Five indicated how campaigning against the EXPO formed an important stage in the evolution of environmental campaigns in Venice: now we can see how the role of the CVN in the EXPO promotion contributed to its construction as the primary villain regarding the mobile dams.

Activists' identification of damage to landscape, vegetation and bird life as basic problems with the dams reflect traditional conservationism, whereas the diagnosis of the CVN's monopoly on power and resources illustrate the ascendancy of radical environmentalism. Our previous assessment of the strength of red subculture in Venice helps us appreciate why emphasizing the radical interpretation of environmental problems has a particularly powerful purchase in the city. The impact of radicalism has allowed the CVN to be depicted by activists as a monstrous figure dominating Venice, as Michele Boato vividly described: "The CVN moved into the lagoon like a godfather, an elephant, like a body full of money that can do everything. They can have newspapers, television, everything, and the political parties" (Boato, M. 2005).

MOTIVATIONAL FRAMING

While diagnostic framing identifies the basic problems and who caused them, motivational framing often creates victim sympathy to help audiences to appreciate why they should care about the problems (Loseke 2003, 59). Motivational framing is usually connected to matters that are perceived to be of importance to target audiences. This section shows how environmentalists' motivational framing of claims about Project MOSE were linked to wider economic and social issues, thus expanding claims beyond conservationist concerns.

Economic Objections – Draining Resources from the City and Its Citizens

"Our judgement of Project MOSE is that it is a money problem. It is just business for big companies and saving Venice is just an excuse," declared Federico Antinori (2005). By promoting the negative financial impact of Project MOSE, environmentalists sought to present themselves as concerned about more than natural environmental matters. Fabrizio Reberschegg also expressed this outlook:

> It is very clear that the worries of the Greens are not only the environment but also the economy. We think that Project MOSE as it will be built will block the port inlets and can bring the death of the Venice port, which is one of the biggest industries in Venice. When the port inlets are closed, ships will not be able to enter and a person with a ship cannot wait a day to enter and eventually will prefer to go to other ports. We risk a crisis for the port and the whole Venetian economy. (Reberschegg 2005)

The creation of shipping locks so that vessels can enter the lagoon when the mobile dams are raised seems to have reduced potential interference with port traffic. We assess changes to the port in Chapter Nine. Here, it is interesting to consider why environmental campaigners prioritized economic over natural impacts as they sought to motivate audiences through the No MOSE permanent assembly. Cesare Scarpa (2005) argued that audiences are not moved by traditional environmental concerns related to protecting nature and this was a mistake made by the previous committee, Save Venice and the Lagoon. He added: "We need to create different motivations. Each of us can choose what we want from the erosion of monuments to economic loss to a lot of other things and alternatives that can convince citizens to stop everything, to stop the machines. Because high waters are not like 1966 and they might start seeing things in a better light."

The focus on the negative economic impact of Project MOSE as a motivational frame developed from environmentalists' diagnosis of the resource question as a basic problem with the Project. This was central to their belief that the CVN is merely serving private economic interests; however, presenting the principal problem with the dams as the dominance of a private monopoly against a weak state may not motivate audiences who do not share the radical environmentalist outlook. So the diagnosis of this basic problem was framed to motivate audiences by emphasizing that the public would be excluded from the benefits of the dams, summed up with the slogan '*Il Mose serve solo a chi lo fa*' ('MOSE is only needed by those who are doing it').

When the No MOSE permanent assembly was founded in 2005, activists decided to promote economic objections as their key motivating claims. The slogan '*Il Mose serve solo a chi lo fa*' was highlighted during the assembly's launch (Gasparetto 2005b), and continually promoted at protest events. For example, on 4 February 2006, the No MOSE permanent assembly organized a protest in St Mark's Square attended by 50 activists and a banner was displayed from a terrace, declaring '*No MOSE perché serve solo a chi lo fa*!' ('No MOSE because it is only needed by those who are doing it'). This slogan was reproduced on banners and T-shirts visible at many demonstrations observed by the author. The slogan helped to single out the CVN as the sole beneficiary of the dams, and assisted the construction of victims through the implication that the CVN's dam project was taking resources away from the public. As previously mentioned, Venice is highly dependent on central government resources that help fund specific projects, including organizations that renovate monuments and housing projects for young local couples who find the city's high property prices expensive. "Funds will not be spent on other things like cultural heritage or money for young couples or for citizens or for some organizations. This is important that all the money is channeled into Project MOSE. It makes no sense," stated Giannandrea Mencini (2005a).

There have been some recent occasions when Venice City Council has set aside money to help young couples meet the cost of high rents or mortgages. But the money allocated "barely covers a small percentage of the needs," said former Venice Mayor Cacciari (Poveledo 2006). Environmentalists argue that the shortfall is due to spending on Project MOSE, despite the suggestion that €1.5 million (US$1.96532 million) was made available to Venetians for renovating their properties in late 2009 while the dams were in an intensive phase of construction (*Il Gazzettino di Venezia,* 18 December 2009). Young couples trying to start new families create feelings of sympathy and victim purity, which are important components in effectively motivating audiences: "Cultural feeling rules emphasize that sympathy should be felt for innocent and morally worthy people, so effective claims therefore emphasize that victims are in no way responsible for the harm they experience" (Loseke 2003, 81).

The apparent victims of the mobile dams were extended to a broad range of groups. "Here with Project MOSE, 10,000 workplaces will disappear, with fishermen, the port and tourism as well, the lagoon tourism," asserted Flavio Cogo (2005). This claim broadens the economic motivation against Project MOSE by constructing almost anyone as a potential victim: a strategy that Loseke (2003, 80) classifies as important to successful claims-making. This strategy prevents narrowing the audience to specific victims and those who sympathize with them, and instead appeals to us all, feeding off a pervasive "culture of victimhood" (Furedi 2002, 100). Motivating audiences against the mobile dams through the construction of economic victims seemed more effective than highlighting damage to nature, because "[w]hile nature or animals can be constructed as a victim, victims most often are constructed as *people"* (Loseke 2003, 77-8. Emphasis in original).

So environmentalists have conjured up images of victims and villains through economic claims against Project MOSE. The principal victims presented by environmentalists are young couples and the primary villain is the CVN. Loseke (2003, 77) suggests that successful motivating frames build on strong local cultural sentiments, and the victims and villains constructed by the campaign against the mobile dams are powerful as motivating frames because they build on core beliefs in Italian culture. One is the strength of the family, and support for the idea of helping young couples to start families in the context of very low fertility rates in Italy (Ginsborg 2001, 68-70). Another is the widely-held sentiment that Italian consortia such as the CVN are created by corrupt and self-interested politicians, who circulate money between themselves without serving the public (Della Porta 1996). "We have a project that is a mistake from the point of view of the public, public money. It's an ancient type of politics. It's stupid finance . . . make a hole and then fill it . . . and the money goes around," was how Fabrizio Reberschegg described Project MOSE as typical of this political characteristic (Reberschegg 2005).

Activists have taken the economic objections against the mobile dams as serving business interests and damaging families, and then presented these problems as reasons for the depopulation of Venice. Marco Favaro explained how local audiences need to be convinced of this claim:

> I think we have to demonstrate to local communities that MOSE is a great business mainly for the building enterprises and the consequence is not only an environmental depopulation of the lagoon but also a social depopulation because MOSE is drinking a lot of money that the government gives to Venice each year. Most of this money is now concentrated on building MOSE. In previous years, the major part was spent to help Venetians to restore a flat or for new families to

> buy a flat or traditional economical activities to achieve the objective of revitalizing the social and economic fabric of Venice. (Favaro 2005)

Thus, Venetian environmentalists built on the claim that Project MOSE will be economically damaging to create the argument that the Project will destroy Venice socially.

Social Objections – Destroying Indigenous City Life

Like the economic claims, social objections about the mobile dams do not dwell on conservationist concerns. The focus is on saving the social fabric of Venice, with direct appeal to the local population and again constructing the CVN as the primary villain, whose monopoly will make the city more dependent on the state. Beppe Caccia claimed that the consortium has already distorted the internal economic and social life of the city:

> There is a risk that we have been living with for almost 20 years: the pollution of the social and political and economic life of the city by a monopoly that can spend enormous resources without being under control. This is a factor in the pollution of civil life that needs to be taken into account. It is said there are too few research and cultural institutions which focus on the various safeguarding problems that have consulting contracts. They must not survive on their economic relations with the CVN. This is an image you can have. There is an imbalance of power in the democratic life of the city. (Caccia 2005)

Caccia thoughtfully developed the economic objection about the victimization of families through the redistribution of resources, and this is worth quoting at length. Failing to assist families financially is here linked to the decreasing population of the city and its deteriorating social fabric:

> With the effect of the direction of resources for other interventions there is a risk that one can feel for the collectivity, the City Council, but also for private people. A part of the significance of the resources from the Special Law is that jobs come to finance the private interventions and institutions. Today, a Venetian family is able, to a degree, to do ordinary and extraordinary maintenance for a proper house if they have enough help and support from the Special Law. In the next few years, these categories of ordinary and unusual maintenance contributions for houses and offices will become impossible, impractical, due to the exhaustion of resources. There is this social risk that contributes to the increase in the process of the triumph of the tourist monoculture. This helps the process of abandoning the city for the mainland by inhabitants, which since the mid-80s we have defined as an exodus for the mainland with an impoverishment of the social, civic life. These elements of relations make a city not a city, but a stage, an un-kept museum to visit. (Caccia 2005)

In this way, the claim is constructed that Venice is being transformed from a city into a museum for tourists through the intervention of the CVN and its concentration of resources on the mobile dams. It is true that the population of Venice has fallen over the past 50 years, and environmentalists have interpreted this trend as leading to a city dominated by tourists with a minimal local population. This claim was promoted during a protest held on Venice's Rialto Bridge in March 2009, at which demonstrators dressed up as American Indians, as if to imply that they were living on a reservation with their cultural identity under threat. They handed out leaflets with the message: "Our cultural identity is at risk of dissolving if Venice becomes a theme park – but we Venetians will not surrender" (Squires 2009). Enrico Sambo suggested that Project MOSE is designed to protect this tourist orientation:

> Behind MOSE there is not only the problem of high water, but also the idea of a city. The idea of the city that Project MOSE has is as a tourist amusement park that should be protected. So if it worked, the MOSE would always be raised. So the idea that it goes up and down with the tides is mad. It will always stay up to defend this theater of tourists. So saying no to MOSE is saying yes to a city that is not this one. It is a city of water that does not need to be kept in this way. (Sambo 2005)

The CVN is thus depicted as leading the destruction of indigenous social life in Venice, while environmental associations are constructed as the defenders of the city's cultural heritage for the local population, as opposed to tourists:

> Today Italia Nostra has a position that sees and understands that the best protection of the cultural and environmental values is to also protect the quality of life of people. Let's say, in a generic but quite an indicative sense, that Italia Nostra wants to save a living Venice made up of a non-tourist, resident population (people that work) and that can better use the cultural and artistic structure; the monuments, the palaces, the very structure of the city. (Gasparetto 2005a)

Tourism is increasingly regarded as a risk for Venice that is comparable to the risks from flooding (Plant 2002, 2), and this claim is analyzed in Chapter Nine. First, we explore how Venetian environmentalists interpret the risks facing the city.

Reversibility, Experimentation and Graduality

When environmentalists were asked during interviews with the author about risks connected to the mobile dams, these three themes emerged: reversibility, experimentation and graduality. These were classified as conditions for safeguarding interventions in the 1984 Special Law for Venice, which committed the Italian state to the design, experimentation and execution of works to protect urban settlements from exceptionally high tides "also via measures at the inlets...with the characteristics of experimentability, reversibility and gradualness" (cited in Fletcher and Spencer 2005, 671. Appendix 2). Giorgio Sarto stated that irreversible work for Project MOSE contradicts conditions in the special laws and introduces risks:

> The special laws of Venice after the disaster of 1966 were...very precise. They said that the Project should be reversible. Instead, the risky aspect of the new Project is that it is not reversible. When we want to take it away because there is a change in scenarios or a change in the climate, we know the most we can do is we can take away the barriers at the top, the submerged steel barriers. But we cannot take away all the 45 meters of foundations, the concrete that is in the foundations. (Sarto 2005)

Reversibility is a requirement enshrined in international and local council laws, according to comments from Fabrizio Reberschegg (2005). Cesare Scarpa (2005) described how the Ministry of Public Works had insisted that the building of Project MOSE should be reversible and experimental, while Michele Boato (2005) highlighted the illegality of the irreversible work on the project. This sentiment connects with some of the statements cited previously by activists who desired more effective state intervention against the private interests of the CVN. Cristiano Gasparetto (2005a) argued: "The Italian law says...that all the work should be reversible and modified and if the work is wrong, it should be eliminated. Project MOSE is not reversible. It is such a tough job that, if it is to be built, it will need this enormous quantity of public money, concrete and years. We will need huge investments, a lot of work and 10 or 15 years to take it away. It is work that has no possibility of reversibility." Similarly, Paolo Perlasca (2005) suggested that Project MOSE is irreversible because it is too costly to remove or change: "MOSE is different from the other irreversible projects because the structure is on the port inlets and it is an enormous structure that is under water. It is really difficult once you have created these foundations of concrete and steel to go and to take them back. That will imply big costs, even to change the Project as well. Nobody would say, okay I'm going to build the MOSE and I can take it away."

The focus on the cost of reversibility connects the dams to the resource question analyzed above. Fabio Cavolo (2005) also held the opinion that the dams could be removed, and stated that the CVN's managers on the Project have commented that they could demolish it with sufficient funds. This implies that reversibility is dependent on the quantity of resources available to remove the dams if this is considered necessary, and that the risk of irreversibility with the dams depends on the circumstances and approach, rather than being an absolute condition.

The conceptualization of risk in relation to experimentation also varied according to different interpretations. The absence of experimentation with Project MOSE is another key claim made by environmentalists, as stressed by Orazio Alberti (2005): "You know that the special law enforces that projects should be reversible, gradual and experimental. The fact is that Project MOSE is not reversible and has not been experimental. It is very important to experiment given that we work in a very unique environment, which is the Venetian environment. It is important to experiment because sometimes the result is not what we expected. Project MOSE is not like this."

As with the claims about irreversibility, the absence of experimentation was referenced to Venice's special laws. "It does not take into consideration the laws that the work should be reversible and experimental. They have not done any experimentation," Maurizio Zanetto (2005) told the author with certainty. However, the MOSE model placed near the Arsenal was designed for experimentation: its name is an acronym for *Modulo Sperimentale Elettromeccanico* (MoSE) (Experimental Electromechanical Module), and the Experimentation Center for Hydraulic Models at Voltabarozzo near Padua was created to model how Project MOSE will function. Orazio Alberti (2005) pointed out that modeling with both these means has been insufficient because it simplifies the lagoon's complexity: "Yes, one part of MOSE was put into a scenario and was left there forever. Now they are doing some mathematical models. They are experimenting with a scale model of the lagoon. However, this is a model, a simplification. The lagoon is much more complicated than the way the model describes it."

Yet the need for proper experimentation with Project MOSE was not perceived as an attempt to eliminate risks. "I don't want to say that we should construct something without risks. Everything has risks. This is a correct fact," explained Alvise Benedetti, a science professor at Venice University as well as President of Italia Nostra's Venice chapter. "However, it is unthinkable that something should be done like this. If you build your own house, you should think about it and try and test it out. This is our approach" (Benedetti 2005). He added that the irreversibility of Project MOSE violates a scientific principle of testing and therefore generates unacceptable risks:

> Our approach should really change and one important question is that the Project should be reversible. Irreversibility is a scientific, not technical, concept. Scientifically, if I go in this direction or the other direction, it should be the same. If I am building a project and I can see it is not going to work, I should go back without doing terrible things. But Project MOSE is not reversible. We cannot go back there. What I say is that on principle this Project is not good. If

> a project is reversible I can test it, I can try it and then I can improve it. It could take me 20 years and at the end I can find a system that I can control. When you put down these big casings, how can you take the casings out? How can you build the foundations again? The risk is connected to the flexibility of the system. (Benedetti 2005)

Federico Antinori similarly connected irreversibility to the failure to pinpoint risks in the early stages of Project MOSE:

> There are terms of some laws that talk not only about irreversibility, but also experimenting and the graduality of intervention. They say, 'let's be cautious about it.' They say, 'if there are some negative sides, let's go back.' However, we will never go back with MOSE. We cannot say there are negative sides to MOSE. We will never be aware of the negative sides of MOSE until the MOSE is there. Maybe we can destroy it, but the damage will be done and then it will be too late. (Antinori 2005)

Antinori thus suggested that the difficulty of identifying the risks presented by the dam project before completion meant that the project should not have been allowed to proceed. This desire to stop projects due to claims about irreversible environmental damage, despite a lack of scientific evidence of risks, is consistent with the precautionary principle, as defined by the 1992 UN Rio Declaration: "In order to protect the environment, the precautionary approach shall be widely applied by States according to their capabilities. Where there are threats of serious or irreversible damage, lack of full scientific certainty shall not be used as a reason for postponing cost-effective measures to prevent environmental degradation" (UNEP 1992). The 1992 definition has been used as a guide to legislation, as when the EC adopted the principle in the year 2000 (Commission of the European Communities 2000, 11). According to the science and technology analyst Indur Goklany (2001), many environmentalists apply the more strongly-worded definition given by the Wingspread Declaration that was produced by the Science and Environmental Health Network in 1998: "When an activity raises threats of harm to human health or the environment, precautionary measures should be taken even if some cause and effect relationships are not established scientifically. In this context the proponent of the activity, rather than the public, should bear the burden of proof" (cited in Goklany 2001, 2).

Environmentalists campaigning against the mobile dams conceptualized the precautionary principle in a rather more flexible fashion. "I think that the precautionary principle is one of those things that can be interpreted in different ways. If the feeling is that the risk is the sinking of Venice, the precautionary principle can take you to say do everything or not. So it depends on what the risk is," elucidated Edoardo Salzano (2005). The perceptions of risks related to the precautionary principle held by the Venetian environmentalists interviewed did not build on international interpretations: rather, their comments suggested that the precautionary principle should be used as a cultural guide for applying caution regarding the implementation of a new technology, in the manner outlined by Furedi (2002, 107). Local cultural influences, especially graduality, were dominant over international definitions. For instance, Giorgio Sarto (2005) believed: "The precautionary principle, this is a principle of steps, of stages. You should go up each step to do the work that you need to do little by little, that are safe and so it is possible to modify them."

References to the precautionary principle thus related to activists' local objections to the dams in terms of the project's lack of reversibility, experimentation and graduality. The elaboration of these concepts in the 1984 Special Law for safeguarding Venice conditioned the subsequent interpretation of the precautionary principle by campaigners. Tony Zamparutti, a member of the board of directors for Italia Nostra's Venice branch, wrote in *The Ecologist* magazine: "In language that presages the precautionary principle, the law calls for all interventions to be 'experimental, gradual and reversible'" (Zamparutti 2003). Many other established figures in Venice have applied the precautionary principle, as the author has explored elsewhere (Standish 2003a, 2003b). In claims made about Project MOSE, the precautionary principle was loosely deployed by environmentalists to identify risks, functioning as a cultural resource backing up claims about the lack of reversibility, experimentation and, especially, graduality. Using a loose interpretation of the precautionary principle for its application to environmental questions is not unusual given the difficulties with adopting a strict definition, as Furedi (2002) notes: "Since the full consequences of change are never known in advance, the full implementation of this principle would prevent any form of scientific or social experimentation" (Furedi 2002, 9). Rather, as Goklany (2001, 3) points out, the precautionary principle is often reduced to the assertion of the cultural saying "better to be safe than sorry."

Although the claims about the mobile dams have built on international trends within sustainable development, local cultural factors were dominant in perceptions of risk. Connecting claims to multiple local and international cultural themes provides activists with great power in motivating audiences. Yet for claims-makers to win audiences away from a proposal, it is also necessary to provide alternative solutions. We now consider how prognostic frames constructing solutions increase the likelihood that audiences will interpret social problems as claims-makers desire.

PROGNOSTIC FRAMING

Diagnostic framing by environmentalists established what they considered to be wrong with Project MOSE; and through motivational framing they sought to encourage audiences to oppose it. But there was awareness among activists that the problem of flooding in Venice demands solutions: "For Venetian citizens, there is another type of provocation. If high water comes, what am I going to do? That is a very particular aspect because we need to explain to these citizens that MOSE is not the answer to their problems," said Edoardo Salzano (2005). Consequently, environmentalists have felt the need to develop alternative solutions to Project MOSE for reducing flooding.

This prognostic framing has occurred over two stages. The first stage consists of various protection measures, including raising pavements, reducing the depth of the lagoon inlets, and improving the condition of saltmarshes. Such measures were advocated as alternatives to Project MOSE predominantly before the year 2000. From 2000 onwards, a range of alternative methods for blocking the three inlets to the lagoon to replace MOSE's mobile dams were proposed, which added to the first stage of protection measures. The objective in examining environmentalists' proposals against flooding is not to assess whether these would be more effective than the mobile dams from a technical perspective, but to ascertain what the various solutions tell us about the campaign.

Activists' Solutions – Stage One

Beppe Caccia listed three examples of alternative interventions to deal with high flooding that he advocated while participating in Venice City Council for the Green Party:

> [T]o only confront the specific question of the high waters, the exceptional high waters above 110 or 120 centimeters, there are a whole series of alternative interventions that cost much less than MOSE. They are easier and quicker to build. And they could definitively resolve the problems that exceptional high water creates for the social and economic life of the city. I can give you some quick examples; they redesign and remodel the three inlets to the port; the re-opening of the fish valleys and the parts of the lagoon that are now closed to the free expansion of the tides. We started all these interventions during the Council administration since 1994-5 with the raising of foundations and ground living levels for economic and productive activities. (Caccia 2005)

These alternatives were linked to the resource problem diagnosed with the dams, through their emphasis on lesser cost. They were also presented as responses to high flooding. According to Scotti of the CVN, re-opening the fish farms in the lagoon would only modestly reduce flooding: "Re-opening of the fish farms would produce an average reduction in tidal levels in Venice of just over 1 cm" (Scotti 2005, 251). Caccia's proposal to reduce the depth of the three port inlets to the lagoon seemed to be particularly important because it was repeated by several environmentalists as a vital measure to decrease flooding: they argued that the deepening of these inlets to allow bigger ships and oil tankers to enter the lagoon has increased the speed and quantity of water coming in, leading to more flooding. "This is a situation that has changed since the Napoleon era. Too much water comes in and too quickly . . . There is no obstacle to water and Project MOSE does not resolve this problem," remarked Fabio Cavolo (2005).

Marco Favaro (2005) predicted that the depth of the lagoon inlets would be made permanent with Project MOSE, leading to sediment loss that could lower the lagoon bed and Venice: "MOSE will also produce indirect impacts regarding the whole of the lagoon, in particular for the loss of sediments from the lagoon to the sea. To stop or reduce this loss we must change the configuration of the inlets, reduce the depth. But with MOSE the actual depth of the inlets will be fixed forever. In this way, we will fix a situation of great disequilibrium." Favaro's comments about the disequilibrium between the sea and the lagoon built on the arguments advanced during the campaign in the 1980s, as outlined in Chapter Five. Moreover, the claim that Project MOSE will set disequilibrium in stone for eternity draws on the motivational framing of irreversible change.

In addition to reducing the flow of water into the lagoon, environmentalists have proposed internal protection measures on the islands in the Venetian lagoon. Internal ('insulae') interventions to raise pavements, protect canals and using pumping and waterproofing have made significant progress with limiting the impact of flooding, as noted in a study conducted by Venice City Council in 1996 according to Caccia (2005). However, environmentalists recognized that these measures would not prevent the highest flooding. "So we can carry on for 10 or 15 years with one or two high waters a year that are not disruptive because the city has adjusted to it. All the electrical cables are higher. All the paths in the city are raised so people can walk around," said Cristiano Gasparetto (2005a). Flavio Cogo (2005) claimed: "In Holland and Russia they have much bigger problems of tides than we have. For what we have in Venice, we just wear our boots. In Holland and Russia they have tides of different dimensions: six meters, etc." These statements by Gasparetto and Cogo accept that some flooding will continue with the internal protection measures, and that these provide an incomplete solution. But is it effective to argue for wearing boots, or to stress that the problem is worse elsewhere? The

Venetian public are unlikely to be convinced on this basis that the internal measures could be a substitute for the mobile dams.

An alternative strategy employed by Stefano Boato was to argue that "Project MOSE is a big mistake because it just makes things worse" (Boato, S. 2005):

> It makes the equilibrium of the lagoon worse, which already in 1973, 32 years ago, the Parliament knew and legislated for because the lagoon and the whole city were not in equilibrium. By making the port inlets deeper and closing and intervening in the fish channels, a lot more water is coming in than before and there is not space to expand. This makes the tides higher. (Boato, S. 2005)

However, it is difficult to appeal to audiences with negative arguments. From this discussion of initial prognostic framing, it is evident that environmentalists had not developed very convincing alternative solutions to Project MOSE during the first stage. This is one of the reasons why alternative solutions for blocking the three inlets to the lagoon were created after 2000. As Stefano Boato (2005) disclosed: "We had to organize the alternatives because we have to say to people; what had to be done and how to do it."

Activists' Solutions – Stage Two

Between 2002 and 2005, 10 proposals for flood control at the three inlets to the Venetian lagoon arrived for review at Venice City Council. In December 2005, the Council held a public meeting to review the MOSE mobile dam project and these proposals, which were set out in a document (Venice City Council 2005).[2] A brief overview of seven flood protection projects was presented in the national newspaper *Corriere della Sera* (Fumagalli 2006).[3] In addition to the MOSE mobile dams, which were outlined in the Introduction to this book, six alternatives were described, which can be summarized as follows:

GRAVITY SLUICE-GATES. Submerged mobile barriers. Instead of rising against the tide, they are turned against it and lifted by it. Their presence is seasonable.

PERLA. A new floating maritime station at the Lido inlet would free it from the restrictions of the port: it could lower the sea tide by 20 centimeters.

ARCA. Closed sea inlets with self-sinking boxes 120 meters long anchored to temporary poles: requires different closing methods according to the sea level.

DOGE. Instead of metal devices, a big rubber sausage that could be inflated with either air or water. It is made from numerous parts and presents extensive closing flexibility.

SHIP GATES. At the beginning of winter these are put across the inlet openings; they rotate around fixed pillars that sink them.

TECHNICAL SERVICES. The Lido inlet closed with barriers and a 10km long dam half way across the lagoon, which would isolate Venice and defend it from the pollution of Marghera Port. (Fumagalli 2006)

A workgroup of experts reviewed all 10 projects for Venice City Council (2005) and narrowed them to five scenarios:

LP I: Closure of the inlets with lifting floating panels, submerged when resting.

LP II: Closure of the inlets with gravity panels, submerged when resting.

LP III: Total closure of the inlets with mobile, inflatable and submerged, resting barriers and the possibility of a partial cross-section blocking of the canals.

LP IV: Reduction of inlet cross-section using fixed elements, with the seasonable possibility of partial or total closure using removable structures (ship gates, auto-sinking or self-propelled vessels), which are resting. (Also including relocation of passenger terminal to the Lido inlet).

LP V: Temporary partition of the lagoon and total closure of the inlets with mechanical mobile barriers, submerged when resting. (Venice City Council 2005, 18)

The Council workgroup assigned values to each of the five scenarios above using the following criteria: economic and social impact, flexibility, engineering factors, sustainability and cost-benefit. The workgroup assessed the fourth scenario (LP IV) as the most effective for blocking high tides at the three inlets to the lagoon. However, the workgroup concluded that "so far no alternative proposal has been developed to the level of a preliminary project" (Venice City Council 2005, 26). The MOSE dam project had been through many phases and years of development: the alternative proposals had not been subject to comparable scrutiny, experimentation and assessment. The Council workgroup recommended that the alternatives should be further investigated with the possibility of providing an integrated defense from high tides: "Some of the solutions examined might not be alternatives, but integrated with each other in synergy. In particular, the solutions for partition and closure might and/or should integrate with each other. They should be in synergy with strategic solutions and with a wider range of systemic approaches" (Venice City Council 2005, 26). The wider range of systemic approaches referred to in this report's conclusion included analyzing the possibility of raising parts of the lagoon and city, which we will address in Chapter Nine. Yet the report was vague about whether the alternatives should be integrated with each other or could complement the mobile dams.

Although the Venice City Council workgroup was vague about the alternative proposals, it clearly favored an alternative scenario to the mobile dams. We should consider the context in which this negative assessment about the mobile dams was made. The document was published by a Venice City Council administration that consistently challenged the MOSE mobile dams. The former Venice Council Mayor, Cacciari, was elected in 2005 after declaring during the election campaign that he would contest Project MOSE; following the election he put together the workgroup to research alternatives to the mobile dams. This is the workgroup that produced the report discussed above, which was published by Venice City Council (2005). The workgroup involved a senior member of Venice City Council, Maurizio Calligaro, and a variety of experts from universities and institutions, including IUAV Professor Andreina Zitelli, a well-known critic of the mobile dams, who had participated in the Environment Ministry Environmental Impact Assessment Commission that produced a negative report on the MOSE EIS in 1998 (see Introduction and Chapter Five). No representatives of the CVN or MAV were included in the workgroup. In addition to selecting members who were already critical of the mobile dams, Venice City Council made its opposition to the dams clear while the workgroup was researching the alternatives. In June 2005, the Council ordered local police to block building work on Project MOSE and to check authorization documents, leading to a brief interruption of the dams' construction.

Venice City Council heavily conditioned the climate in which the workgroup made its assessments of the mobile dams and the alternatives. But what do environmental activists in Venice think about the alternative proposals? Most of the environmentalists interviewed favored the alternative proposal named 'ARCA', which was listed on the Venice chapter website of Italia Nostra (2007c) as the alternative method to MOSE for closing the lagoon's inlets. The gravity sluice-gates, proposed by Vincenzo Di Tella, were also discussed in positive terms by several activists. For instance, in June 2010 Cristiano Gasparetto favorably compared the gravity sluice-gates to MOSE, following a report by Principia R.D., a French modeling company, which raised concerns about a possible design defect with MOSE's barriers under steep wave conditions (Gasparetto 2010; Pirazzoli 2009). Principia's report advised that the "MoSE gate requires an active control system of the water ballast to maintain the design working condition" (Principia R.D 2009, 10). This may be a positive suggestion that engineers should examine, although the report also pointed out many problems with modeling the mobile dams.

While some MOSE critics made positive comments about the gravity sluice-gate alternative to MOSE, Michele Boato of the Eco Institute explained how the alternative projects could be used in different ways at various stages: "It is better to start with ARCA because in six months with little money it is possible to show that the problem can be solved. Then we could start on experiments with Di Tella, but taking it easy" (Boato, M. 2005). Boato thus linked these alternatives to the resource problem diagnosed with the mobile dams, and drew on the idea of the alternatives being moveable and experimental. Giorgio Sarto (2005) highlighted this advantage compared to the inflexibility of Project MOSE, and Paolo Perlasca (2005) suggested that the flexibility of the alternative projects such as ARCA meant that they could be integrated into the building work that has already been done in preparation for Project MOSE.

The ARCA proposal, with floating moveable barriers, is attractive to environmentalists in favor of reversibility because these barriers could be replaced by another tide-blocking system. Enrico Sambo (2005) described how the ARCA floating barriers could be placed at the three inlets of the lagoon during winter when there is more flooding, thereby complementing the Stage One solutions examined above; Cristiano Gasparetto (2005a) explained how ARCA would be experimental, drawing on this aspect of motivational framing analyzed earlier in this chapter. None of the environmentalists interviewed advised that the preparatory work for Project MOSE should be scrapped: they claimed that alternative projects would develop work already done for the mobile dams. By mixing the modern techniques of the mobile dams with alternative technologies, environmentalists' proposals for blocking the inlets to the Venetian lagoon build on interest in alternative technologies that dates back to the 1970s, when Ivan Illich (1973) explored alternative ecological engineering technologies, which he called "tools for conviviality", and Ernst Friedrich Schumacher (1973) recommended alternative technologies that would mix modern and traditional techniques in a creative combination. Yet it was not until 2001 that Venetian environmentalists advocated integrating Project MOSE's traditional, modern approach with what they considered as alternative, environmentally-friendly technologies.

Combining existing work done for the mobile dams with alternative technologies appears to have public appeal for several reasons. Firstly, integrating alternative technologies with preparatory construction for the dams has been pitched as environmentally-friendly because resources already used would not be wasted. This linked the alternative proposals to the resource issue. Secondly, the alternative proposals for intervening at the lagoon inlets would directly address the problem of the highest flooding. Together with some of the proposals from Stage One, they seem to present an effective solution for flooding at different heights, requiring no need to resort to suggestions that Venetians should just wear boots. The alternative methods for blocking the three inlets to the lagoon developed the local framing of risk in the campaign by following the principles of reversibility, experimentation and graduality. Beppe Caccia (2005) stated that all the alternative dam proposals were compatible with the 1984 Special Law requirement of reversibility, and Fabrizio Reberschegg (2005) argued that the absence of fixed foundation casings in the alternative projects like ARCA makes them reversible.

Another aspect of the alternative projects for blocking the lagoon inlets that environmentalists welcomed was their complete rejection by the CVN. In December 2005, the CVN and MAV refused to take part in the high-profile public meeting organized by Venice City Council to review the alternatives to MOSE. The CVN defended this refusal by documenting that a wide range of alternatives had already been assessed during the conceptual design of the project in 1989 (CVN website 2009b). The rejection of environmental campaigners' preferred alternatives helped to develop the CVN as the prime villain in the campaign. Marina Corrier (2005) said that the CVN and MAV were trying to avoid assessments of Project MOSE and the alternatives; Michele Boato (2005) commented that the alternative projects had not been considered because they were not backed by the key politicians, yet Edoardo Salzano (2005) stated that he believed Romano Prodi supported the ARCA Project. When Salzano was interviewed in December 2005, Prodi was preparing for the general election in 2006, which he won.

With the election of a new government, Venice City Council also promoted the alternatives to MOSE. In April 2006, Venice City Council voted to approve a document on modifying the blocking of the three lagoon inlets with alternative methods, and sent the document to the government. But in November 2006, Prodi's government gave its final approval to the MOSE works, as did the Comitatone; and in December 2006, the Minister of Infrastructure, Antonio Di Pietro, rejected alternative proposals to MOSE. "The Council presented alternative projects but it was decided to only finance MOSE, even though it is the most costly project," bemoaned former Venice Mayor Cacciari (cited in Dina 2009b). Whatever the possible merits of the alternative systems, the Prodi government never seriously entertained the possibility of replacing MOSE. Alternative systems had already been scrutinized in the 'competition of ideas' that was initiated in 1970, as well as at the conceptual design stage in 1989, and the consideration of alternative systems came after preparatory construction work for the MOSE system had begun. None of the alternatives to MOSE had been developed to the level of a preliminary project (Venice City Council 2005, 26). Stefano Boato (2005) explained how he helped bring engineers together in weekly meetings to develop the alternative proposals from 2001, but the proposals were not integrated into the campaign and reviewed by Venice City Council until 2005: reflecting, as Orazio Alberti's comments in Chapter Five indicate, that environmentalists' prioritization of the campaign against MOSE came late in the game (Alberti 2005). The re-election of Cacciari as Mayor of Venice in the City Council election of 2005, and the election of Prodi as Prime Minister in 2006, created political opportunities for environmentalists, but the lateness of presenting alternatives to MOSE at this stage and the lack of engagement with them at a national political level meant they were never seriously considered. Nevertheless, when the author interviewed Flavio Cogo for the second time in February 2010, Cogo believed that the dams could still be adapted using the alternatives: "We intend, as always, to change the direction of operation, block the work and to recover all that is possible. Then, in time, we would apply the alternative projects, which are on the Council website" (Cogo 2010). At this stage, the dams were 60 percent built: but Cogo held out hope that the alternatives could still be implemented so long as the barriers were not in place.

MYTH AND REALITY IN ENVIRONMENTAL CLAIMS

Diagnostic claims about damage caused to the natural environment during the mobile dams' construction have a real basis. According to the OECD (2010, 85), the dams are the largest infrastructure project in Italy and the following work was completed by March 2010: 1,400 hectares (3,459.475 acres) of tidal mud flats, saltmarshes and sandbars had been reconstructed; 35km (21.75 miles) of industrial channels and five former landfills had been sealed to prevent leakage into the lagoon; 100km (62 miles) of embankments had been rebuilt; 45km (28 miles) of beaches had been created; and 10km (6.2 miles) of wharfs had been reconstructed. Some of these measures had improved the lagoon environment, but the excavation of beaches has undoubtedly disturbed birds, other wildlife and vegetation, especially on the Lido Island. The construction of breakwaters and the placing of the casing foundations for the dams have also disrupted marine life and introduced some pollutants into the water. It is virtually impossible for a large-scale engineering project to be constructed in an area containing a significant number of wildlife habitats without causing environmental damage. It is also true some pollution will occur during the operation and maintenance of the dams, although the extent is currently unclear and should be carefully monitored.

The environmental costs of the dams are often traded against the benefits of reducing flooding for Venice, other islands in the lagoon and the people, who live, study, work and visit the islands. The Commission (*Collegio*) of International Experts (1998) judged the benefits as outweighing the costs. Environmentalists disagree with this assessment, and this is because the benefits cannot be compared with the costs technically. Whether the environmental damage during the construction of the dams is worth the protection they will bring against flooding is a political, not technical, question. It depends on whether we prioritize protecting Venice or the natural habitats near the building sites. Ultimately, this is a political choice between defending nature and developing human society: an argument expanded in Chapter Nine.

The diagnostic claims about the aesthetic impact of the dam project raise similar political questions. Now it is well known that the dams will be submerged under water, hugely reducing their impact on the landscape compared with projects like London's Thames flood barrier. Nevertheless, structures have been built that affect the landscape, such as the artificial island at the Lido inlet. The principal complaint made against this island by Italia Nostra was its 20-meter (65.6-foot) tower: but as discussed, this tower was the subject of negative aesthetic claims while numerous similarly sized lighthouses near the lagoon inlets, including those that are inactive, have been ignored. This tells us that the environmentalists' claims about the tower are motivated more by the need to construct claims against the dams than by obstructions of the view. The building sites, cranes and erected structures have disrupted the landscape; but again, whether these have a negative aesthetic impact is a subjective matter. Are the human and environmental benefits of the mobile dams worth significant short-term and minimal long-term changes to the landscape? This is a matter of opinion governed by whether natural or human needs are considered priorities.

The claims about the financial costs of the mobile dams also depend on what we prioritize. Keahey (2002) points out that the $1.85 billion cost of the mobile dams should be compared to $15 billion spent on the Channel Tunnel between the UK and France, $14 billion spent on the 'Big Dig' interstate road in Boston (Massachusetts) and $1.75 billion spent on refurbishing Rome for the 2000 Holy Year Jubilee (Keahey 2002, 223-225). The cost of building the dams has increased since 2002, and the CVN website put the total cost in November 2010 at €4.678 billion (US$6.12723bn) (Consorzio Venezia Nuova 2010a). This is still not a huge figure in comparison with other contemporary infrastructure projects, as we outline in Chapter Nine.

There is some truth in the claim by environmentalists that the dams have meant Venice has been left with less money for other projects. Yet we should bear in mind that Venice receives significant resources from Rome that other cities in Italy do not enjoy. The special laws for Venice have brought substantial funding to the city since 1973. Successive governments have also provided extra funding to support the construction of the dams. Additional money for the dams has been offered from other sources, including €1.5 billion (US$1.96532bn) from the European Bank of Investments. Overall, the dams have attracted funding offers for the city that otherwise would not have been forthcoming; thus the environmentalists' claim that the dams have generally drained resources from the city is over-simplistic. The specific claim that young families struggle with housing costs in Venice is based on a real problem (Poveledo 2006). But the reasons for this problem are complex, including property prices, incomes and state support. To claim that the financial problems of young families are solely due to the dam project lacks justification, and the families' presentation as victims of the project distorts a more complex reality.

Regarding the principal villain in environmentalists' claims about the dams, it is true that the CVN has largely monopolized their planning and construction. This is the legacy of the political circumstances in which the CVN was created, and it has been difficult to change as the dam project has progressed. Over time, some aspects of the project have been put out to tender and carried out by other companies. The claim made by several environmentalists that the CVN is not subject to state control, however, does not stand up to scrutiny. If this were the case, how could we explain that the project was blocked by the Ronchi-Melandri government decree in 1998? The overview of the project's development offered in this book's Introduction gives numerous examples of how the CVN was subject to impositions by successive Italian governments, and the CVN is governed by the decisions made by the Comitatone, which is chaired by the prime minister. The CVN has become powerful in Venice, which is not surprising given the impact of the dam project on the city, and City Council administrations that have opposed the dams have occasionally come into conflict with that power, finding it difficult to affect policy through the Comitatone. But the CVN does not have a monopoly on power: as illustrated when Venice City Council brought in local police briefly to stop construction work on the dam project in June 2005. In reality, the local state has some influence over the dam project, while the central state dominates the CVN and the project through the Comitatone and the Objective Law.

It is also difficult to justify the environmentalists' claim that the CVN has distorted the social fabric of Venice. Undoubtedly, Venice has altered since 1984, when the CVN was created, and as the dam project has progressed. It is correct that the indigenous population of the city has decreased and the proportion of non-Venetian inhabitants has increased. This has been driven by many companies moving premises out of Venice, changes in employment in the Marghera industrial area, property prices and multiple socioeconomic factors. The fluctuating composition of the city's population cannot be attributed to the impact of the CVN and the dams.

Three themes connected to risks presented by the dams were promoted by environmentalists with the aim of motivating audiences. The irreversibility of the dams was strongly emphasized, building on requirements in special laws for Venice. Irreversibility also connected with the international application of this theme in environmental management, especially through the UN definition of the precautionary principle. But Venetian environmentalists themselves explained how it would be difficult - rather than physically impossible - to remove the dams. While it is hard to accept the claim that the dams are irreversible, perhaps it is even more important to consider why we should focus on removing a project before we have observed its effectiveness. Moreover, the question of removing the dams would depend on a future situation that we cannot evaluate now. With technological developments, climate change and economic shifts, we can be confident that in 20 years' time we will be in a different position to make this assessment. It might be possible to adapt the dams by replacing the moveable barriers with different barriers or using the foundation casings in another way. However, we cannot address the question of whether we should remove the whole project now. The irreversibility claim is currently inappropriate and, according to statements by environmentalists themselves, lacks substance.

Venetian environmentalists claimed that the lack of experimentation with the dams has generated risks. Alberti's (2005) comment that modeling has over-simplified the complexity of the lagoon holds some truth: by their very nature, models simplify reality. However, in addition to modeling the dams in the Voltabarozzo Experimentation Center, the MOSE experimental model did test a mobile gate in the lagoon. What else could have been done to experiment with the dams before building them? As Benedetti (2005) remarked, not all risks can be eliminated before proceeding with construction: this would violate basic science.

This brings us to the claims about risk and graduality. Venetian environmental claims about the lack of graduality with the dams were not based on the UN or Wingspread definitions of the precautionary principle; rather, they were linked to the special laws for Venice and operated as a cultural guide for applying caution. One wonders just how gradual the design and construction of the dams should have been to satisfy environmental demands. As outlined in the Introduction, the idea for the dams emerged in 1970 and the design was modified many times before construction began in 2003. Delays were caused by many factors and were not simply because caution was being applied: nonetheless, the claim that the dam project has not been gradual does not stand up.

Finally, we need to assess the balance of myth and reality in environmentalists' claims about solutions to flooding. Most of the solutions put forward in the first stage of their proposals seem to have real advantages. Raising pavements and reinforcing local defenses are positive initiatives, and have been implemented with success over the past two decades. There is some evidence of benefits for flood protection by regenerating saltmarshes and re-opening fishing valleys in the lagoon. Scotti (2005) notes the modest benefits of re-opening fish farms, and describes how the CVN has helped reconstruct significant areas of saltmarshes and mudflats (Scotti 2005, 250 and 255). But these measures have a limited impact on high flooding, as confirmed by Rinaldo (2009, 167-170). The claim that raising the height of the three lagoon inlets would reduce flooding is more controversial. This is because this claim is directly linked to the lowering of the inlets to build the mobile dams. Environmentalists have claimed that lowering the inlets increases the speed and quantity of water flowing into the lagoon, as well as sediment loss from it. These claims appear credible if we assume the dams are absent: in the past, lowering the height of the inlets for bigger ships to enter the lagoon may have increased flooding. In early 2010 Paolo Canestrelli, the Director of the Center for Tide Forecasting, did record a small increase in the speed of some currents entering the lagoon during the preparatory work for the dams, although he regarded the effect on flooding as negligible (Testa 2010). Yet once the dams are completed, lowering the inlets comes with the possibility of also closing them. Dam closures would stop water from flowing into the lagoon and could be used to regulate sediment loss. Therefore, the claim that deepening the inlets for the mobile dams will increase flooding does not take the advantages of dam closures into account.

Some of the environmentalists' first stage solutions had real advantages, although others lacked supporting evidence. The principal problem with all their initial proposals was that they only aimed at addressing low-level flooding. As sea levels have risen, these solutions have become less appealing to Venetians who want protection. The second stage of environmentalists' solutions supported the promotion of various alternative projects to the mobile dams for blocking the lagoon inlets. But these were proposed too late to be considered as alternatives to the mobile dams. Politically, the momentum at a national level was behind the MOSE dams. Practically, alternatives had been examined long ago and preparatory construction for the dams was proceeding. This is not to imply that the alternatives to MOSE put forward during the Venice City Council review in 2005-6 did not contain new or useful ideas: they were seriously put together based on engineering expertise, and contained real innovations that could be beneficial in the future. However, as the Venice City Council workgroup report concluded, the proposed alternatives to MOSE had not even developed into preliminary projects and need further investigation.

Dividing Venetian environmentalists' claims into diagnostic, motivational and prognostic framing helps to assess the balance of myth and reality. Overall, there is more reality in their diagnostic than their motivational claims. The building of the dams has caused locally-concentrated environmental damage. We could regard this damage as a reason not to build the dams; or we could treat this damage as a small, though inevitable, drawback in creating a defense against high tides, which should improve life for people in Venice. The motivational framing included some real claims, such as the monopoly of the dam project by the CVN. Yet developing this monopoly status into presenting the CVN as a monstrous body is akin to creating a modern mythical demon: the CVN does not dictate policy to the Italian state and has not depopulated Venice. The construction of young couples and others as victims of the dams has similarly distorted reality. Regarding prognostic framing, the first stage contained some valid solutions, although only for low-level flooding. The alternative solutions to the mobile dams for high flooding may provide some possible improvements in the future, but they are currently insufficiently developed to offer real alternatives to the mobile dams.

NOTES

1. All currency conversions were correct on 21 December 2010.
2. In explanations of blocking the inlets, "submerged" refers to structures that are below the surface of the water and "resting" means structures are on the seabed.
3. A more detailed explanation of all the alternative projects can be found in section 6 of this website: http://www2.comune.venezia.it/mose-doc-prg/.

Chapter Seven

Myths about Venice's Mobile Dams

"All social movements construct myths. By myths, I refer not to the popular use of the term as a false belief, but rather to the anthropological notion of a myth as 'a sacred narrative explaining how the world and people came to be in their present form' (Dundes 1976, 279)."

Robert Benford (2002, 57), Professor and Chair of the Department of Sociology at Southern Illinois University.

This chapter explores the interplay between myth and reality about the mobile dams to assess what this reveals about contemporary claims and perceptions of Venice. In the above quotation, Benford reminds us that myths should not be treated as lies to be dismissed because they are not supported by evidence. On the contrary, an appreciation of how myths shape claims is vital to understanding current issues of modernization and sustainability in Venice. Judging the balance of myth and reality in these claims and establishing the truth is necessary to guide policy decisions.

MYTHS AND CULTURAL NARRATIVES

Cultural narratives are stories that often build on past myths and can also help create new myths. They are "general or canonical narrative models that exist within any cultural setting, functioning to pattern and constrain the types of stories regarded as plausible and acceptable within that setting" (Davis 2002, 23). It is also important to distinguish between individual and movement cultural narratives. In contrast to individual narratives, "*[m]ovement narratives*...refer to the various myths, legends, and folk tales, collectively constructed by participants about the movement and the domains of the world the movement seeks to change" (Benford 2002, 54. Emphasis in original). Movement narratives are different from collective narratives. The American sociologist John Steadman Rice (2002, 81) explains that collective narratives are what we conventionally refer to as history, which is not exclusively constructed by movements.

This chapter concentrates on cultural movement narratives, rather than individual or collective narratives. Cultural narratives play special roles in linking past and present events that may otherwise be perceived as distinct. This point is crucial for our analysis of how the history of the Venetian Republic is connected to contemporary environmental campaigns. When the author began the process of interviewing Venetian environmentalists about their campaigns against the mobile dams, the telling of stories about the Venetian Republic initially came as a surprise. As the interviews progressed, it became clear that these narratives have played a critical role in environmental claims-making. This insight was produced through the narratives told by Venetian activists and their interpretation by the author: as Davis (2002) explains, "Stories involve two parties, a teller (or narrator) and an interpretive audience (listeners/readers)" (Davis 2002, 16). Interpreting how environmentalists linked the Venetian Republic to claims about the mobile dams requires a detailed deconstruction of their narratives.

CULTURAL NARRATIVES ABOUT THE VENETIAN REPUBLIC AND THE MOBILE DAMS

Parts of an unpublished letter by Lidia Fersuoch (2004), who was Vice-President of the Italia Nostra Venice chapter in 2004, highlight the principal debates about the Venetian Republic and the MOSE dams:

> [A]ll the interventions were reversible. This approach guaranteed the survival of the Lagoon and of Venice. Those who claim instead that the survival of the Lagoon depended on the courage [of] those who forced through the start and completion of gigantic works evidently do not know well the history, nor the approach of the Republic's leaders. The Venetians were anything but courageous, at least in the banal sense of the word. This can be proved for the two works that most frequently are invoked by the Mo.S.E.'s supporters: the re-routing of rivers flowing into the Lagoon (to stop sediment from filling it in) and the construction of the *Murazzi*, the sea wall along the Lido and other barrier islands, to protect against storms. (Fersuoch 2004)

The letter introduces how both opponents and proponents of the MOSE dams have constructed cultural narratives about interventions during the Venetian

Republic to provide weight for their contemporary claims. The proponents believe these interventions prove that the Venetian Republic aggressively curbed and controlled the impact of nature, which is also necessary with the mobile dams; while opponents argue that these ancient interventions respectfully guided nature in a manner ignored during the construction of Project MOSE. During their interviews with the author, campaigners told narratives that formed the myth of an environmentally-friendly Venetian Republic. Fabrizio Reberschegg (2005) identified the Republic as having made a special contribution to the lagoon's ecosystem: "The lagoon ecosystem was created over time, thanks to the Republic of Venice that had some specific connotations of flora, fauna and birds. Now these are in discussion." According to various narratives, major interventions during the Venetian Republic, especially river diversions and the building of sea walls, are framed as environmentally-friendly in contrast with many modern interventions. These cultural narratives about the Venetian Republic and their deployment as an argument against the mobile dams will be referred to as 'Venecoism.'

Venecoism appeals to an ideal past condition of equilibrium between man and nature. Chapters Five and Six explained how Venetian environmentalists campaigned to restore the lost equilibrium between the city, the lagoon and the sea, especially during the 1980s. Framing the history of an ecological problem in this way is not uncommon, as Hajer (1998) suggests: "The historical account is framed around the sudden recognition of nature's fragility and the subsequent quasi-religious wish to 'return' to a balanced relationship with nature" (Hajer 1998, 251). In the context of Venice, this insistence on a harmony between society and nature has led to a contemporary emphasis on recreating equilibrium between the city, the lagoon and the sea. The construction of the Venice problem in the aftermath of the 1966 high floods was predicated on perceptions of increasing fragility and disequilibrium, which in turn stimulated the politics of safeguarding the city. Safeguarding shares many assumptions with the cultural narrative of Venecoism, in combining the contemporary protection of Venice with the myth that the Venetian Republic achieved equilibrium. Beppe Caccia (2005) described this relationship:

> The other theme that is a current and historical initiative for the Green Party is the safeguarding of the city. It is a theme that, from one point of view, comes from the history of the Serene Republic and with the difficult and delicate relationship of the equilibrium from 800 or 900 years after Christ until now. It is a millennial theme and was constructed between human activities and natural phenomena in the Venetian lagoon. There is a long history of the relationship between anthropic presence and the equilibrium of the lagoon environment with the sea on one side and the earth on the other.

Venecoism is countered by cultural narratives advocating anthropic intervention in the lagoon. Positive interpretations of man-made projects during the Venetian Republic have been discussed to support the case for contemporary interventions, such as the mobile dams. These cultural narratives will be referred to as 'Venanthropocentrism.' Cristiano Gasparetto (2005a) argued that there was human intervention in the lagoon during the Venetian Republic, although this was controlled to prevent nature from being abused: "The difference in the past, in all the history of the Venetian Republic, is that anthropocentrism, which is the use and modifications of the lagoon by human beings, was always studied because the natural system could get metabolized. So some transformations were wanted by man, but were compatible with the natural system." These remarks build on a wider critique of anthropocentrism within green thinking: as Bramwell (1994) notes, "One of the key elements of ecologism is its opposition to anthropocentrism" (Bramwell 1994, 182). In the cultural narrative constructed by Venetian environmentalists, the Venetian Republic guarded against the anthropic exploitation of nature whereas now, nature is neglected. By comparison, Venanthropocentrism welcomes aggressive human intervention in the lagoon during the Venetian Republic and today to restrain the negative impact of natural forces.

Supporters of the mobile dams claim that the dams follow the tradition of significant projects by the ancient Venetians to protect the city, especially the diversion of rivers and the building of dam walls. This cultural narrative has been used by Rinaldo (2001) to bolster his argument for interventions in the lagoon, including the mobile dams: "throughout the centuries, Venice's rulers have had to continually intervene and tamper with the lagoon in order to fashion it into a liveable and vital environment in accordance with the demands of the city's economic and social development" (Rinaldo 2001, 66). But positive claims about the Venetian Republic's use of anthropic intervention are contested by Venetian environmentalists, as indicated by Fersuoch's (2004) letter cited above.

Competing Interpretations of the Venetian Republic's Interventions in the Environment

Fersuoch (2004) identifies the two main points of conflict in the debate about the mobile dams and how the Venetian Republic intervened in the environment: the diversion of rivers, and the building of sea walls (*murazzi*). The walls were constructed during the eighteenth century, measuring four kilometers (two and a half miles) long and consisting of stone blocks, fourteen meters (45 feet) wide at the base. The sea walls were a "tremendous bulwark" (Norwich 1983, 604), which held back the sea until the storm of 1966. This project was a considerable feat, given the limited technology of the time, and was fundamental to maintaining the lagoon. Likewise, the diversion of the Piave and Sile rivers between 1324 and 1683 was necessary to prevent the lagoon from becoming dry land due to silt from these rivers (see Figure 1.1). The river diversions also limited the replacement of salty water with fresh water in the lagoon. More fresh water could have exaggerated mosquito breeding and fatal malaria, and Venetians were mindful of this threat after malaria lead to multiple deaths at the early Venetian settlement of Torcello (Lane 1973, 16). Maintaining the lagoon was crucial for the city's trading port and defense against invaders, who found its shallow waters and hidden underwater channels almost impossible to navigate.

"The river diversions and seawall constructions were incredible engineering feats—as incredible in their time as the modern proposal to put giant hydraulic steel gates at the lagoon's three mouths to protect Venice from widespread flooding," writes Keahey (2002, 71). While Fersuoch (2004) argues that these protection measures were not examples of courageous, anthropic intervention, Fabio Cavolo (2005) acknowledged that they defined the existence of the lagoon: "The work involved a lot of expenditure by the Serene Republic. They created the defense against the sea and the diversion of the rivers and now we have this lagoon. Now we have this lagoon since 1791. This great work made the lagoon in Venice." This sense of grand achievement was also described in a mid-sixteenth century sonnet attributed to Cristoforo Sabbadino, the chief hydraulic engineer of the time:

> How great your walls were, you know
> Venice, now see how you find them.
> and, if you don't provide against their ruin,
> deserted and wall-less you will remain.
> The rivers, the sea, and men you have
> as enemies, and for proof and belief
> don't delay, open your eyes and get moving,
> (otherwise) what you wish to do, you will find
> impossible.
> Drive the rivers away from you, and check the greedy
> desires of men and then, only the sea
> remaining, it has always obeyed you.
>
> (Cited in Caniato 2005, 11)

The river diversions and sea wall constructions were artificial projects that transformed the lagoon to curb the destructive impact of natural changes. On this basis, Rinaldo (2009) argues "It is a mistaken idea that the lagoon of Venice was in a magical state of hydraulic equilibrium and morphology during the times of the most Serene Republic" (Rinaldo 2009, 172). This is a formidable argument against Venecoism, which assumes a past harmonious equilibrium between man and nature during the Venetian Republic. Marco Favaro of WWF Italy recognized that there cannot be a static interpretation of the state of the Venetian lagoon, and that river diversions can arrest the natural evolution of a lagoon to dry land: "You know a lagoon is considered a temporary state from a naturalistic point of view. The natural evolution of a lagoon is to become dry land because rivers put the sediments into the lagoon year after year and transform the lagoon into dry land. For this reason, Venetians, starting in the sixteenth century, decided to put the rivers outside the lagoon" (Favaro 2005). Despite this evidence, Venetian environmentalists whom the author interviewed asserted that the Venetian Republic followed nature, not intervening aggressively but guiding it gently. For example, Edoardo Salzano (2005) claimed:

> All the intervention should be gradual, experimental and reversible, mainly. Everything should have been done in a correct way. The logic of the Venice Republic was to follow nature and guide it. They used to do a little bit of channel, carefully. They could check what would happen. If it was okay, they would dig it and finish building. If it had gone badly, they would change direction. So this is the experimental way of following nature and its evolution without ever violating it. When there was the risk that too much sand would come into the lagoon, they intervened and moved the two rivers so that less water would come into the lagoon.

The three principles of experimentation, reversibility and graduality have frequently been deployed by environmentalists to negatively assess the MOSE dams, as outlined in Chapter Six. Here we consider how these principles were grounded in the cultural narratives constructed by Venetian environmentalists to provide their contemporary claims with additional legitimacy.

Narratives about Experimentation, Reversibility and Graduality

"When there was the Venetian Republic, they always tried to use the reversibility criteria; graduality, experimentation and reversibility. Then they were taken into consideration by Italian law. The Italian law put them as conditions for Project MOSE. But Project MOSE did not take these into consideration," stated Edoardo Salzano (2005) in an interview with the author. Fersuoch (2004) contends that respect for these principles of experimentation, reversibility and graduality during the Venetian Republic and their presence in Italian law meant environmental associations were previously unnecessary:

> [A]t the time of the Republic of Venice there was no need for either environmentalists or for associations like ours, because the criteria that inspired the

> various magistrates and councils in charge of the waters then were perfectly in line with those that we invoke now (and are also invoked by Italy's special laws for Venice): gradual, testable, reversible actions. Reading the thousands of documents produced by the Venetian authorities, one can see that when it came to managing the Republic's waters, public prudence guided all choices. Each project, decided after a long debate with experts, engineers, 'politicians' and citizens with practical knowledge...was tested over a long period...Frequently, after observing closely and over a long period the effects of a project, they decided to scrap it and return to prior conditions. This was always possible because all the interventions were reversible. (Fersuoch 2004)

Examining whether these environmental principles really were followed during the Venetian Republic helps us to assess the balance of myth and reality in such claims. There is historical evidence of conservationist sentiments towards construction during the Venetian Republic, especially after the fourteenth century. As Crouzet-Pavan (2002) notes, when selecting plans for the reconstruction of a market after the fire of 1514 at Rialto, "a persuasive reason for accepting Scarpagnino's project was that he respected the history and the nature 'of the place.' Venetians deliberately chose to maintain their ties with the architectural past and rebuilt in coherence with it" (Crouzet-Pavan 2002, 26). She also observes astutely that the desire for political continuity translated into construction focusing on restoration rather than transformation: "[A]n ideology of government further reinforced Venice's concern for continuity. Duration, seen as a desideratum, and an explicit respect for previous modes of governance explain the Venetian attachment to tradition and to the culture of the city's origins. Conservation, restoration, and refashioning were long the rule. The Byzantine portion of the Ducal Palace did not disappear until a fire gutted it in the fifteenth century." (Crouzet-Pavan 2002, 26)

Yet these conservationist desires in the fifteenth century Republic need to be balanced alongside the century's "building boom" (Crouzet-Pavan 2002, 30), which included some 200 palaces along the Grand Canal and created one of the biggest urban complexes in the medieval West. In addition, there had been previous periods of rapid transformation and land reclamation. As noted in Chapter Three, land reclamation has been associated with modernization in the nineteenth and twentieth centuries and environmentalists have called for reopening of reclaimed land to reduce the lagoon's water level and flooding by increasing the size of the lagoon (Italia Nostra Venice 2007c). But in the early fourteenth century the Venetian state took direct command over the creation of the Giudecca Nuova Island and there is plenty of historical evidence that tracts of land were reclaimed during the Venetian Republic. The frantic pace of land reclamation and construction during the thirteenth and early fourteenth centuries raises questions about whether concerns about graduality and the environment were respected:

> Expansion continued throughout the thirteenth century, but the pace was particularly brisk up to the early 1340s. In spite of the large number of victims of epidemics, the first demographic declines, in 1307 and in 1320, did not yet affect the rising population curve. Pressure continued, and building activity must have been intense. Day after day, at the edges of Venice, at the foot of hundreds of gardens on the Giudecca, at Santa Croce, or at Cannaregio, Venetians planted pilings and set planks into place to shore up a few square meters of spongy soil. Householders dumped their rubbish there, along with a bit of soil or mud, slowly nibbling outward to gain ground. Flotillas of boats loaded with dirt and refuse circulated among the various quarters of the city. Mud from canal dredging, gravel from construction sites, sweepings from the market or the streets—everything contributed to filling in one more pond or reclaiming one more parcel of land. (Crouzet-Pavan 2002, 12-13)

Some projects designed to protect the lagoon during the Venetian Republic took many years to execute, and this fact adds weight to the Venecoist emphasis on graduality. For example, construction of the sea walls along the shores of the Lido continued throughout much of the eighteenth century. Giovanni Caniato (2005), who has worked at the Venice State Archive, notes that delays were not necessarily due to lengthy testing: they were often caused by economic and political constraints, including border disputes with barbarians on the mainland (Caniato 2005, 10-11). It should also be borne in mind that other interventions during the Venetian Republic were carried out remarkably quickly:

> Sometimes, when the government finally decided to realize important hydraulic enterprises, such as massive excavations or river diversions, from the moment of the decision to the final execution a very short time elapsed. Just think—for example—of the diversion of the lower branch of the Po, the greatest of Italian rivers. Once it became clear that the easterly flows from the delta contributed to the silting of the southern portion of the lagoon, it was quickly decided to divert it to the south. The Senate took the final decision in the year 1600 and a huge excavation was completed in 1604, in spite of the delicate diplomatic problems that arose, the lower portion of the Po river being a political boundary between the Republic and the still powerful and certainly not always friendly State of the Roman Church. (Caniato 2005, 10)

The environmental interpretation that work in the lagoon during the Venetian Republic was based on the principle of graduality thus needs to be assessed against the reality of rapid interventions, and the fact that when projects did progress gradually, this was for reasons often unrelated to conservationist concerns.

Regarding the environmental principle of experimentation, Orazio Alberti commented on the use of a prototype for the building of the sea dam walls during the Venetian Republic, and contrasted this with the absence of experimentation for the MOSE dams: "So, pay attention, what did the Venetian Republic do? They did not then construct dam walls. They made a prototype and they put it in front of the sea by some meters. When they saw that it was durable and it did not let water in, then they built it. First, they experimented with it and they saw that it was working, then they built it" (Alberti 2005). On the other hand, Alberto Scotti (2005), a leading engineer on the mobile dam project, stresses the experimental evolution of Project MOSE, highlighting models of the barriers at the Voltabarozzo Experimental Center and the prototype placed at Venice's Arsenal in 1988:

> Experiments in the laboratory and on site have been undertaken to indicate the possible future trend...at the Ministry of Public Works—Venice Water Authority Voltabarozzo Experimental Centre for Hydraulic Models, near Padua, studies and tests relating to implementation of the safeguarding measures have been carried out using physical and mathematical models...A life-sized prototype of a gate has been constructed (*Modulo Sperimentale Elettromeccanico*—MoSE) and subjected to four years of testing. (Scotti 2005, 247-55)

Experimentation for the dams is linked to the Venetian Republic through cultural narratives, which add support to contemporary claims. For example, Marina Corrier (2005) constructed a narrative about how, hypothetically, the Venetian Republic would have rejected an intervention like Project MOSE because it is not experimental: "I think that Project MOSE is work that the Serene Republic would have totally refused because to do anything they used to experiment with everything. Now they are continuing with the work without experimentation. Locally, they do not know what the result will be at the end of the work." But others argue that the dams do, in fact, follow the principle of experimentation.

On the question of whether the principle of reversibility was followed during the Venetian Republic, Enrico Sambo (2005) argued: "I think the dam walls are not reversible but they have served their function of stopping the sea in a way that reduces environmental impact. The landscaping impact of the dam walls is terrible. But the environmental impact, the pollution of the environment by the dam walls, is small." These remarks raise an interesting point about how the impact on the landscape is considered a criterion for assessing a project by environmentalists. As we now examine, this illustrates that the ancient Venetians did not have priorities that are comparable with those of contemporary Venetian environmentalists.

The Myth of an Environmentally-Friendly Venetian Republic

It has been established that there are significant gaps between the environmentalists' cultural narratives and the reality of ancient and contemporary interventions in the lagoon. These gaps are evident regarding the principles of experimentation, reversibility and graduality. Yet these cultural narratives helped to construct the myth that the Venetian Republic was environmentally-friendly. In order to assess fully the myth that the ancient Venetians respected the environment, we now investigate other aspects of how nature, the conservation of buildings, and the lagoon were treated during the Venetian Republic.

Firstly, let us consider how the ancient Venetians regarded environmental landscaping. Maurizio Zanetto of Italia Nostra pointed out that the sea dam walls demonstrate that the ancient Venetians did not share modern concerns about environmental landscaping: "Now we can talk about the dam walls. They had a big impact on the morphological aspects, with the environment and the landscape. Obviously, they did not have a concept of landscape then" (Zanetto 2005). Some environmentalists therefore recognize the building of the sea dam walls violated aesthetic environmental principles. If we compare the sea dam walls built during the Venetian Republic with the construction of the mobile dams using criteria of environmental landscaping, the latter project has a much more positive impact on the environment. Even Stefano Boato, one of the most prominent environmental critics of the mobile dams, acknowledged that the placement of the barriers under the water avoids blocking the view: "Nobody else ever built dams under the water that have to be maintained under water...In this way, everything is under the water to avoid a clash with the environmental and landscape aspects because one does not see it" (Boato, S. 2005). Conversely, environmental aesthetic concerns were largely ignored by the ancient Venetians relative to their priorities regarding wider strategic interests. "Venetians were not interested in such 'aesthetic' questions concerning the lagoon where they lived: their primary, if not their only, concern, as we shall see, was with strategic security, maritime navigation and sanitary matters. The lagoon was modified over the centuries as the permanent site for a once Mediterranean-wide civilization," writes Caniato (2005, 7). Rinaldo (2009) draws similar conclusions from his analysis of archives from the Venetian Republic:

> From the description it comes out that the uninterrupted and untidy action of the state to safeguard the lagoon environment only functioned for the city's interests, threatened as they were by sedimentation created from muddy rivers and by health and military problems. Such action does not show any consideration for servicing the ecosystem or for conservation. The actions of the State were constantly conducted by all the organs that determined the political will, but almost never in a systematic and coordinated manner. (Rinaldo 2009, 213-14)

This antithesis between the environmental priorities of modern green activists and the ancient Venetians is also illustrated by the way in which buildings were treated. Rather than preserving buildings for their aesthetic qualities or to conserve the past, the ancient Venetians often preferred to destroy and rebuild: "The attitude was that they did not intervene in damage to buildings that they thought was serious. They used to destroy and build again. Who would do that now?...With the problems of high water, if a palace and the doors were too low, they used to just destroy it and build it again. They used to raise the pavement... If a house was left with a low entrance, they used to either lower the head or demolish the house. They did not think much about it." (Zanetto 2005)

This interpretation is supported by the demolition of the Church of San Teodoro in 1071, and the destruction and relocation of San Geminiano Church from the middle to the end of St Mark's Square just over a century later. The demolition of structures continued until the last days of the Venetian Republic: several buildings were demolished for the construction of the Fenice Theater, which opened in 1792. Crouzet-Pavan (2002) argues the need to acknowledge that ambitious interventions meant social and environmental destruction: "We cannot deny the violence inherent in the grandiose public-works projects of the thirteenth and fourteenth centuries. Expropriations and sweeping abrogations of ancient rights were required in order to open up passageways and knock down walls" (Crouzet-Pavan 2002, 28). The demolition and construction of buildings by the ancient Venetians prompts Keahey (2002) to contrast their attitudes with conservationist thinking nowadays: "Ancient Venetians dealt with rising water simply by raising the pavements around the buildings, or they demolished the buildings, built up the ground and constructed new structures—a process that today's preservationists would find unthinkable" (Keahey 2002, 68).

The treatment of the lagoon during the Venetian Republic raises doubts about green endorsements of ancient attitudes towards the environment. Crouzet-Pavan (2002) refers to historical proof that the ancient Venetians over-exploited the lagoon for fishing, violated the lagoon's equilibrium by over-using resources, and created pollution (2002, 43 and 105). Caniato (2005) documents how one of the river diversions by the ancient Venetians caused substantial environmental damage to the lagoon and its banks:

> In 1324, they decided to build a long earth bank in front of the Brenta lagoon delta, in order to divert the river to the south-west, as far as possible from the site of the capital. This enterprise soon caused serious environmental problems, both to the nearby mainland, where fresh water lakes and marshes expanded and forced peasants to abandon their lands and houses, and to the main sea-channels, which shallowed with the loss of the previously adequate and continuous river flow that had kept them naturally deep. (Caniato 2005, 12)

The evidence suggests that the Venetian Republic paid little attention to environmental issues as they are now framed. Despite this evidence, environmental campaigners criticize the construction of the mobile dams because they claim that environmental safeguards are not respected as they were by the ancient Venetians.

Modern Attitudes towards Nature

In green cultural narratives, the mobile dams are only one illustration of how modern projects have failed to follow the Venetian Republic's environmental principles. Former Green Party Senator Giorgio Sarto pointed out that environmental principles were initially maintained after the fall of the Republic, but by the early twentieth century environmental traditions had been abandoned due to modernization projects: "The great ability of the Venetian Republic continued through the Italian reign of the Water Authority that could check the water and could legislate. That stopped with an operation in 1917 to intervene in the saltmarshes, then to build Marghera Port and all the drains, etc. There was an ecological disaster and an interruption there compared with all the old traditions that today we could call ecological." (Sarto 2005)

Modernization at the end of the nineteenth century and start of the twentieth century represents a midpoint in environmentalists' cultural narratives, which begin with the Venetian Republic and continue with the mobile dams. It may appear strange that another midpoint is identified as Napoleon's victory over the Venetian Republic in 1797. Yet as Davis (2002) writes, in addition to a beginning "[t]ypically, at least two middles and two endings are portrayed in movement narratives" (Davis 2002, 54). Characters like Napoleon become dynamic elements in narratives: for Venetian environmentalist Fabio Cavolo (2005), Napoleon's victory was the start of the modernizing approach to protecting Venice that would culminate in the completion of Project MOSE unless activists could stop it. Edoardo Salzano (2005) compared the reticence of the ancient Venetians with the modern belief that nature can be transformed by man, as represented by Project MOSE:

> The Venetian Republic tested interventions to ensure they were compatible with nature, whereas MOSE symbolizes the belief that man can change nature. This belief has led to widespread damage...They were very careful with natural evolution and they followed it. Whereas here, the 1800 concept that man can do everything to nature is in Project MOSE. Nature is a material thing that can be changed or canceled. This concept has prevailed on everything that has created damage everywhere.

The reason identified by environmentalists for the ecological degradation created by modern projects is the belief that man can modify nature for human benefit. Project MOSE's dams are judged as an imposition on nature by man, while the ancient Venetians were allegedly guided by nature:

> If you think about how the ancient Venetians defended the lagoon, they did a lot of hydraulic projects like the deviation of rivers...However, the natural element was the force behind human activities. They always wanted to go back if something wasn't working, so everything was reversible. The MOSE system is not reversible. It is a project that modifies the hydrodynamics and the natural equilibrium. It is imposed by human beings. It cannot enter this laboratory. It is something that stands on its own and does not integrate with Venice. It risks ruining this relationship with man and the environment that has been delicately established over the centuries. (Mencini 2005a)

These comments typify the environmentalist belief that the ancient Venetians integrated engineering interventions into a balanced relationship between man and nature, while Project MOSE fails to do this. The Green Party's Beppe Caccia argued that well-established cultural knowledge about the Venetian Republic shows how engineering solutions need to be linked to social relations within the city and cannot be considered in isolation:

> A city is such that there is a civil institution of relations - if there is a social link, if there is an idea that is not monocultural about the development of the city. Either these two questions stand together or there can be no engineering solution that could stand on its own. This is cultural knowledge, not what we have invented. It is the cultural knowledge that has allowed the city to live for centuries. It is even how the city was historically thought of during the Republic. (Caccia 2005)

By linking engineering solutions to Venetian society, green cultural narratives draw upon the agenda of sustainable development that was discussed in Chapter Four, claiming that the ancient Venetians integrated engineering interventions with social and environmental considerations to demonstrate respect for sustainability. Flavio Cogo (2005) argued that the ancient Venetians even incorporated concerns about climate change into engineering projects: "We should remember that the Venetian Republic constantly studied the hydraulic works that were changing with climate change. So from the Middle Ages to the 1700s this was done. So why isn't this done now?" Connecting powerful international debates about sustainable development and climate change to the traditions established during the Venetian Republic adds historical authority to activists' contemporary claims.

THE MOBILE DAMS AND MYTHS ABOUT CLIMATE CHANGE

During the 1980s and 1990s, a range of new environmental issues emerged internationally. Sustainable development became prominent during the early

1980s; climate change became a major debate during the 1990s (Jamison 2001, 93). Media commentators began to promote flooding in Venice as a leading example of the problems with sea level rises associated with global warming. For example, the *International Herald Tribune* ran a series of four articles about global warming in 2002, the first of which was titled "A global threat laps at the gates of Venice" (James 2002) and began on the newspaper's front page.

The international discussion about global warming allowed the Venice problem to be reframed. The Venice problem, as defined in the Introduction, originally focused on the threats of flooding and sinking, and these threats acquired new meaning as concerns about global warming grew. Public audiences became influenced by the perception that global warming represents a danger, and journalists and experts frequently linked climate danger to Venice. This enabled local claims-makers to develop this link to add weight and urgency to their claims. Expanding the domain of claims that can be made about a problem and building on claims made by others are typically useful strategies for claims-makers, as Loseke (2003) explains: "[P]iggybacking and domain expansion are effective strategies because claims-makers can build upon the successes of previous claims-makers by linking a problem constructed as new to a problem that already has achieved some level of audience acceptance" (Loseke 2003, 62). Our objective here is to analyze how the debate about Venice's mobile dams was redefined through global warming.

We recognize that global temperatures increased during the twentieth century and worldwide sea levels have risen over the past 40 years, adding to the other causes of flooding in Venice: the reality of these climate changes was discussed in the Introduction. Our aim in this section is to look at how claims-makers have drawn upon debates about these changes to construct new myths about the mobile dams.

Like the cultural narratives about the mobile dams and the Venetian Republic, there are two sides in this debate. Although both proponents and opponents of the mobile dams have developed their claims through the discussion of climate change, they usually disagree about what this problem is. They certainly do not have a shared understanding of how to respond. These are common features of how claims are diffused and translated into actions, as Best (2001a) explains: "Just as claimsmakers may not agree about all of a problem's characteristics, they may advocate various actions" (Best 2001a, 10). Those in favor of the mobile dams have stated that climate change will make the project more urgent and useful, while critics have argued that global warming will render the project useless. "[A]ccording to the best climatic models, there is a clear risk that the Experimental Electromechanical Module gates will become obsolete within a few decades and that they may have to be demolished shortly after their construction to separate the lagoon from the sea more effectively," suggests Paolo Antonio Pirazzoli of the French National Center for Scientific Research (Pirazzoli 2002, 217-223). On the other side, Rafael Bras of the Massachusetts Institute of Technology (MIT) claims that the mobile dams were designed to account for sea level rises. Bras and a team of experts from MIT and the University of Padua were hired by the Venice Water Authority in 1995 to oversee the environmental impact study of the mobile dams. Bras argues: "The bottom line is that the gates work...To argue that the design of the barriers did not consider sea-level rise is just wrong... The barriers, as designed, separate the lagoon from the sea in an effective, efficient and flexible way, considering present and foreseeable scenarios" (cited in Conor 2002).

It is difficult to evaluate these conflicting claims about the mobile dams and climate change on the basis of empirical evidence, because they discuss future possibilities. Having said this, the passing of time since these claims were made allow us to make some provisional assessments. Predictions that global warming will lead to higher sea levels and that this will make the dams useless have often been referenced to forecasts by the Intergovernmental Panel on Climate Change (IPCC). For instance, American archaeologist Albert Ammerman and his geologist colleague Charles McClennen, both from Colgate University in New York, argued in the journal *Science*: "If the new Intergovernmental Panel on Climate Change (IPCC) report that is forthcoming sustains its previous position on global warming, then the handwriting could be on the wall regarding the project" (Ammerman and McClennen 2000, 1302). Since this claim was made, data have been released indicating that sea levels have risen. The IPCC (2007, 2) establishes that global sea levels rose from 1961 by an average rate of 1.8mm/year (0.07 inches/year), and it has been estimated that rising global sea levels added 5cm (2 inches) to the water level in Venice between 1970 and 2005 (Istituzione Centro Previsioni e Segnalazioni Maree 2009).

However, local circumstances mean global sea level rises have less impact upon Venice's Adriatic Sea than many other seas. In a paper published in 2009, Carbognin et al (2009) analyzed how global sea level rises are predicted to affect the Adriatic Sea, which is an arm of the Mediterranean Sea. Compared with nearby seas, "[i]n the Mediterranean, a semi-enclosed basin, sea level trend rates are smaller than in the neighbouring (*sic*). In particular the Adriatic Sea assumes very peculiar and different characteristics due its shape and low depth" (Carbognin et al. 2009, 7). These authors refer to research by Holgate (2007) showing that the eustatic rate in the northern Adriatic Sea is consistently lower (approximately 35 percent) than the global mean. Similarly, the OECD (2010) notes that, "since the Mediterranean has registered stationary and even falling sea levels in recent decades, its take-up of global average sea level rise could be lower than in other places" (OECD 2010, 158). For Venice and the nearby region, the OECD further suggests: "Although climate change is not the sole or even primary cause of flooding, it implies greater uncertainty and probability of catastrophic events, intensifying the challenges in water governance facing the city-region" (OECD 2010, 148).

The predominant concern is that rising sea levels will lead to excessive closure of the dams: as explained in the Introduction, the mobile dams were designed to rise and close the inlets between the sea and the lagoon when tides reaching 110cm (43.3 inches) above the tide gauge at Punta della Salute are predicted. But there is so much uncertainty regarding climate change forecasts that it is impossible to predict dam closures accurately. Carbognin et al (2009, 7) emphasize the uncertainty of predictions and that climatic changes such as meteorological storms could substantially alter water dynamic perspectives. Despite the uncertainty inherent in predicting the future number of dam closures, we can document how often the dams would have been closed since the start of this century. Data for past high tides in Venice allows us to assess what recent implications rising sea levels would have had, if the mobile dams had been in place.

Past high tide events are shown in Table 1.2. If the dam's barriers had been closed to block the tides above 110cm, they would not have been used at all during 2003 or 2007. In 2005 and 2006, they would have only been closed once. During 2002, the dams would have closed the lagoon from the sea on 12 occasions. The years 2000, 2001, 2004 and 2008 were more typical, with between four and six extreme tides above 110cm, which would have led to dam closures. Fletcher and Da Mosto (2004) note that "[a]nalysis of the past 30 years' data shows fewer than four such 'extreme' episodes per year" (Fletcher and Da Mosto 2004, 41). Taking an even longer time span, between 1960 and 2008, the dams would have been raised on average three and half times per year, as noted by Pierpaolo Campostrini, Director of the Venetian research association CORILA (Campostrini 2010). When there was a lot of flooding during 2009, the dams would have closed the lagoon on 16 occasions (Campostrini 2010). Even the relatively high number of dam closures in 2009 or 2002 would not have made pollution and the water quality in the lagoon considerably worse.

Yet claims about excessive dam closures due to climate change are depicted as leading to the Venetian lagoon becoming little more than a stagnant pond. This is then linked to exaggerating the infamous stench from the city's canals. Such fears were reported in the British newspaper *The Guardian* by Paul Brown (2002): "Once built, the barrier will be raised when a high tide threatens to engulf the city. But there are fears about how this might affect the Venice lagoon, particularly the possibility that it could further restrict the flushing of the city's waterways by the tide, making the famous stinking canals more stagnant...With global warming expected to add at least another half metre to the sea level this century, the city's problems are bound to get worse" (Brown 2002). The connection between Venice's smelly canals and the raising of the dams during high tides, which predominantly occur during the winter, is not as straightforward as it may appear. The stench from Venetian canals is usually caused by low tides and the exposure of algae on canal walls to the air and hot sunlight.

Other claims about excessive dam closures discuss the negative impact on the whole lagoon. Ammerman and McClennen (2000) predict problems due to rising sea levels during the later years of the dam's functioning, especially in winter: "As such a high concentration of gate closure will limit the circulation of water that is essential to biological life in the lagoon, this could have negative impacts on levels of water pollution and the ecology of the lagoon" (Ammerman and McClennen 2000, 1302). It is true that a greater number and duration of gate closures is likely during the winter months, when worse weather conditions are more common. Nevertheless, pollution is usually a bigger problem in Venice during the summer, according to Giovanni Cecconi, director of engineering for the CVN (Cecconi 1997b, 5). A report by experts connected with the CVN claims that closing the mobile gates could even be used to reduce pollution and benefit marine life: "Once storm surge gates have been installed, they can also be used to induce lagoon flushing. By preventing water leaving through the central tidal opening, a net circulation flux through this opening of 2000m(3)/s will result. This will flush the tidal flats south of Venice and immediate benefits can be expected in terms of the oxygen content during periods of fast *Ulva* biomass decay" (Runca et al. 1996, 16).

We cannot simply assume that a longer duration or greater number of gate closures would decrease or increase pollution in the lagoon. These claims are further complicated by the multiple causes of pollution, including the quantity of sewage released into the lagoon, agricultural run-off, algae, industrial waste and emissions from water traffic. Reducing pollution from these sources would mean that longer and more frequent gate closures could be less problematic. On the other hand, increased pollution in the lagoon could mean that shorter and less frequent gate closures could have a more negative impact. So isolating gate closures as either a problem or a remedy for pollution is unsatisfactory, and does not enable us to assess these conflicting claims. Rather, it is more useful to treat the different opinions about dam closures and pollution as claims constructed to support the positions of supporters and critics of the dams on the question of climate change. In reality, the impact of the mobile dams on pollution will depend on how many different aspects of Venice and the lagoon are managed: a question that we address in Chapter Nine.

We have observed how journalists, academic experts and people involved with the building and assessment of the mobile dams have related their claims to climate change. But how were Venetian environmentalists influenced by the international discussion of global warming? This discussion, and the way it was widely linked to Venice's dams, provided an opportunity for Venetian environmentalists to expand the domain of their claims against the dams. During the 1980s, environmentalists' campaign claims focused on the disequilibrium that the dams could create between Venice and the sea, as described in Chapter Five. The emergence of the global warming discussion in the 1990s allowed Venetian environmentalists to develop their claims against the dams at a time when they needed strengthening. During the 1990s, environmental protest events steadily declined in Italy (Diani and Forno 2003, 138-140), and by the early twenty-first century, the national demobilization of environmental protest was affecting Venetian activists. The campaign against the mobile dams hit a low ebb between 2003 and 2005, when the government was advancing the construction of the dams and environmental activists participating in the Save Venice and the Lagoon committee expressed a lack of direction in their campaign. Venetian environmentalists were able to find new momentum by building on the connections that were being made between Venice and the global warming issue internationally.

The founding document of Chioggia's No MOSE permanent assembly highlighted the damage the lagoon environment would suffer from the dams due to rising global sea levels (*Assemblea Permanente No MOSE* 2005). Fabrizio Reberschegg, a leading figure in the Venice Green Party, was adamant that the city must reflect on the debate about the greenhouse effect[1] and the way that rising sea levels would impact upon Project MOSE: "The greenhouse effect is the response for the protection of our planet against our unacceptable situation...We maintain that we need, absolutely, here in 50 years, to imagine the work of MOSE and if it will function little or badly against the sea rise" (Reberschegg 2005).

By developing new claims around global warming and the potential redundancy of the mobile dams, Venetian environmentalists connected with pervasive risk consciousness in Western societies. The German sociologist Ulrich Beck has theorized that the public's heightened risk consciousness will often be sufficient for the manufacture and acceptance of specific risks (Beck 1995, 91), and pointed to public perceptions of mounting risks generated by modernization (Beck 2000, 70). British sociologist Anthony Giddens (1999, 26) has elaborated on the way in which risks are manufactured by our modern impact on world. In this vein, Project MOSE is often perceived as a modernization project that could protect against global warming, yet one that also generates new risks. Giorgio Sarto (2005) thus linked the greenhouse effect and protective responses to extensive risks for Venice: "So the greenhouse effect is a very high risk because the measures to stop it are really precarious. So the scenario for Venice is of very high risks."

The advantage for claims-makers of linking the mobile dams and climate change is that the risks are derived from future possible scenarios: for example, forecasts of future sea level rises, rather than data collected from events of the recent past. Deriving risks from the future makes it difficult to prove if such risks are real or mythical, as Beck (1992) writes: "the actual social impetus of risks lies in the *projected dangers of the future*" (Beck 1992, 34. Emphasis in original). Beck also suggests that the risks caused by pollutants and toxins seem more imaginary than real (Beck 1992, 73). That is why we need to take risk consciousness and perceptions into account, instead of adopting a technical approach used in many texts about Venice's protection. In their examination of numerous megaprojects, Nils Bruzelius, Bent Flyvbjerg and Werner Rothengatter (2005, 137) demonstrate how technical assessments of risk rarely correspond to actual risks.

This is not to imply the risks associated with the mobile dams due to climate change are imaginary. The claims made about the risks with the dam project and global warming are based on the broadening of an objective risk, rather than the construction of a fictitious risk. The objective risks created by subsidence and flooding have been redefined as problems connected with climate change by journalists, experts and environmental campaigners. This understanding of risk has a great deal in common with Giddens' (1990) "manufactured uncertainties" and the way in which claims-makers broaden the awareness of risks and dangers (Beck 1999, 77).

Danger has become a salient theme in the risks connected with global warming, which was discussed by Venetian environmentalists in their interviews with the author. "I think that regarding the greenhouse effect and eventually the rise of the sea level, Project MOSE will not be able to do anything because what will the mobile dams be able to stop regarding the rise of the sea? The sea will go into the islands. All the coastlines are not high enough to maintain a sea that rises more and more. That is dangerous," commented Marina Corrier (2005). Cesare Scarpa (2005) predicted: "The greenhouse effect has a big influence because it creates some heating of some areas that creates emissions in the circle of the seas and oceans and portions of polar icebergs that have been shrinking for a long time. This will not only create a rising of the quantity of water present everywhere. It will also create some modification of heating elements for the entire environment that are very dangerous."

Reference to the greenhouse effect has provided environmentalists with formidable power for their claims, as Bramwell (1994, 167) remarks. Moreover, she observes that the uncertainty surrounding the greenhouse effect does not mean it can be neglected as a consideration in the discourse of environmental disasters (Bramwell 1994, 167-8). In green thinking, speculative disasters represent sufficient reason for action: "The ecologists' argument is that even if these disasters lie in the realm of speculation, the situation is so threatening that action should be taken to avert them" (Bramwell 1994, 166). Yet it is not only environmentalists who have piggybacked on speculation about the greenhouse effect for the domain expansion of claims regarding Venice's dams. The CVN uses IPCC predictions to claim that the mobile dams will mean that Venice is better protected than other Italian coastal cities from sea level rises:

> The system to regulate tidal floods, able to withstand a difference between sea level and lagoon level of as much as two metres, is extremely useful today, but will be indispensable in the future if the city is to survive a significant increase in sea level. Mose has been designed on the basis of a precautionary criterion to cope with an increase of up to 60 cm in sea level, in other words, higher even than the latest estimates from the IPCC (International Panel on Climate Change) which predict an increase in sea level of between 18 and 59 cm during the next 100 years...Thanks to the system, Venice would be the only city protected, as the coasts of Italy could be invaded by water and the sea would represent a concrete threat. (Consorzio Venezia Nuova 2009c)

ASSESSING MYTHS, NARRATIVES AND REALITY

The evidence presented in the second part of this chapter has shown that there are serious disagreements about whether sea level rises will make the dams more necessary, or render them useless. At the center of this debate is whether extreme high tides will lead to excessive dam closures. So far, modest sea level rises would not have led to a problematic number of closures if the dams had been in place. There are also disagreements about the impact of dam closures on pollution within the lagoon. In the long term, it is hard to predict whether rising sea levels would mean the mobile dams will be useless. Carbognin et al (2009, 7) emphasize the uncertainty of predictions, and that climatic changes could alter water dynamic perspectives. We agree it is difficult to predict the long-term impact of sea levels on the dams, but we should bear in mind that the dams were designed to protect Venice for a period of 100 years, and were not proposed as a permanent solution for high tides. Although it cannot be guaranteed that the dams will function effectively in 100 years' time, their design suggests they will be able to cope with predicted sea level rises in the short term. The Commission of International Experts (1998) has pointed out that by the mid-twenty-first century, increased knowledge and experience will inform decisions about protecting Venice (Commission of International Experts 1998, 6). If climate change and subsidence bring very substantial sea level rises, the MOSE dams would not necessarily be redundant. One of the benefits of the dams being mobile is that they can respond flexibly to separate the lagoon from the sea as conditions vary, and that they could be used in combination with another protective measure.

Nevertheless, the claims made by Pirazzoli and Corrier that the mobile dams could be useless in the long-term due to sea level rises cannot be ruled out. Both critics and proponents of the dam project have proposed that a permanent division between the Venetian lagoon and the sea might be necessary in the future. Cecconi (1997b), an engineer, advises that the lagoon could be separated into several basins (Cecconi 1997b, 34). Fabrizio Reberschegg, of Venice's Green Party, argued against the mobile dams in favor of a more permanent division between the sea and the lagoon: "We should have more important pieces of work that have an ability to create a relationship between the sea and the lagoon. So some projects that go under the sea with fixed casings or like in the port of Rotterdam" (Reberschegg 2005).

Although it is difficult to be definitive about the future impact of climate change on the mobile dams, there is no doubt that this topic has enabled both proponents and opponents of the dams to help redefine the problem of flooding in Venice as an example of global warming. Through the narratives of various claims-makers, the Venice problem has been transformed into a global warming myth. Using the definition of myths at the beginning of this chapter by Benford (2002, 57), this is not to imply that flooding in Venice is unconnected to climate change: rather, it is to understand this belief as a narrative that claims-makers have constructed according to changing cultural moods.

Similarly, the cultural narratives about the mobile dams and the Venetian Republic should not be ignored, even if we doubt their validity. During the first part of this chapter, fundamental questions were raised about the cultural narratives used to construct the myth that the Venetian Republic was environmentally-friendly: in particular, the narrative about how the Venetian Republic achieved an environmental equilibrium between society and nature that has now been lost. Advocates of anthropic intervention claim that this environmental equilibrium has never existed in Venice or, indeed, elsewhere. Competing interpretations of interventions during the Venetian Republic, especially river diversions and the building of sea dam walls, were scripted into cultural narratives to argue for and against the contemporary mobile dams. Narratives about the Venetian Republic were constructed to back up claims against the mobile dams and were conceptualized as Venecoism. This was counterposed to Venanthropocentrism, which is a positive interpretation of anthropic intervention during the Venetian Republic and in modern projects such as the mobile dams. Venecoism was judged to be based more on myth than reality compared with Venanthropocentrism, because numerous interventions by the ancient Venetians did not conform to the environmentalists' criteria of graduality, experimentation and reversibility.

Several statements by Venetian environmentalists revealed how the ancient Venetians tended to prioritize political, economic and military concerns over conservationist sentiments. As Benford (2002, 72) describes, movement participants sometimes contradict movement myths and create discrepancies, especially when movement narratives are not effectively shared among the whole group. Moreover, the myth that the ancient Venetians respected the environment was often not supported by reality, as documented by historians. As Davis (2002) notes: "In constructing stories, tellers select some events for inclusion, while excluding other actions and details that do not serve to make the endpoint" (Davis 2002, 14).

Davis makes a further astute observation that provides an insight into the myths analyzed in this chapter. The myths about the Venetian Republic as environmentally-friendly and Venice as a problem of global warming both reinterpret past events while anticipating future scenarios. "[N]arrative explanation operates retrospectively, since the events earlier in time take their meaning and act as causes only because of how things turned out later or are anticipated to turn out in the future," he writes (Davis 2002, 12). The past is typically reinterpreted in narratives that can appear diverse, yet share underlying assumptions. American sociologists Joshua Yates and James Hunter (2002, 145) conducted research revealing how various fundamentalist movement narratives share a world-historical account of perverted modernity, despite theological, cultural and historical differences. The myths about the Venetian Republic as environmentally-friendly and Venice as a problem of global warming are about different periods in history and appear to have diverse frames of reference, but they share underlying doubts about modernity.

NOTE

1. It should be noted that the Venetian environmentalists interviewed often referred to global warming as the "greenhouse effect". An explanation of the greenhouse effect is provided here: http://www.ipcc.ch/pdf/glossary/ar4-wg1.pdf.

Part Four

MYTHS ABOUT MODERNIZATION

Chapter Eight

Modernization and Environmentalism

"The Green Party, like the rest of the environmental associations, considers the actual situation as the result of anthropic impacts started in particular from the start of the last century until our present time. Man has radically changed the physical situation of the lagoon during the last century, especially the hydraulic configuration for the harbor. From these changes new natural processes started in the lagoon, natural but induced by the anthropic transformation. Now we want to really fix with MOSE the present assets of the lagoon. We say that if we really want to solve the problem of high water and the lagoon that year by year looks more like the sea than the lagoon, we have to go back to the situation at the beginning of the twentieth century."

Marco Favaro (2005), Venice Green Party Provincial Councilor and WWF member.

For Marco Favaro, modern intervention in the Venetian lagoon has created many problems. Although Project MOSE is supposed to address these problems, especially high flooding, the risk is that the dams will merely reinforce them. Like his environmental associates, Favaro would like the lagoon to return to how it was at the beginning of the twentieth century. This desire is not simply about embracing the past: it expresses the rejection of modern engineering interventions.

This chapter explores environmentalists' reticence about various human interventions that they characterize as modernizing attempts to transform the lagoon. It is shown that the claims made by environmentalists in Venice about the mobile dams have been able to draw upon a particular reaction against human engineering as part of a general cultural unease with modernization. In Chapters Three and Four, doubts about industrialization, urbanization and human intervention in nature were identified as themes in conservationism and environmentalism. Here we examine the reaction against the belief that man should use science to intervene in nature for human benefit, and discuss how this reaction was affecting many international organizations and institutions by the 1970s. A holistic environmental culture has emerged based on integrating nature and society, which has assisted environmentalists in presenting engineering interventions like the mobile dams as epitomizing an outdated modernizing approach.

GROWING UNEASE WITH MODERNITY

Bramwell (1994, 10) notes the contemporary disillusionment with modernity, especially among intellectuals. Unease with humans engineering the future is part of a transition to an age of post-modernity. Giddens (1990, 2) writes: "[P]ost-modernity refers to a shift away . . . from faith in humanly engineered progress." Post-modernity is usefully distinguished from post-modernism by the British scholar Terry Eagleton (1996), who defines post-modernism as "a form of contemporary culture" and post-modernity as "allud[ing] to a specific historical period" (Eagleton 1996, vii). Sociologist Krishan Kumar (1999) observes that "post-modernism can be understood to some extent as a reaction against the kind of modernism represented by modernist architecture" (Kumar 1999, 98-9). However, although post-modernism originated in the cultural sphere, the concept has spread to encompass philosophy, politics and the economy: "The suggestion is that industrial societies have undergone a transformation so fundamental and wide-ranging as to deserve a new name. The question then becomes, are we living not simply in a post-modern culture but in an increasingly post-modern society" (Kumar 1999, 112).

Given the depth of this reaction against modernity, it is imperative to define what modernity is and how it is interpreted by environmentalists. For many environmentalists, modernity is defined by the belief that man could progress through the transformation of nature. This is traced back to the early seventeenth century and the Enlightenment philosopher Francis Bacon. Donald Worster (1977), a leading figure in the American environmental organization the Sierra Club, accuses the "imperialist" Bacon of promoting the belief that natural resources could be used for human benefits: "Bacon promised to the world a manmade paradise, to be rendered astonishingly fertile by science and human management. In that utopia, he predicted, man would recover a place of dignity and honor, as well as the authority over all the other creatures he once enjoyed in the Garden of Eden" (Worster 1977, 30). The traditional beliefs and institutions of Western science are generally associated with the emergence of modernity in the Enlightenment of the seventeenth and eighteenth centuries. "It is fairly obvious that what is commonly referred to as science and technology, and which we might characterize, somewhat more precisely, as the institutions of Western science, emerged as an integral part of a much broader cultural transformation (Huff 1993). For Karl Polanyi, it was simply the 'great transformation', while for others it has been termed the 'birth of modernity,'" writes Jamison (2001, 47).

Several scholars have questioned whether there has ever been a true modernity in terms of the application of objective, rational scientific knowledge. This was the explicit message put forward by sociologist Bruno Latour (1993), and was implicit in the earlier writings of science historian Thomas Kuhn (1962) and Michael Polanyi (1998 [1958]), the eminent scientist and philosopher. These writers stress that modernity was culturally imbued with tradition. There is general agreement that Enlightenment thinkers including Immanuel Kant, A.R.J. Turgot, Francis Bacon and Marquis de Condorcet gave birth to the idea of human-inspired progress, and that the industrial revolution in Britain that gave modernity its material form. Kumar (1999) explains that "modernity and industrialism are closely if not intrinsically linked" (Kumar 1999, 83) and defines modernization as "the social and economic processes of modernity" (Kumar 1999, 93). The idea of modernity was established by the end of the eighteenth century, and scientific and political thought during the eighteenth and nineteenth centuries was dominated by the belief that nature was to be used for human benefit (Kumar 1999, 85; Jamison 2001, 75-6). Yet many movements reacted against modern technologies and the application of science for industrial purposes, including the Luddite revolts in early nineteenth century England. Romantic Movement intellectuals like William Blake, William Wordsworth, John Keats, Mary Shelley and Robert Owen expressed doubts about modern science. By the end of the nineteenth century, complex reactions to modernity had emerged. This was the cultural movement of modernism, which contained currents that both affirmed modernity and denied it: "On the one side science, reason, progress, industrialism; on the other a passionate denial and rejection of these, in favour of sentiment, intuition and the free play of the imagination" (Kumar 1999, 85).

The nineteenth-century elevation of sentiment over Enlightenment rationality led to new interpretations of Venice. No longer were the decline and fall of the once-dominant Venetian Republic mourned - the decay of Venice became a source of interest and attraction. Pemble (1996) contrasts the perceptions of the city held by philosopher Jean-Jacques Rousseau during his visit in the 1740s with those of naturalist writers Jules and Edmond de Goncourt a century later: "In Rousseau's time Venice was an independent republic whose past and present reputation precluded sentimental rhapsody. By the time of the Goncourts the city had become a phantom, a relic, whose political power was extinct and whose history had been rewritten. Tyranny had gone; and in its place was the pathos of merit traduced and majesty dethroned" (Pemble 1996, 3). This changing perception of Venice was affected by a shift away from Enlightenment universalism towards embracing relativity. Pemble's lucid explanation of this development is worth quoting at length:

> The nineteenth-century historians brought about this change in perception by looking at Venetian history not from the outside, but from the inside...They judged it not by the standards of a notional universal morality, but by the standards of the period they were discussing. Eighteenth-century historians—Gibbon, Voltaire, Turgot, Condorcet, Hume—had used the comparative method in order to discover an invariable. They had looked, amidst the accidentals of time and place, for human nature—something they assumed to have been always and everywhere the same . . . To the historians of the following century the results of such inquiry seemed highly suspect. The older historians were accused of having found what they wanted to find—which was a universal man who was rational, benevolent, and happy. They had used history to validate Rousseau's assumption about the natural goodness of man and the corrupting influence of society. The new historiography rejected Hume in favour of the Italian philosopher Giambattista Vico. By adopting Vico's view that human nature was subject to psychological and moral development, it introduced relativity into historical judgement. (Pemble 1996, 3)

Embracing relativity and sentiment over Enlightenment universalism and rationality encouraged nineteenth-century fascination with Venice. This made the city ripe for the flourishing of conservationism, as explored in Chapter Three. The rejection of Enlightenment traditions assisted the growth of conservationist organizations in Venice and elsewhere. Jamison (2001) details how conservationist organizations, such as the Sierra Club in the United States, were formed in the late nineteenth century as a reaction to scientific and industrial intervention in nature. The Sierra Club has been compared with Italia Nostra as a model of early conservationism (Bramwell 1994, 41). In the post-Second World War period, conservationism became institutionalized into the United Nations. Julian Huxley, a naturalist, was the first director of UNESCO and helped create the International Union for Protection of Nature. We have discussed the leading role played by UNESCO in framing the Venice problem after the high floods of 1966.

Conservationism was pervasive in Venice and internationally by the 1960s, encouraging widespread questioning of Enlightenment ideals and modernity. As the twentieth century progressed, doubts emerged about the very separation of nature and society in Enlightenment thought. In the early twentieth century, American architectural critic Lewis Mumford (1922, 1934) outlined a "human ecological" perspective, which challenged the traditions of natural science and sought to integrate natural and social knowledge. Although a human-centered approach to the environment was developed by radicals like Murray Bookchin (1963), Jamison (2001) shows how Mumford's ideas became highly influential in academic and activist thought after the Second World War. The influence of Rachel Carson's best-selling book *Silent Spring* (1962) illustrates how prominent the reaction against the human application of science in nature became after the Second World War. *Silent Spring*, which investigated the environmental impact of chemical insecticides, had a major impact on the emerging environmental movement in the 1960s (Jamison and Eyerman 1994, 92), and is widely credited with having inspired environmental consciousness and the emergence of organizations that helped form the environmental movement. The WWF began in 1961, Friends of the Earth was created in 1970, and Greenpeace originated in North America during the 1970s. Environmental organizations and protests that they organized grew significantly in the 1970s, and by the 1980s, Italy had a formidable environmental movement. As environmental campaigns developed in Venice, activists were able to build on the wider questioning of modern intervention in nature.

Questioning Modern Intervention

During the author's interviews with environmentalists in Venice, the problematic use of intervention by man was often associated with modernity. This association was used to back up claims against the mobile dams. Fabio Cavolo (2005) identified Napoleon's reign as the period when intervention based on modernity began to violate the natural equilibrium in the lagoon, representing a modernizing approach that culminated in Project MOSE: "The history of MOSE started with Napoleon at the end of the 1700s. He wanted to make a big port of Venice. The disequilibrium in the lagoon started with the arrival of Napoleon in Venice. The port was not compatible for Venice in terms of where it is now. With an understanding based on industrial modernity, it became a necessity. So they started to make a big port in the city of Venice."

Modernization of the port situated within the Venetian lagoon continued during the nineteenth century. This included changes to the three inlets between the sea and the lagoon to enable larger ships to reach and leave the port. Between the mid-nineteenth century and mid-twentieth century, breakwaters, dams and jetties were built at the three inlets to the lagoon. The jetties accentuated tidal current, which reduced sediments blocking up navigation channels that were dug for ships to pass through the shallow lagoon.

More recently, there have been significant modifications to the lagoon inlets for the construction of the mobile dams. In Marina Corrier's view, these alterations to the lagoon entrances have created higher tides: "The dam walls and jetties of the 1800s created passage for water that was more open than it was in the past...Through the jetties, the foundations have been deepened at the sea entrances to the lagoon. So that brought a bigger influx of water into the lagoon. But they are continuing to dig them. They are accelerating and helping the entrance of the water into the lagoon. So the high waters are higher because the depth helps the water to come into the lagoon." (Corrier 2005). Venetian environmentalists linked the changes to the lagoon inlets to the combination of modernity and industrialization. For example, Cristiano Gasparetto (2005a) argued:

> Beginning with modernity, so halfway through two centuries ago, halfway through the 1800s, they started, in the name of modernity and industrialization, some physical transformation of the lagoon that compromised life in the lagoon. The main ones were the transformation of the three inlets, the ones of San Nicolò, Alberoni and Chioggia, which used to be inlets to the sea without dams. They created these dams and made the inlets of the ports deeper to allow the ships to come in. This was the first great transformation. They started in 1860/1870 and Venice was ruled by Austria then.

Cultural narratives outlined in Chapter Seven constructed the myth that the Venetian Republic was environmentally-friendly, and that industrialization eroded this harmonious tradition. Several Venetian environmentalists distinguished the twentieth century from previous historical periods: it was allegedly at this time that intervention in the lagoon generated the main environmental risks. "There are two great problems of environmental themes we focus on connected to the history of this city, the distant past and the history of the last century, the 1900s. The history of the 1900s has in many ways conditioned the development of the city and represents the productive development of the 1900s. It is the principal risk factor for the future for the city," warned Beppe Caccia (2005). The connection between twentieth century modernity and mounting environmental risks is well-established within sociological theories of "risk society" by Beck (1992, 2000) and Giddens (1999). Caccia gave some examples of industrial interventions that created the specific risk of reducing tidal expansion in the lagoon, which could exaggerate flooding: "The progressive grounding of various areas of the lagoon that have reduced by 30 percent the surface of the free expansion of the tide activities in the course of the 1900s; the closure of the fish channels, the achievement of the first, second and preparation of third industrial areas, the creation of the international airport Marco Polo. A series of interventions have reduced by a third the surface for the free expansion of the tides in the lagoon." (Caccia 2005)

Scientists have made observations that support Caccia's comments about the reduction of the space for tidal expansion in the lagoon. Fletcher and Da Mosto (2004) state that "[t]he natural extent of the lagoon has shrunk by over 20 percent," although they do not define "natural extent" or the time period over which this reduction has taken place (Fletcher and Da Mosto 2004, 23). The shrinking of the lagoon is significant because environmentalists attribute it to modern interventions. However, it should be noted that the size of the lagoon was also decreased for commercial reasons during the Venetian Republic (Crouzet-Pavan 2002, 141), and that large areas of the lagoon were reclaimed by the ancient Venetians to provide land for agricultural cultivation (Lauritzen 1986, 158). Interventions by the French and Austrian administrations of the nineteenth century reduced the total area of the lagoon. "The French first and then the Austrians had filled in the channel between the monastery islands of San Cristoforo and San Michele to create the city's cemetery island. The Austrians had enlarged the island of Sant'Elena tenfold to provide parade grounds and barrack space for the cavalry regiments billeted in the city," writes Lauritzen (1986, 152). In spite of these pre-industrial historical facts, Cristiano Gasparetto (2005a) blamed the reduction in the size of the lagoon on industrial development and the second major phase of change: "The second main transformation was the invention of the industrial port of Marghera, which from a capitalist, productive point of view was a big invention and a big success...But all of Marghera Port was obtained by occupying an area of saltmarshes that used to be an area where the tide could come in and out. So a part of the lagoon was taken away from the lagoon and became an industrial port." In addition to inadequate recent maintenance of the lagoon, Gasparetto argued that various interventions since industrialization have increased flooding in Venice:

> To put together these phenomena that are physical transformations, we can add 30-40 years of no maintenance on the physical elements of the protection of the lagoon that are essential—the dam walls that were partly built by men to protect the lagoon from the sea—without maintenance of the natural channels of the lagoon. These physical transformations and this lack of maintenance have made the lagoon system very fragile. Not only this. It is absolutely the cause of the increasing frequency of high tides and the height of high tides. (Gasparetto 2005a)

There is scientific evidence to link flooding and sinking in Venice with modern industry, in that subsidence due to the extraction of groundwater for the industrial zones has been established as the principal reason for the rise in the relative sea level during the twentieth century (Ghetti 1988). But it is far more difficult to maintain that other modernizing interventions in the Venetian lagoon over the past 300 years have exaggerated flooding, while previous changes did not.

Another tension in the claims put forward by Venetian environmentalists has emerged regarding the question of intervention. Gianfranco Bettin, a leading Venice Green Party politician, argues, using the examples of the Marghera refinery and the fishing industry, that paralysis in Venice has been caused by doing too much, not too little (Bettin 1991, 12-13). This could imply that the problem is perceived as the degree of intervention in the lagoon, with the logical solution being that intervention should slow down or cease altogether. However, Venetian environmentalists whom I interviewed expressed a more subtle attitude towards interventionism. Despite unease with modernizing interventions in the lagoon in general, and opposition to the mobile dams in particular, many acknowledged the need for something to be done to protect the city from flooding. For example, as Giannandrea Mencini said:

> There is no doubt that we must defend Venice against high water. There was the huge flood that transformed the lagoon into an arm of the sea. This was grave. It is necessary to intervene. There are lots of little interventions that should be done before thinking about closing the port inlets. In the last 20 years they always thought about closing the port inlets without giving priorities to other more efficient things that are more important for the lives of citizens. I am accusing them of thinking only of the big work of MOSE without comparing it to another work like they do in Europe. This kind of work, even if it compared to the last centuries, is a very worrying intervention. It is an intervention that modifies the past philosophy of interventions. (Mencini 2005a)

These comments suggest that Project MOSE goes against past traditions of minimalist technology. Even though the mobile dam project was conceived in the 1970s, it did not incorporate the perspective on technology of "small is beautiful" famously proposed by Schumacher (1973), who recommended that large-scale, environmentally-destructive projects should be stopped. In Chapter Seven, several Venetian environmentalists described how there was a similar tradition of cautious, small-scale intervention based on graduality during the Venetian Republic. Yet Crouzet-Pavan (2002) argues that the Venetian Republic was exceptionally interventionist against nature: "In a lagoon environment in constant evolution, Venetians confronted singularly active and always hostile natural forces. Mastery of those forces was a necessity: life in the lagoons was only possible at the price of constant labors and incessant maintenance operations. Hence my hypothesis is that the public authorities in Venice were undoubtedly more interventionist than their counterparts elsewhere" (Crouzet-Pavan 2002, 27). Conflicting interpretations of intervention are central to the tension between historical reality and environmentalists' cultural narratives about the Venetian Republic. As Fabio Cavolo suggested: "We know that it is necessary to do something, as the Venetians of 500 years ago were aware. So the problem is what to do" (Cavolo 2005). Some Venetian environmentalists feel they have been accused of rejecting intervention *per se*, because they oppose the MOSE dams. Mencini was highly sensitive to this allegation, arguing that intervention should follow the ancient Venetian traditions based on adaptation to the needs of the lagoon:

> It is not that I am against work, especially technological work. It is the people that are against us who accuse us of being against technology. It's absolutely untrue. There is work that we need and work that we don't need. What I think is that the way MOSE has been projected is not useful. It is a wrong way of intervening. In the past when the Venetians intervened in the lagoon in the wrong way and they were aware of it, they used to stop it. Now nobody is aware of it and they just want to carry on with this Project without realizing how much damage for the lagoon it is. (Mencini 2005a)

Recognizing the need for intervention against flooding led Paolo Perlasca to acknowledge that Project MOSE could address this: "Project MOSE obviously would stop high water and that would be positive. But for Project MOSE they need big foundations and the foundations of the port inlets need to be done forever to a depth that is not compatible with the lagoon. That means creating disequilibrium in the lagoon" (Perlasca 2005). Fabrizio Reberschegg (2005) commented that the problem was not intervening against flooding, but the outdated modernizing approach of the mobile dams: "Venice is at risk. We are not against intervention for safeguarding Venice with some instruments. We are against Project MOSE because we think it is old and useless." Marco Favaro (2005) suggested:

> In the last century the level of Venice with respect to the sea is lower by 27/28 centimeters because the sea level is growing. But also because of subsidence because the industries of Marghera Port used underground freshwater wells and the affect was subsidence. For this problem MOSE is completely useless and we have to study other solutions. We have to create and insulate. In some buildings they are experimenting with a layer in the walls that is impermeable to the salinity. It is probably the only way to solve this problem. But as you can understand, it is another answer. MOSE is not able to answer this problem.

Although Venetian environmentalists have challenged the mobile dams, they do not oppose intervention to deal with flooding: rather, it is the specific approach to intervention based on engineering that they are against. It is worth examining this concern in detail.

ALTERNATIVES TO ENGINEERING INTERVENTION

As with many of the activists interviewed for this book, leading environmental thinker Jeremy Rifkin (1983) favors ecological solutions in harmony with nature over engineering intervention: "We can choose to engineer the life of the planet, creating a second nature in our image, or we can choose to participate with the rest of the living kingdom. Two futures, two choices. An engineering approach and an ecological approach" (Rifkin 1983, 252). Giannandrea Mencini (2003) explains how an ecological approach differs from an engineering approach for Venice and its lagoon:

> [T]he physical and environmental safeguarding of Venice, of its land and its lagoon, cannot be exclusively orientated with an engineering approach, focusing only on internal lagoon actions or at the port inlets, but should be the result of ecosystem planning and management of the territory that interacts in a complex and articulated way with all the hydraulics of the basin and also the height of the lagoon, making environmental politics the future planning and the socioeconomic considerations of the territory. (Mencini 2003, 8)

It appears that the problem with engineering is the tendency to concentrate on a single intervention, which fails to take wider environmental complexity into account. To understand better this environmental critique of engineering intervention, the author asked Mencini the following question during an interview: "You wrote that for the safeguarding of the lagoon, it is not possible to be exclusively orientated with an engineering point of view. Why?" Mencini responded:

> Because it is not only an engineering problem. It is also a planning problem. Now the problem is that they have not got any permission with respect to regulating plans. It is a problem of the landscape, of ecological, geological problems. It is also a historical problem. It is a social problem. This is what I mean when I say it a multi-disciplinary problem. It's all done from the engineering and hydraulic approach. But it's not only this problem. It is architectural, cultural, environmental, social, about tradition of history, everything. It is what I meant when I said that it is not only an engineering, mathematical problem or a technical problem. The Project also has some social problems that are not only technical. (Mencini 2005a)

An environmental approach thus goes beyond the narrow hydraulic considerations of engineering intervention to incorporate social and historical factors. It is this failure by engineers to consider the whole environment when they are planning projects like the mobile dams that supposedly creates future problems. "When the engineers are drawing they do not think about the whole environment. When we start to see the execution of the work a lot of problems will come. So we do not think that in 10 years it will be possible. We think that when they will have finished Project MOSE it will be useless because they do not take these facts into consideration," predicted Fabio Cavolo (2005). Gustavo De Filippo (2005) argued that engineering planning for projects like the mobile dams is too limited because it is based on statistics and financial assessments, whereas the environment does not conform to numerical considerations:

> We start to destroy this eco-system. If you want to prevent high water in the lagoon, if you are an engineer, you think I need to stop water. But the main problem is not to stop water. It's to decrease the speed of the water and the water that can come in. There are so many natural possibilities if you want to do it. There is a better system all around the lagoon. Unfortunately, they did not consider it in a serious way. They talk about it only in a general way. Basically they agree with the engineering system because it is the only one that can produce numbers for the time period and the money. All the environmental models don't have numbers. They probably have statistics for a period. But they have been there for over 2,000 years. Unfortunately, with all the environmental problems, we don't know what they will do.

Here De Filippo introduces the existence of environmental models that offer alternatives to engineering models. Cristiano Gasparetto (2005a) provided some specific suggestions for environmental alternatives to mechanical or technical interventions, including reducing the size of the lagoon inlets and directing ships to a new port situated in the sea outside the lagoon:

> You don't need to always give a technical answer to needs. You should give an answer of a different nature. For example, instead of making the channels deeper and the dams longer, the bigger ships could not enter and we could find a more technically modern port in the sea. We could do smaller entrances and reduce the size of the dams and raise the foundations. You should find an environmentally compatible way to answer questions, not a mechanical way.

Gasparetto's proposals involve engineering work, but he framed them in terms of environmental compatibility. Gasparetto recognized the additional need to block high tides from coming into the lagoon, acknowledging that the alternative methods for closing the lagoon entrances, discussed in Chapter Six, would not block the highest tides. He hypothesized the alternatives would stop the medium tides that Project MOSE was not designed to prevent:

> When they build the MOSE dams we would avoid one or two, maximum three high waters in one year and all the other 70 or 80 high waters would be there

> because Project MOSE can close only when the water is 110 centimeters above the sea level. We environmentalists have done alternative projects that can be made in three or four years, which can be experimented with to see if they are functional before completing them. They would only leave one or two high waters a year when they are very high. This is the point of taking the lagoon and the city of Venice and its islands to the situation that existed before the transformation of modernity. At the beginning of the 1800s, Venice used to have one or two high waters in one year. (Gasparetto 2005a)

Gasparetto indicated that a merit of the environmentalists' preferred systems would be that Venice would be returned to pre-modern flooding levels. An attachment to pre-modernity appears to be a strong motivating factor for environmentalists in rejecting engineering interventions like the mobile dams. Modernizing intervention is presented as the cause of the flooding problems. That another modern intervention, the mobile dam project, could address these problems is derided as likely to make the problem worse.

A pivotal claim against the dam project is that it is outdated. Five of the environmentalists interviewed (Caccia, Cogo, Gasparetto, Mencini and Sarto) criticized Project MOSE on the grounds that it was initially proposed more than 30 years ago. But the problem is perceived not so much as the passage of time between the mobile dam project's initial design and actual construction, but its conceptualization with an industrial twentieth-century mentality that is overwhelming and oppressive. "This Project is very old. It is from the early eighties and it was conceived at the end of the sixties. The very name says what it is (MOSE). So the name is like industrial archaeology. It is a Project of industrial archaeology. It is a mentality from the twenties and the thirties and they think only about a pharaonic, costly and forceful act that goes against any logic," declared Flavio Cogo (2005).

SCIENCE AND ENVIRONMENTAL CLAIMS

Despite the reservations expressed by environmentalists whom the author interviewed about scientific engineering, their comments did not include any hostility towards science itself. On the contrary, science was used to support their claims. This is unsurprising: the American sociologist John Hannigan (2006) points out that environmental social problems tend to be tied more directly to scientific findings than other public social problems, which include a wider range of personal troubles. Environmental problems such as pesticide poisoning or global warming tend to be rooted in scientific findings (Yearley 1992, 117), and the political scientist Charles Rubin (1994) demonstrates how prominent thinkers including Carson, Barry Commoner, Paul Ehrlich and Garrett Hardin have helped to popularize scientific findings as environmental problems.

Many Venetian environmentalists are formally enthusiastic about science; Federico Antinori talked about his fascination with science; Alvise Benedetti, a senior figure in the Venice chapter of Italia Nostra, is Professor of Chemistry and Physics at Venice University; Fabrizio Reberschegg (2005) argued that scientists are generally sympathetic to the arguments against Project MOSE, and that it was only finance from the CVN that led to scientific backing for the mobile dams; Giannandrea Mencini (2005a) pointed out that recent scientific opinion on climate change has meant most scientists agree that Project MOSE is outdated.

The environmental movement's use of scientific claims illustrates its failure to make a decisive break from modernity. "The environmental movement, while in many ways hostile to modern science, has at the same time depended for much of its social authority on scientific, seemingly morally neutral, claims about the physical threats produced by present social trajectories," writes Szerszynski (1998, 112). A tension exists at the heart of environmental thinking between reacting against modern science, and relying on science to back up environmental claims. Ecologists have addressed this tension by developing a particular interpretation of science based on taking values from nature. Bramwell (1994) claims that ecology does contradict the principles of modern science, but is not a rejection of science *per se* because a holistic version of science has been constructed: "Ecologism promotes holistic science and simultaneously attacks mechanistic or analytical forms of science" (Bramwell 1994, 16).

The holistic environmental worldview was famously presented in the book *Only One Earth* by René Dubos and Barbara Ward for the 1972 UN Conference in Stockholm. Jamison (1998) writes, of the book's influence: "From the early 1970s, environmentalists argued that nature is a whole, the earth is one, and that natural processes that take place in one part of the world have an impact on the environment thousands of miles away" (Jamison 1998, 241). Douglas Macdonald, a professor of politics, ascertains that the holistic perspective of ecology gave the environmental movement some authority within science and attracted devotees of expanded consciousness, such as Zen Buddhists (Macdonald 1991, 89). However, holism has been attacked as alien to the scientific method based on the province of reason, of hypotheses made, evidence produced and arguments proved or disproved (Popper 1982, 137). "[T]he problem is whether the Greens should indeed be seen as a reaction to humanistic and Enlightenment values, as their past might suggest, because of their opposition to what they see as the dominant mechanistic and Cartesian paradigm of the last 300 years," writes Bramwell (1994, 16-7). Some sociological scholars have even located the birth of the environmental movement in the reaction to the Enlightenment emphasis on humans controlling nature: "To the extent that this idiom of the Enlightenment presumed subsumption of the object under the weight of the subject, it also legitimated man's domination of nature, itself conceived as apodictically knowable object. The ecology movement is born in the *crisis* of these discourses of identity and technology—in the legitimation crisis of determinate and propositional knowledge." (Szerszynski, Lash and Wynne 1998, 11. Emphasis in original)

The end of the dichotomy between nature and society, celebrated in the holistic approach, has important implications for the investigation of environmentalism. Giddens (1991) describes how an "invasion of the natural world by abstract systems...brings nature to an end as a domain external to human knowledge and involvement" (Giddens 1991, 224). As outlined in Chapter Three, the distinction between man and nature underpinned conservationism when it emerged in the late nineteenth and early twentieth centuries. Conservationism was based on man protecting the natural world, and presumed the separation of the subject (man) and the knowable object (nature) (Szerszynski, Lash and Wynne 1998, 11). In the process characterized in Chapter Four as the transformation of conservationism into environmentalism, radical input from the social movements of the late 1960s and 1970s led to reframing the environment with an emphasis on integrating nature and society, and environmental concerns expanded from defending nature to wider social and economic questions. As sustainable development became the dominant environmental discourse during the 1980s for activists and many institutions, the distinction between nature and society eroded. A holistic environmental culture evolved whereby nature is treated as part of society's history, rather than separate from society. "At the end of the twentieth century", writes Beck (1992), "nature is *neither* given *nor* ascribed, but has instead become a historical product" (Beck 1992, 80. Emphasis in original). The erosion of the culture of modernity, which separated nature and society, and the emergence of a holistic culture has helped environmentalists in their presentation of modernizing projects like Venice's mobile dams as 'outdated.'

In contrast to an engineering mentality, which is based on human intervention to restrain nature and protect a society from flooding, environmentalists have proposed minimal forms of intervention that appear to integrate natural, social and historical factors. In this regard, Venetian environmentalists have been able to draw on the contemporary unease with modernization as a cultural resource to support their claims. However, these claims have been only partially successful, in that the culture of modernity has not been completely subsumed by environmental culture. Many scientists, engineers, politicians and claims-makers still advocate that man needs to limit the impact of disruptive natural forces like floods and hurricanes with engineering and technology. Environmentalists thus face competing claims-makers.

Chapter Nine

Modernizing or Sustaining Venice?

"The Venetian Republic saved the islands and citizens' social structure to maintain a living Venice. The new economy must succeed a nearly dead one, but cannot be based on tourism."

Cristiano Gasparetto (Italia Nostra 2010b).

At a national press conference in Milan in March 2010, Cristiano Gasparetto deployed the myth of social and political harmony during the Venetian Republic to criticize the growth of Venice's tourist economy. Tourism now permeates so many aspects of Venetian life that it is fundamental to decisions about sustaining or modernizing the city. It is frequently claimed that tourists are swamping Venice and that Venetians are fleeing, leading to the death of the city. This chapter challenges these claims. As the late novelist and essayist Mary McCarthy wrote in the 1960s, "there is no use pretending that the tourist Venice is not the real Venice, which is possible with other cities—Rome or Florence or Naples. The tourist Venice *is* Venice" (McCarthy 1963, 8. Emphasis in original).

In their promotion of sustainable policies, environmentalists have contributed to claims about tourism leading to the death of Venice. This chapter argues the case for modernization and development, which are used interchangeably for Venice in general. For specific projects, modernization implies restructuring whereas development refers to transformative projects. Development means constructing the mobile dams, a sewage system, a subway, new bridges and transforming the mainland with new industries, hotels and accommodation for others. This is a not a proposition for building skyscrapers on St Mark's Square or demolishing ancient structures that constitute a vital part of Venice's heritage. In contrast to the 'sustainable' environmental approach, a developmental approach would improve life for people, conserve ancient buildings and protect the long-term condition of Venice's environment.

We begin by focusing on the relationship between past and present trends regarding tourism and population changes, and move on to consider more recent changes and assess options for the future. We also recognize the difference between the ways that the concepts 'conserve' and 'sustain' are used. Conservation refers to the protection of nature, monuments and buildings, whereas sustainability is more widely applied to social, economic and political policy-making. Protective traditions and regulations prior to the 1960s were governed by conservation, whereas sustainability has come to be used for an array of issues since the 1980s and is ubiquitous in debates about tourism.

TOURISM AND THE DEATH OF VENICE

An advertisement for UNESCO recently declared: "Tourism is yet another issue that could help send the vulnerable Venice to a watery grave" (UNESCO 2009). At first sight, it may seem improbable that tourism could sink Venice: no one is suggesting that trampling tourists are lowering the city's land level. But this apparent threat comes from Venetians moving out of the city and masses of tourists outnumbering those who remain, making Venice "not a city, but a stage, an un-kept museum to visit" (Caccia 2005). This belief that Venice is little more than a museum is not a new. Mary McCarthy wrote in 1963: "It has been part museum, part amusement park, living off the entrance fees of tourists, ever since the early eighteenth century, when its former sources of revenue ran dry. The carnival that lasted half a year was not just a spontaneous expression of Venetian license; it was a calculated tourist attraction" (McCarthy 1963, 7). To understand claims about how tourism is suffocating Venice, we need to appreciate how the mythology of death in Venice has evolved.

Myths about Venice's death have found different focuses over time. Death lurked as an emergent theme early in Venetian history as the city was threatened by plagues, floods, silting and fires. After the fall of the Venetian Republic, the myth of Venice dying became culturally elevated, and, by the twentieth century, dominant. The slow character of Venice's death has been repeated in this mythology: threats that could have destroyed Venice in a short space of time have not taken hold, and there has been a reluctance to imagine military destruction, including by the French in 1797, Austrian bombing during the First World War, and German bombing following Venetian resistance in the Second World War. Myths about the slow death of Venice have been particularly effective because they express cultural pessimism rather than objective dangers. Intellectuals have consistently wallowed in the death of Venice, and they have typically refrained from contemplating its fortification: "I pity Venice insofar as the centuries have abandoned it; but I would not wish my pity to revive it," remarked author and politician Maurice Barrès in 1903 (cited in Pemble 1994, 143, footnote).

New claims about the problem of tourism have emphasized the gradual demise of the city. This was illustrated by a mock funeral that was performed on gondolas sailing down the Grand Canal on 14 November 2009, staged by a group of residents who formed an organization called *Venessia*. "We're going to turn into a city of ghosts if something isn't done soon," stated Matteo Secchi, a spokesperson for *Venessia* and a local hotelier. "In 30 years there might be zero Venetians left," added Secchi (Donadio 2009). One gondola in the procession carried a wooden coffin; at the end of the 'funeral', the coffin was broken open and a flag with a phoenix was pulled out, signifying hope for the city's rebirth. The coverage of this event by *Newsweek* (Nadeau 2009), the *International Herald Tribune* (Donadio 2009) and newspapers across the world suggests the slow death of Venice touches international cultural sensibilities. The mock funeral attracted considerable interest from Venetians, tourists and the world's media, even though this was an unoriginal event; Venice has a long history of mock funerals. In order to assess what is new in claims about tourism today, it is necessary to examine how tourism has affected the city in the past.

Tourism in History

The British sociologist John Urry (1990) defines the tourist as a sightseer who is drawn to an unordinary gaze: "the tourist gaze is directed to features of landscape and townscape which separate them off from everyday experience. Such aspects are views because they are taken to be in some sense out of the ordinary" (Urry 1990, 3; cited in Davis and Marvin 2004, 13). Using this definition, pilgrims were the first mass tourists to Venice. Between the 1380s and the 1530s, Venice dominated pilgrimage transit between Europe and the Holy Land. Pilgrims often waited months in the city for connecting ships. By the fifteenth century, Venice had become an attraction in its own right for pilgrims, who were keen to see the wide collection of religious relics. Most pilgrims combined sacred and secular tourism while in the city, and even bought trinket souvenirs. Venice created an infrastructure of lodging, food, and entertainment to serve pilgrims, which increasingly provided for secular tourists too. In the fourteenth century, "[t]he authorities invented a season of festivals and fairs, from the end of April to the beginning of June, which could be used to inveigle more visitors," writes Ackroyd (2010, 289).

After Venice's pilgrim trade collapsed in the mid-1500s due to war between Venice and the Ottoman Empire, Grand Tourists were attracted to the city. "[I]n the year 1775 the number of those who arrived on the eve of the Ascension day, amounted to 42,480, exclusive of the preceding days," wrote Johann Wilhelm Von Archenholtz in 1785 (Von Archenholtz 1785, 37; cited in Davis and Marvin 2004, 36). "Even half that many foreigners, coming in such a short time, could make news in the Venice of today," note Davis and Marvin (2004, 36). During the nineteenth century, the Grand Tour to Venice became mass 'package' tourism.

The famous Thomas Cook's tour to Venice started in 1864. The transatlantic connection with the USA grew rapidly between 1851, when 200 Americans visited the city, and 1894. "[B]y the late 1800s, five thousand or more package tourists were coming to Venice daily, just by train, during the high season," write Davis and Marvin (2004, 216). Working people as well as elites were able to visit the city: a group of 60 artisans, clerks, salesmen and schoolteachers from Toynbee Hall social settlement in East London visited in 1889. Venice had a profound impact on some visitors: a notable example was Thomas Okey, a basket weaver from Spitalfields in London, who traveled to Venice from Toynbee Hall in 1892. He gained an education in his spare time, wrote two books on Venice, translated Dante and St Francis, and became a professor of Italian at Cambridge University (Pemble 1994, 176). Although many factors undoubtedly assisted Okey's success, the impact of his Venetian experience is a reminder of the contribution tourism can make to people's lives. Despite such success stories, it was the addition of working-class masses to the Venetian Grand Tour that provoked growing criticism. After numerous visits to Venice during the second half of the nineteenth century, Henry James deplored the tourist masses that had overwhelmed the city:

> The barbarians are in full possession, and you tremble for what they may do. You are reminded from the moment of your arrival that Venice scarcely exists any more as a city at all; that she exists only as a battered peep-show and bazaar. There was a horde of savage Germans encamped in the Piazza, and they filled the Ducal Palace and the Academy with their uproar. The English and Americans came a little later. They came in good time, with a great many French. (James 1909, 7-8)

Paradoxically, as Venetian tourism became more popular, elite tourists including James came to defend the city from the "barbarians". As discussed in Chapter Three, elite conservationists including Ruskin challenged the development of bridges, the port, hotels, steam and motor boats, the railway and road link to the mainland. The opposition to development in Venice by elite conservationists was intertwined with their dislike of mass tourism; this infrastructure supported the growing tourist industry.

Contemporary Tourism

As vacations have become commonly affordable and tourism to Venice has grown, it is the mass character of tourism that has been increasingly criticized. Distinctions are repeatedly made between mass tourists on package holidays and independent travelers. In a perceptive critique, British lecturer Jim Butcher (2003) demonstrates how moral distinctions are created between mass package holidays and "New Moral Tourists," who plan their holidays individually with the environment in mind. Venice is often held up as an example of a place to visit for an environmentally-sensitive vacation due to its lack of cars and modern infrastructure, as depicted by Susan Sontag's 1983 film *Unguided Tour*. In a study advocating sustainable tourism in Venice, Jan van der Borg and Antonio Paolo Russo of Rotterdam University recommend "special itineraries of eco-environmental interest" around the lagoon (van der Borg and Russo 2001, 184). Mass tourists are negatively compared with independent travelers carefully exploring Venice's environment.

Tourists who do not stay overnight in the city face extra criticism. "In 1951, 1,128,699 travellers came to Venice; in 2007, it was around 16.5 million...12.5 million come just for the day," document Anna Somers Cocks and Thierry Morel (2009, 31-33). Day tourists include a variety of people, who stay outside the city, come by bus, boat, train or car. "[T]he day-trippers' contribution to the various social costs is much greater than that caused by residential tourists. In other words, the economy earns more from tourists who stay overnight than it does from day-trippers," write van der Borg and Russo (2001, 164). The likely conclusion from this analysis seems to be that people who desire (or need) to spend less money in Venice are less welcome. Is this just a contemporary version of the nineteenth century elite dislike of working-class people visiting Venice? Or is it true that the city is suffering due to mass tourism in general and day tourists in particular? The day tourist supposedly has a disproportionately negative impact on Venice's environment and society, especially through transport use: but it is difficult to prove that day tourists have a more negative environmental impact than visitors who stay overnight. Moreover, analyzing tourists simply in terms of their environmental impact obscures their positive economic and cultural contribution to the city.

Although distinctions are made between types of tourists, the claim that Venice is suffering from mass tourism usually compares the overall number of tourists to residents. "Number of Venetian residents in 2007: 60,000.

Number of visitors in 2007: 21 million. In May 2008, for example, on a holiday weekend, 80,000 tourists descended on the city like locusts on the fields of Egypt," writes Cathy Newman in *National Geographic Magazine* (2009). This vivid disdain for mass tourism is created using an inappropriate comparison, or what Best (2001b, 97) describes as "comparing apples and oranges." Comparing the total number of residents with the annual number of visitors creates a huge contrast, while mystifying daily reality: a more accurate picture of who occupies the city on a daily basis is required. Calculating the daily effective population, which includes an array of people who inhabit the city on a typical day, is a better sociological measure. This measure has been applied by Venice City Council since 1994 and by COSES (Consortium for Research and Educational Training) in reports for the Council. In 2007, COSES assessed the daily effective population in Venice and its islands as follows: residents 88,519, students 4,174, owners of second homes 15,224, commuters for work 15,181, commuters for study 6,894, tourists 54,003 (Da Mosto 2009, 20). The conspicuous statistic is that the number of residents in Venice far outweighs the number of tourists. These statistics are supported by an OECD (2010, 75) assessment of Venice that "[t]ourists account for approximately 30% of the daily city users." The daily effective population measure helps reveal the reality that contradicts the myth of tourists outnumbering residents.

The daily effective population measure does, however, include statistical gaps. It ignores residents' vacations and those who commute out of Venice for work or study. The most recent figures for commuting are from the Census of 2001, which showed that 17,414 people commuted out of Venice every day, with 62,222 people commuting into the city daily. This led to a daily daytime net increase of 44,408 people in Venice. Including holiday periods, the average daily presence of commuters in the city was 36,728 (Da Mosto 2009, 18-19). Secondly, the daily effective population measure does not take seasonal variations into account. This is a significant variable: for example, Venetians tend to go on vacations in August when many tourists visit the city. Seasonal variations mean that, in 2007, day-trippers and other tourists outnumbered residents in July, August and September within the historical city. On the other hand, there were more residents than day-trippers and other tourists within the historical city during all other months (except for June, when the numbers of residents and tourists were approximately equal) (COSES 2008, cited in Somers Cocks and Morel 2009, 42). Thirdly, the daily effective population measure does not calculate the time different people spend in the city: day-tripping tourists spend less of each day in Venice than most residents.

The claim that Venetians are being suffocated by tourists is typically constructed using illegitimate comparisons, and is difficult to sustain. Some claims highlight tourists' impact on central parts of the city. "When a tourist tsunami hits town, the central areas of Venice are completely swamped," warn Davis and Marvin (2004, 109). Yet even when we consider tourists in the most popular area to visit in Venice, St Mark's Square, their numbers have not surpassed saturation point. Somers Cocks and Morel (2009, 41) calculate that the maximum number of people present in St Mark's Square a day is 134,000, taking into account an average stay in the square of 50 minutes and six peak hours of visiting. St Mark's Square undoubtedly gets very busy at times: but Somers Cocks and Morel also note that the main routes to the Square can carry 140,000 to 150,000 people (Somers Cocks and Morel 2009, 41). It is easy to avoid St Mark's Square and routes to it when it is crowded. It is also easy to move around in the Square, in comparison with periods in the past. Davis and Marvin (2004) describe how St Mark's Square in the eighteenth century was often littered with moneychangers, preachers, jugglers, cluttered stalls, chicken coops, garbage and even urine (Davis and Marvin 2004, 61-2). Since the early nineteenth century, most commercial activity and mess has been cleared from the Square, and all souvenir stalls, food sellers and portrait artists are now banned. Even pigeons are less common, since seed vendors were prohibited in 2008.

Venice is now considerably easier to navigate and more pleasant for tourists and residents than it was before the early twentieth century. "From the first days of the pilgrim tourists until the time of Henry James, one took a gondola for any jaunt of more than a few hundred meters. The city was not only confusing and complex for outsiders; it could also be dangerous, or least annoying: anyone who ventured into an unknown alley could expect, at the very least, to be assaulted by beggar children, whores, or even robbers," write Davis and Marvin (2004, 88).

Current Population Changes

"Venice isn't sinking as much as it is shrinking — demographers predict that by 2030, there won't be a single full-time Venetian resident left," announced Barbie Nadeau in a 2009 *Newsweek* article (2009). This prediction assumes that recent falls in the Venetian population will continue, based on projecting the current situation forwards without significant changes. In fact, the city's population shifts indicate that it is not dying, although it is certainly changing. "In March 2009, the number of registered residents of the historic city was 60,208, with a further 30,362 living in Venice's islands and estuary. This gives a total of 90,570 people designated as permanent residents of the 'lagoon city', a term to embrace the historic city, the Lido, Pellestrina and the islands of the lagoon," records Da Mosto (2009, 12). This compares with a peak number of 218,991 residents of the lagoon city in 1952. The resident population of the historic city has fallen even further, from 174,808 in 1951 to 60,208 in March 2008 (Da Mosto 2009, 12). Why has the number of residents fallen? One reason is the higher number of deaths than births per year: "Since 2000, the average number of births per year has been 705 for the 'lagoon city' (472 in the historic city and 233 on the islands) and 2,113 for the municipality. Over the same period, the annual death rate is 1,387 for the 'lagoon city' (970 and 417 in the historic city and the islands respectively) and 3,329 for the municipality" (Da Mosto 2009, 14).

The longer-term decline in the number of Venetian residents since the 1950s has been caused by Venetians leaving the city. Chapter Three explained how the failure to modernize Venetian housing and infrastructure encouraged Venetians to move elsewhere. Howard (2005) writes: "Lower-class housing in the city is still well below acceptable standards, and the desperate shortage of amenities such as parks, sports facilities and children's playgrounds adds to the loss of morale. Modernization of the more modest housing has tended to raise the rents beyond the means of the original inhabitants, leading to further emigration. At the heart of the city, the historic architecture has become increasingly inviolate" (Howard 2005, 294). It is entirely understandable that many Venetians have left the city due to insufficient modernization and development, which are fundamental to maintaining a residential population. In any case, the reduction of the city's population is not entirely negative. Roderick Conway Morris (2000), a British journalist living in Venice, observes the ironic nostalgia for overcrowded, poor quality housing: "[A]t the time of the last population boom between the late 1940s and early 1960s, much accommodation was, by today's western European standards, overcrowded. Accordingly, if the city were to accommodate the population size that Venetians look back on with positive nostalgia, its inhabitants would have to live in conditions to which most of them are unaccustomed and would now consider unacceptable" (Morris 2000).

Changing employment opportunities also motivated Venetians to leave the city, as businesses have progressively relocated premises outside of Venice's center. "[I]n 1971, businesses located within the municipality, the Comune of Venice (which includes large tracts of the mainland), accounted for 61% of jobs in the province and 12% of those in the Veneto region. Both of these figures had fallen, to 42% and 7% respectively by 2001," register Renato Gibin and Stefania Tonin (2009, 51). *Assicurazioni Generali*, a market-leading Italian insurance company, is a prominent example of a major employer that moved out of Venice between the late 1980s and 2001: its headquarters had been located in the city since 1832. The daily newspaper *Il Gazzettino* and state TV organization *RAI* have also deserted the city. Ironically, it is tourism that has expanded job opportunities and is helping to prevent the death of the city for working Venetians, as Gibin and Tonin (2009) observe: "The local economy has recently come to specialise in providing services for people who use the city but are not resident. The only employment sectors that showed increases between 1991 and 2001 were professional services and the hotel, bar and restaurant trades. Moreover, in 2001 the latter became the most important private-sector activity for jobs" (Gibin and Tonin 2009, 54).

The belief that Venice's population shifts are leading to the death of the city reveals a lack of historical awareness and an incapacity to embrace change. Historically, the population of Venice has varied significantly. "The 117 mud-flat islands became a metropolis of two hundred thousand inhabitants at a time when medieval Paris had a population of forty thousand, London was inhabited by twenty thousand and Rome a mere fifteen thousand," notes Lauritzen (1986, 36). Venice's population was nearly 190,000 in the third quarter of the sixteenth century, despite plagues that had reduced its number of residents by three-fifths in the fourteenth century. Contemporary population variations analyzed above are relatively small, although immigration still needs to be taken into account. Recent immigration is much lower than in many past periods. Following plague epidemics in the Middle Ages, Venice's population was replenished by waves of immigrants from the mainland, Greece and Dalmatia. Between 2000 and 2009, there was a net average loss of 1,046 people from Venice city to the province of Venice and 458 to the Veneto region per year. Against this, there was a net annual inflow into Venice of 1,769 people from abroad and 253 people from other regions of Italy during the same period (Da Mosto 2009, 15). Even though these are small changes, more people coming to live in the city than leaving it has reduced the fall in Venice's population since 2000. There has been a growing influx of people buying second homes in Venice and staying occasionally: in 2004, 15,224 people had second homes in the 'lagoon city', which was double the number for a decade earlier (Da Mosto 2009, 17). Venice has historically coped with vast population swings and hordes of visitors, including pilgrims, artists and traders. "Cultural diversity and a changing population have always been integral to the history of Venice. Its resident population has been marked by sharp oscillations yet has always remained resilient. In the past, parts of the city were left empty when the population went into decline. Today, the picture is very different," writes Da Mosto (2009, 21). Proclaiming the contemporary death of Venice indicates a loss of historical imagination, more than it reflects the emergence of real threats.

Venice's population is in a dynamic state of change. Is there sufficient imagination to manage change? Answering this question requires addressing two more specific questions. Firstly, can tourism be better managed to improve the city? Secondly, can the city be modernized? In the remainder of this chapter, the author makes ten recommendations for an imaginative approach to developing Venice.

A TEN-POINT PROPOSAL TO DEVELOP VENICE

1) Modernizing Accommodation

We begin with the question of modernizing housing and accommodation — for tourists, students and residents alike.

TOURIST ACCOMMODATION. A key driver of Venice's changing accommodation has been the liberalization of planning laws, which has allowed private homes to be turned into Bed and Breakfast (B&B) businesses and rooms to be rented. Implementation of this liberalization has been gradual since 1996, when the municipal Councilor Roberto d'Agostino lifted restrictions on altering the use of residential property. The liberalization has given Venetians more opportunities to make money from tourism and provided tourists with better accommodation options. Between 2000 and 2004, the total number of beds for visitors in the historic center increased from 28,000 to 32,000 and the number of B&B establishments doubled (Somers Cocks and Morel 2009, 38-9). There was a 30 percent increase in hotel beds available in the historic center between 2006 and 2007, and non-hotel accommodation establishments within the historic city proliferated from 142 to 1408 between 2000 and 2007 (Gibin and Tonin 2009, 65). A survey published by Venice City Council in 2008 recorded increases in three types of non-hotel accommodation between 2004 and 2008 (Gibin and Tonin 2009, 66). Many properties have been internally refurbished, thereby improving the quality of Venetian housing stock: 44 percent of the 952 B&B establishments opened between 2001 and 2007 were refurbished internally (Somers Cocks and Morel 2009, 38).

It is positive that the liberalization of planning regulations in Venice has made it easier for residential property to be transformed into tourist accommodation. Yet when Michele Boato stood as the Ecological and Radical candidate in the April 2010 Venice Mayoral election, he called for an end to transforming residential properties into tourist accommodation (Boato, M. 2010). There have been some new accommodation problems, which have been attributed to the liberalization of planning laws: between 2000 and 2009, property prices rose sharply and fewer dwellings became available for residents and students (Gibin and Tonin 2009, 59). However, these problems cannot be attributed solely to the liberalization of planning restrictions - they have been caused by multiple factors, including changing employment and population patterns analyzed above. Indeed, property prices were assessed to have decreased by 2.4 percent in Venice between 2009 and 2010 (Nomisma consulting group, cited in Poveledo 2010). Residential property prices were partly inflated in 2001-2 by the introduction of the single European currency (Gibin and Tonin 2009, 59).

A more significant factor in long-term rising property prices has been the failure to build new and modern residential and student accommodation. This failure is due to the historical dominance of conservationism in planning, as reviewed in Chapter Three. While the liberalization of planning restrictions was a positive step, this was insufficient to meet demand. Conservationist traditions and laws have meant not enough accommodation has been built, and these same traditions and laws are adding to pressure on companies to transform Venetian palaces into hotels. The following palaces have been sold for this purpose: Palazzo Garzoni, Palazzo Genovese, Palazzo Nani Mocenigo, Palazzo Rava Guistinian, Palazzo Ruzzini Priuli, Palazzo Sagredo and Palazzo Soranzo Piovene (Leo Schubert cited in Agnoletto et al. 2009, 69-85). Most of these palaces are situated on the Grand Canal. To become hotels, they require major internal restructuring, including the construction of bedrooms with *en suite* bathrooms, meaning that part of Venice's heritage will be lost forever. It is the author's view that the restructuring of palaces to become hotels is a form of modernization that should be rejected. If individuals cannot afford the upkeep of palaces, the state or consortia of investors could fund their use as offices, museums, galleries, libraries or educational institutions. New hotels are already being built on the mainland and the Lido, and from there tourism could be spread from central Venice to islands throughout the lagoon, along the waterways of the 600km (373 miles) Litoranea Veneta, to Treviso, Vicenza, Padua and Palladian villas across the countryside.

It is a sad irony that conservationist opposition to building new hotels is leading to the loss of ancient palaces and the concentration of tourism in Venice's historical center. A developmental approach favors maintaining ancient heritage, while constructing new hotels and accommodation for residents and students.

STUDENT ACCOMMODATION. Da Mosto notes: "With around 30,000 students and university staff — equivalent to half of the entire resident population of the historic centre — attached to institutions of further education that are mostly in the historic centre, they make a very significant contribution to the effective population" (Da Mosto 2009, 18). Many educational staff and students are also residents. In the academic year 2008-9, approximately 30 percent of all students at Ca' Foscari University were resident in the province of Venice (Da Mosto 2009, 18). Students who do not stay in the city commute in: according to a major study in 2004, "[l]ess than 60% of the 8,000 university students wanting accommodation in Venice were able to find it" (Da Mosto 2009, 18). The rise in property prices over the past decade is likely to have discouraged some students from enrolling at Venice's educational institutions. Although Venice expanded its educational institutions and student populations during the twentieth century, the total number of students has been falling during the twenty-first century. The total number of students at Ca' Foscari University was 4.6 percent lower for the academic year 2007-8 compared with 1992-3. Over the same period, the number of students at IUAV fell by 51 percent (Da Mosto 2009, 17).

Building decent and affordable student accommodation should help to reverse this decline. A welcome example is the joint project between Venice City Council and IUAV to restore the Ex-caserma Manin building in the historical center. By the end of 2011, this will provide 174 rooms and apartments for students, as well as 37 flats for public housing (Palazzo 2009). Making use of premises available due to the falling number of schoolchildren could present another opportunity for the expansion of higher educational facilities. More student accommodation could be built on the mainland bordering Venice, near the university site being constructed in Mestre, where a new hospital and science park will stimulate demand for dwellings for various kinds of people.

RESIDENTIAL ACCOMMODATION. Despite the construction of a few apartment complexes since the 1980s, housing provision for Venetians is still inadequate. "Although the modern apartment blocks on Sant'Elena or Sacca Fisola are comparable in their comforts to similar working-class dwellings in Mestre, many of the dwellings on the Giudecca and far Castello are simply squalid," write Davis and Marvin (2004, 101). It is the author's view that conservationist restrictions on property restructuring should be lifted to allow the modernization of residences, providing nearby ancient buildings are not damaged. The refurbishment of many Venetian properties for tourist use since 1996 has demonstrated that internal changes to buildings can be made without destroying the city's ancient fabric. External planning restrictions should also be canceled to allow the transformation of properties in residential areas, especially Castello.

An opportunity to expand existing residential accommodation will be created by the mobile dams. After the high floods of 1966, approximately 6,000 ground-floor apartments in Venetian houses were abandoned due to flooding fears; by the 1990s, it was estimated that 4,700 of these apartments remained empty (Tamburrini 2009). When the dams have begun operating against high flooding, it should be clear which ground-floor flats are sufficiently protected to be renovated and reoccupied. Other opportunities exist for expanding accommodation, including on the mainland, in the Arsenal and its neighborhood.

2) Modernizing the Arsenal

Venice's Arsenal of maritime factory production used to be the largest industrial enterprise in the world and could make 30 galleys in ten days (Ackroyd 2010, 219-20). It was closed down in 1957 when the Command Center of the Maritime Department of the Upper Adriatic was moved down the coast to Ancona. The Arsenal area covers 48 hectares (118.6 acres), which is one fifteenth of the total area of the historic city. It is also situated close to the key tourist zone of San Marco and has great access to the lagoon. This means the Arsenal has a lot of potential to support tourism, as well as fulfilling other needs for the city. The Italian Ministry of Defense suggested in April 2010 that military buildings in the Arsenal could be replaced by infrastructure that serves tourism (Fullin 2010).

There have been very limited attempts to restore parts of the Arsenal, including the old rope making workshops, the armory, the entrance gates, the small island of Isolotto, and the Tese delle Vergini warehouse. Some dilapidated buildings have been restored through a partnership between IUAV and MAV. The northern parts of the Arsenal and its dry docks have been leased to the company in charge of water buses (ACTV), which has repaired boats there. Thetis S.p.A., a marine technology research company, restored three buildings for company premises. Warehouses are temporarily occupied by Biennale exhibitions. Offices for the National Research Council and for the Museums' Foundation are being constructed in the northern Arsenal. To date, these activities are falling far short of the Arsenal's potential, as indicated by the levels of employment: in 2009, approximately 300 people were employed in the Arsenal, excluding seasonal staff for the Biennale (Panzeri 2009a, 58).

One reason for the failure to realize the potential of the Arsenal is that planning has been governed by sustainable criteria, which were defined in the 1998 Program for Urban Rehabilitation and Sustainable Development of the Territory (PRUSST). PRUSST was approved by Venice City Council in 2001, with a specific plan for the northern part of the Arsenal agreed in 2003. These plans did not even consider how to develop the Arsenal water docking and port facilities: "Oddly, there is no plan at all for the waters of the Arsenale," notes Lidia Panzeri (2009a, 57). Since these plans, the CVN has been given permission to restructure 125,000m^2 (1,345,488.802 feet^2) of the Arsenal, including six warehouses, for monitoring and maintaining the mobile dams.

Modernization of the Arsenal should go beyond this. Warehouses and workshops need to be restructured so that they can be used for permanent exhibitions, museums and educational facilities. Given the historical significance of the Arsenal buildings, it would be a shame to lose their outer structural forms. Internal restructuring could provide a public space to enhance

Venice as a cultural and educational center. Considered by many as an arts capital, Venice does not have a permanent and publicly-funded museum of contemporary art (Palazzo Grassi and the Dogana host private collections). Plans for a museum and a headquarters for the Institute of Naval Studies have been stalled due to funding shortfalls. The Porta Nuova tower has been earmarked as an exhibition center. Restructured warehouses could host guest Biennale exhibitions when required. Currently, placing exhibitions in Arsenal buildings periodically for the Biennale is an inefficient use of the space. Even though these Arsenal buildings are also occupied for other occasional events, such as during Carnival, they are underemployed in a city where land is limited. These proposals would require restoring the basic infrastructure of the Arsenal, which is affordable: "According to a reliable, though not particularly detailed, estimate, it would cost €503 million to rehabilitate the entire Arsenal complex completely... This sum breaks down as follows: €381 million to rehabilitate and restore the buildings and €122 million to bring mains services up to standard and rebuild the paving and embankments" (Panzeri 2009a, 56).

The Arsenal and its neighborhood present opportunities for expanding accommodation. If some of the Arsenal warehouses remain vacant, they could be used for tourist or student accommodation. It is not uncommon for old buildings to be internally transformed to create hotels: Oxford prison in the UK is a creative example. It would also make sense to build new residential properties near the Arsenal. Following the Arsenal's closure, 11,000 families were moved out from the neighborhood in 1960 (Ackroyd 2010, 220-21), and much of the remaining nearby residential property is in a poor state. Restructuring could provide attractive dwellings.

3) Port Development and Maritime Services

Venice port consists of three parts: the maritime passenger station on the edge of the city; the Marghera industrial goods zone; and the San Leonardo oil terminal. All three parts are already being transformed due to economic changes and additional recommendations are proposed here. The recent economic downturn has adversely affected overall business for Venice's port: in 2009, total traffic through Venice port declined by 16.7 percent compared to 2008, with an EU average decline of 11.9 percent. Most of this decline came from a reduction in cargo: Venice's cruise passengers increased by 16.9 percent in 2009 compared to 2008 (Pietrobelli 2010). This improvement followed a rise in the number of cruise ships sailing through Venice from 200 to 510 a year between 2000 and 2007 (Agnoletto et al. 2009, 9). The cruise business for Venice has prospered both in tough and favorable economic periods. It has prospects for further expansion, as do other services.

Passenger port services are located at the large maritime station docking facilities and the smaller San Basilio and Santa Marta quays (see Figure 1.2.). Venice's combined passenger port is the thirteenth largest in the world and the third biggest in the Mediterranean, and the city's maritime services provide for ferries, yachts and cruise ships. In 2007, Venice University estimated that direct expenditure by cruise ships constituted nearly 10 percent of Venice's tourist economy (Somers Cocks and Morel 2009, 45). The construction of cruise ships in Venice has also boosted the local economy, with 20 completed in the *Fincantieri* Marghera shipyard over the past 20 years, although orders were reported as scarce in 2010 (Favarato 2010). There has also been an increase in luxury yachts mooring in Venice. During 2007, 201 luxury boats tied up - a rise of 1,000 percent on 2002 - and stayed an average of four days, spending between €800 (US$1,055)[1] and €2,000 (US$2,367) a day for various port services (Somers Cocks and Morel 2009, 37). Better facilities for luxury yachts could be built, such as in the Arsenal, which hosts many Biennale exhibitions that attract numerous yacht owners.

New docking for larger cruise ships (those 350 meters–1,148 feet long) is planned by the Venice Port Authority near Fusina on the mainland edge of the lagoon. This will be linked by road and train to Venice railway station for transferring passengers. The Port Authority President, Paolo Costa, confirmed a €200 million (US$261.938m) investment for these docking facilities (Crema 2009). In April 2010, it was reported that these new docks would be ready in 2015 (Favaro 2010). The Venetian lawyer Alessio Vianello proposed that two million passengers per year could be received at Marghera, reducing the volume of tourists arriving closer to the ancient city center (Alessio Vianello 2009). But environmental campaigners have challenged the expansion of cruise ship services. Italia Nostra demanded the exclusion of cruise ships from the lagoon tract between Lido Island and St Mark's Square, while other local campaigners, including the group *Amici di Venezia,* presented petitions against air pollution, noise pollution and damage to canal sides from the undertow caused by the movement of water as cruise ships sail. Countering these claims, "Attilio Adami of Protecno srl concluded that due to their slow speed and the shape of their hulls, the cruise ships caused fewer waves than the vaporetti and lighters. In 2003, Mr Adami addressed the question of the effect of the undertow on the canals opening off the Giudecca and concluded that the action of the water was no stronger than a tide" (Somers Cocks and Morel 2009, 44). These findings had been confirmed, according to Adami, by a 2004 study from Venice's Center for Study of Tides, the National Research Center and the City Council Commissioner for Wave Action (Somers Cocks and Morel 2009, 44).

Regarding polluting emissions from cruise ships, Admiral Stefano Vignani, Maritime Director of the Veneto, told *The Venice Report* (Somers

Cocks and Morel 2009, 44) that shipping companies had agreed to limit the sulphur content of the smoke from cruise ships entering Venice to 2.5 percent, and further decreases to 1.5 percent are planned: this is in relation to an international limit of 4.5 percent. Fifty spot checks between 2008 and 2009 found that limits were generally being respected (Somers Cocks and Morel 2009, 44). Where cruise ships dock, "[w]ater quality, particularly near the city, is considered to be improving": the main concerns for lagoon water quality are industrial pollution, agricultural run-off and algae (OECD 2010, 155). Reductions in pollution from these sources have improved water quality, while cruise ships have little impact.

Additional aesthetic complaints have been raised about huge cruise ships blocking the view or being unsightly. Many of these complaints are a matter of opinion. Some people enjoy seeing impressive cruise ships next to historic buildings as a contrast of the ancient and modern, and the view is only momentarily blocked by moving cruise ships. However, more consideration should be given to where cruise ships dock so that they do not block people's view of important events, such as the Redentore Festival, and this is another positive reason for the new docking facilities near Fusina.

Further monitoring of cruise ships and more independent studies of their environmental impact should be conducted. It is the author's opinion that in the absence of evidence that cruise ships cause significant environmental damage, they should continue coming into the lagoon. If evidence of environmental damage is found, docking facilities could be constructed elsewhere with transport links to Venice, as explained below.

4) Transforming Marghera

Various plans have emerged to transform the 30km (18.6 miles) stretch of land along the mainland bordering the lagoon. This is primarily being driven by the decline of the chemical and oil refining businesses. "Chemicals at Marghera Port are finished," announced Renato Brunetta, Minister of Public Administration and Innovation, in March 2009 (Fullin 2009). Shifting methods of energy production have already been explored through the opening of the world's first hydrogen power plant at Fusina in July 2010 (*Corriere del Veneto*, 13 July 2010): a step towards creating a larger Hydrogen Park at Marghera. The first hydrogen plant cost €50 million (US$65.4844m), has the potential to generate 12 to 16 megawatts of electricity and supplies enough electricity for 20,000 households a year. But hydrogen is a poor choice for energy distribution to homes and this hydrogen plant's output compares to approximately 1,000 megawatts from a nuclear power station. The British writers James Woudhuysen and Joe Kaplinsky (2009) explain the advantages of nuclear power given the huge quantities of power generated, while noting the limitations of hydrogen (Woudhuysen and Kaplinsky 2009, 324). Marghera would be a more appropriate site for a new nuclear plant as part of the plan to restart domestic nuclear energy production (Standish 2009).

In addition, Marghera has been identified for expanding the use of waste incinerators, which environmentalists and other activists have challenged (Bertasi 2010a). Waste from Venice and elsewhere needs to be dealt with somewhere. Strong winds in this coastal area make this is a suitable location for activities that may create air pollution, including waste processing and power generation. There has also been environmental opposition to expanding the goods-handling services of Marghera Port. Large container ships would need to pass through the Malamocco inlet into the lagoon, continuing to Marghera Port. This requires deeper dredging of the channel between the Malamocco inlet and Marghera Port. Dredging this channel began in 2004 and was predicted to be completed in 2011. In May 2009, the Port Authority submitted a project to expand the goods-handling services at Marghera Port to the Italian Senate. More dredging could be needed.

Concerns about environmental risks caused by similar dredging in the past have been raised by Dr Tom Spencer, Director of Cambridge University Coastal Research Unit. “The excavation of the navigation channels, largely in the 1960s, altered water and sediment circulation patterns in the lagoon. Increased tidal currents and increased wave heights in deeper water have been used to explain the changes in the lagoon bathymetry, particularly during strong NE (bora) wind events,” he states (Spencer 2009, 26). Rinaldo’s analysis of these navigation channels, which were mainly deepened for petrol tankers, found that they had virtually no impact on exceptional high tides: “It is evident that the petrol channels are not the cause of increasing high water in Venice. The primary causes of the increasing frequency of high tides in the historical center are the combined phenomena of subsidence and the increase in the average sea level” (Rinaldo 2009, 157). Like Spencer, Rinaldo (2009, 161) did find that these channels changed the morphology of the lagoon. Spencer advocates a program of environmental monitoring over a number of years to evaluate the navigation channels, followed by further monitoring after the mobile dams begin operation. “Only then would it be possible to come to an informed decision as to whether or not the navigation channels might be increased in depth,” he concludes (Spencer 2009, 26). These proposals for lengthy monitoring programs risk preventing port transformation by implementing the precautionary principle discussed in Chapter Seven. While the need for definitive data is important for effective environmental management, monitoring should not postpone the expansion of goods-handling.

Some alternatives to deeper dredging in the lagoon are being explored. One proposal is to build a new offshore platform, which would reduce large ship traffic within the lagoon. A plan for this platform to be positioned 14km (8 miles) from the mainland was submitted in October 2010 with an estimated cost of €1.3 billion (US$1.72875bn). This platform could provide a terminal for ships waiting to enter the port when the mobile dams are closed. Even when the dam barriers are not raised, larger vessels would dock at the offshore platform to load or unload goods. Then smaller vessels would transport goods between the platform and mainland. They could also travel along inland canals to port facilities and roads. The Venice to Padua canal could be improved as a route, although developing road links would mean faster deliveries. A different alternative to dredging deeper channels in the lagoon is to avoid it altogether: this could be facilitated by increasing access to the sea at the Port of Levante, which is south of Venice (see Figure 1.1). This port and transport links would need improving; the main road (SS 309—E55) between Venice, Levante Port, Marghera and Mestre would benefit from additional lanes, and new railway lines could be created between Levante Port, Marghera, Mestre and Venice. Existing railway lines would also need modernizing for connections with the cities of Rovigo and Padua.

For the expansion of cargo-handling at Marghera Port and deeper dredging of channels, a staged approach is recommended. In the short-term, deeper dredging of channels and expanding goods facilities should continue. Although deeper dredging will affect the lagoon, Rinaldo's (2009) analysis indicates this will not increase high tides. The impact of deeper dredging and waves created by big ships should be further studied to assess whether they cause serious environmental damage. Deeper dredging and the passage of larger vessels through the lagoon should be reviewed on the basis of compelling scientific evidence, rather than the precautionary principle. Plans for the offshore platform should be investigated as a medium-term alternative. If the offshore platform is not feasible, ships can be routed to a developed Levante Port.

Another plan for the Marghera area will develop the existing Venice Gateway (VEGA) science and technology park, as journalist Enrico Tantucci (2009) describes: "Construction has already begun in the Vega area for the Science and Technology Park that the Comune launched more than 10 years ago with the universities and other bodies such as ENI, the Italian energy company, the Region and the Province to be an 'incubator' for businesses interested in research and innovation. Now, next to Vega 1, there will be Vega 2, Vega 3 and Vega 4, all for service industries, commerce and transport" (Tantucci 2009, 29). As with many of the proposals to transform Venice's port, the expansion of VEGA is connected to other mainland projects.

5) Constructing Tessera City

Tessera City has been devised for the mainland near Venice's Marco Polo airport, and its creation would be recompense for missing the opportunity to build a multi-function center known as the 'Magnet' near the airport as part of EXPO 2000. As explained in Chapter Five, EXPO 2000 and this plan for the airport area were blocked after a successful campaign by Venetian activists. According to Italian sociologist Giuseppe De Rita, holding EXPO 2000 would have sorted out Venice's infrastructure for the next 100 years (cited in Rinaldo 2009, 216). Rebuffing EXPO 2000 "was rejecting a certain idea of modernity," stated Cesare De Piccoli, Vice-Mayor of Venice between 1987 and 1990 (Scalzotto 2008). New plans for Tessera City were developed by Massimo Cacciari, previous Mayor of Venice, and the former Veneto Regional Governor, Giancarlo Galan, although both were not re-elected in April 2010.

If political momentum behind Tessera City is revived, it is envisaged that the city will include a huge hotel, casino, congress center, shopping malls and sports stadium. Nearby Marco Polo airport would have a second runway, enlarged terminal and new cargo area. SAVE, the company that manages Marco Polo and Treviso airports, has been planning Tessera City with Venice City Council, Venice Casino (which is owned by the City Council), Venice Province and the Veneto Regional government. Venice Casino already has a branch on the mainland at Ca' Noghera and has proposed a large casino next to the airport with an adjacent hotel. Tessera City would include a business park and a multi-storey car park. SAVE plans to build 100,000m^3 (3,531,466 feet^3) of office space, while Venice City Council volunteered to manage an area of woodland. The idea of a new 30,000-seat stadium suffered a setback after Venice soccer team was relegated to division D due to lack of funds (Tantucci 2009, 29). So this idea requires new business backing and a better soccer team.

Tessera City could be a place to accommodate and manage larger groups of tourists visiting Venice from here. Their journeys would be assisted by a new airport train station, linked to the regional railway system, the high speed European train (TAV), a subway for Venice and the People Mover (see below). Predictably, however, moves to block Tessera City have been made. Venice's Safeguarding Commission declared in May 2010 that environmental experimentation and observation on numerous aspects of the project need to be discussed before it proceeds (Tantucci 2010). Stefano Boato of Italia Nostra also criticized the Tessera development plan as an idea developed just to generate cash (Bertasi 2010b), and Italia Nostra opposes the proposals along with their associated transportation improvements (Piran 2010).

6) Improving Transportation

Recent and future transportation improvements are fundamental to most development recommendations in this chapter.

TRAINS AND TRAMS INTO VENICE. The People Mover is an aboveground train service that will eventually be linked to other train routes. Since its opening in April 2009, the People Mover has provided transportation over 857 meters (2,812 feet) between the multi-storey car park and ferry terminal at Tronchetto and the principal car/bus arrival point in Venice at Piazzale Roma. The final spending on this project was estimated at €23 million (US$30.1242m) (De Rossi 2010). It provides another welcome example of development on the outskirts of Venice's city center, and provoked no significant opposition. In November 2010 Venice City Council launched a similar proposal for a tram service through Marghera and Piazzale Roma to the San Basilio quay. It remains to be seen whether this proposal will provoke protests.

MAINLAND TRAIN LINKS. Progress has been less impressive here. Neither of the two regional airports of Marco Polo and Treviso is connected to railways, even though 8.5 million passengers used these airports in 2009, representing more than a fivefold increase since 1990 (OECD 2010, 72). Goods-handling has also been hampered by inadequate train services for Venice's airports and port. Over the past decade the number of trains servicing Venice Port has fallen from sixty to twenty. This has resulted in greater reliance on trucks and more road congestion (OECD 2010, 71), despite the opening of the Mestre bypass in 2009. The majority of container traffic originating locally now passes through the ports of Genoa and Trieste: of the 1.3 million containers generated in north-eastern Italy, 25 percent are shipped through Venice Port (OECD 2010, 130). As trade is increasing with Eastern Europe, the Balkans and across the Mediterranean (OECD 2010, 74), poor train services between the port and Venice's hinterland are constraining business opportunities.

The 2005 Veneto Transport Plan initiated a new metropolitan regional railway system. A high-speed link between Mestre, Venice and Padua is established and a route between Venice and Marco Polo airport has been approved. The approval and financing of a route to Treviso Airport, improved port services and high-speed links with Verona, Vicenza and Milan would boost Venice's development and help spread Venice's tourism around the mainland. But high-speed train services across northern Italy have been partly hindered by activists' campaigns against them since 1990 (Della Porta and Piazza 2008, 13-22).

SUBWAY. A subway for Venice has been debated for many years, but at the time of writing, there was still no definitive design. A plan was prepared by Venice City Council in December 2008 to connect Lido, Tessera City and Murano Island with the Arsenal and Fondamente Nuove near the city center (Vanzan 2008a). Ideas for 12 subway stations were exhibited in May 2010 at Venice's Santa Marta; the exhibition was inaugurated by IUAV Professor Gianni Fabbri while IUAV students protested against the plan (Vitucci 2010). Opposition to the subway has been consistently voiced by Venice's leading environmental associations, including local chapters of the WWF and Italia Nostra - Gherardo Ortalli of Italia Nostra went so far as to call the subway "a dangerous project" (cited in Vitucci 2010). The *Nosublagunare* committee collected 12,000 signatures against the subway with proposals for small hovercrafts and more motorboats instead, while former Venice Mayor Cacciari was reported to be in favor of the subway (Zanon 2009). The subway could take center stage in the politicization of Venice's environment as debate about the mobile dams recedes, yet the tram proposal may turn out to be framed as an alternative to the subway. We favor the subway over the tram proposal, although there may be advantages to building both. Construction of the subway would lead to environmental disruption in the short term around the construction sites. But in the long term, the subway would reduce water traffic and environmental damage from boats in the city center, unlike the proposed tram service that would stop at the outskirts of Venice.

BOATS. Various types of motorized boats in Venice create diesel exhaust pollution and waves that erode buildings. Propellers also churn the waters disturbing rotting sediments on canal beds. According to a survey by COSES in April 2000, on an average day 25,000 boat trips are made around the city center and another 5,000 in the wider lagoon (Davis and Marvin 2004, 199). As an alternative to the subway, Italia Nostra has proposed the introduction of mini water taxis that carry only four or five passengers. This proposal (Italia Nostra Venice chapter 2010a) would be completely inadequate to service multiple types of travelers around Venice. If Italia Nostra's proposal for mini water taxis is introduced, transportation problems would be compounded by the association's demand for the exclusion of water taxis from the inner canals (except for emergencies). Insisting on smaller water taxis than are currently operating would increase the number of vessels affecting Venice's environment.

Several improvements have the scope to reduce boat traffic. The public water bus service needs better planning. In recent years, competition has led to rival water bus services, increasing the number of vessels competing for passengers: this service should be run by one public authority. Similarly, the transport of cargo by water boats should be regulated through better use of the warehouse storage center at the Tronchetto to reduce traffic. The subway system could also be used to transport cargo around the city. Separate services and pricing for tourists and residents now operate in Venice: a divisive measure, which sets residents against tourists and compounds the negative attitude towards tourists discussed above. There should be an integrated water bus and subway system that treats all travelers in Venice equally.

7) Island Regeneration

Inevitably, most discussion of policy tends to focus on the islands that make up the center of Venice. Creating a better environment and a prosperous future will also require attention to the other islands in the lagoon. The Venetian brothers Giorgio and Maurizio Crovato (2008, 14-16) usefully categorize the islands as follows: absorbed into the city or mainland, conjoined, vanished, drastically reduced, lost their former role, private (not abandoned), entirely new and abandoned. Many of these islands are neglected and not fulfilling their potential as part of a wider strategy for the area (see Figure 1.2 for the lagoon's islands).

Most islands need to be better connected to central Venice to benefit from its economy. Murano is a good example of how islands can prosper through tourism and strong links with the city. Its glass industry was of considerable interest to tourists from the fifteenth to eighteenth centuries, but went into decline during the early nineteenth century; it has been partly revived through the expansion of modern tourism. Numerous tourists take boats to observe examples of glass manufacturing and buy glass objects. "[M]ore than 60 percent of Murano's workforce, or around eighteen hundred individuals, are involved in some way with glass: in production, supply, sales, or packaging and shipping," note Davis and Marvin (2004, 171). Although space and labor costs limit production on Murano, as Venice's transport links improve the island's trading role could be increased by outsourcing more production to parts of the world with lower labor costs. Being situated close to Venice's central islands, Venice airport and Marghera's container services helps Murano link to businesses elsewhere.

Despite the relative success of Murano, specialization for different islands is more effective than attempting to replicate this model. Remote islands have potential as bird sanctuaries and for environmental tourism. Natural parks on the islands of Certosa, Campalto, Lazzaretto Nuovo and San Giuliano have received some finance and deserve more. These islands should be better served by water buses, which would be easier to achieve if the subway is constructed and boats become available. Lazzaretto Nuovo has attracted regular tourist interest since national, local and UNESCO cultivation of its archaeological facilities, and the island was added as a water bus stop that can be specially requested. San Francesco del Deserto Island, with its ancient churches, has similar potential. Additional channels in the lagoon may need to be dug to serve new water boat routes and mooring needs to be improved.

Regenerating Venice's islands beyond the historical center should not be limited to the less developed islands. Significant changes and construction plans are already underway on the Lido Littoral Island, including a new cinema, hotels and apartments. The company Est Capital SGR has invested heavily to re-establish Lido as an upmarket vacation destination, and is restructuring the iconic Hotel Excelsior and Hotel Des Baines into luxury rooms and apartments. The company also spent €50 million (US$65.4844m) to create Malamocco Bio Island, a wellness center for the Lido. It is awaiting approval to convert a former state hospital into luxury apartments, a hotel and marina. A new publicly-funded convention center is being built on the Lido seafront for the Venice International Film Festival and conferences. But these developments have attracted environmental protests.

Other islands should also be targeted for regeneration by specializing in a way that complements wider development. For instance, the lace-making industry on Burano Island could grow using its lace museum and, like Murano Island, by appealing to tourists and outsourcing production. Modern housing was constructed on the island of Mazzorbo, and more facilities could be added. Lazzaretto Vecchio Island is creating a sports complex. Islands with some infrastructure could improve by hosting restaurants and hotels that offer quieter alternatives to those on the central islands. Private purchases in the lagoon have shown potential, especially on Crevan Island and Tessera Island. San Clemente Island has become a five-star hotel resort. The large combined space on Sant'Erasmo and Vignole Islands have comparable potential within striking distance of Venice's principal islands. Torcello Island has built on interest in its cathedral and archaeological museum by constructing the five-star Locanda Cipriani restaurant. Torcello's quay was recently modernized, despite objections from Italia Nostra.

San Servolo Island has developed through promoting education and culture. The island hosts a craft center, conference hall, museum, the Academy of Fine Arts, and Venice International University (VIU). This university is associated with Venice University, IUAV and international universities. VIU could grow into an educational hub with the provision of more student accommodation, possibly on the nearby Lido Island where there are modern facilities. San Servolo was formerly a monastery, a nunnery, and a retreat for plague victims and then mental patients. It provides a great example of adaptation to change in Venice, as Crovato and Crovato (2008) observe: "There are islands that are no longer 'abandoned' and that have found a viable role (public or private); others that were on the verge of being abandoned which have enjoyed a decisive change of direction (San Servolo most obviously).

If it is true that Venetian life in general has changed markedly, it is no less the case that a number of the islands have now rediscovered a positive part to play, cultural or touristic, within the lagoon context" (Crovato and Crovato 2008, 11).

8) Environmental Defenses

Much of this book has been devoted to defense from the highest flooding using the mobile dams. This penultimate chapter addresses the dams' completion, which continues to be contested. In June 2010, Italia Nostra's Cristiano Gasparetto had hopes of stopping the installation of the barriers and adapting them with other methods: "Still today we can block these and can recover work already done and use it in other ways" (Gasparetto 2010). But the alternative proposals to the dams have been insufficiently developed and scrutinized, as explained in Chapter Six. It is the author's opinion that the barriers should be installed and the dams completed as designed.

Since the financial crises of 2008 and 2009, further challenges have questioned the dams' cost. In May 2010, the final construction phase was estimated at approximately €1.434 billion (US$1.87825bn). Of the total €4.678 billion (US$6.12723bn) cost for the dams, €3.244 billion (US$4.24905bn) had been assigned and €2.409 billion (US$3.20092bn) committed; roughly 63 percent of the work for the dams was done with the remaining 37 percent to be finished by 2014 (Consorzio Venezia Nuova 2010a). "Two thirds of the work has been completed now and we're waiting for remaining 25 percent of the money to do the remaining," stated Giovanni Cecconi of the CVN in October 2010 (cited in Di Marino 2010). In July 2010, the Italian government approved a 10 percent cut in spending for all ministries during 2011 and 2012, which could squeeze finance for the remaining work.

Despite claims about excessive cost, comparisons with other contemporary projects reveal that the dams are not particularly expensive. For the EXPO 2015 in Milan, the city is putting up €3.5 billion (US$4.65243bn) with an additional €4.5 billion (US$5.98170bn) coming from central government (*ANSA* 2010). It is noticeable that Milan alone is providing money for the EXPO that is close to the total cost of Venice's dams, and the additional finance from central government means the total state provision for EXPO 2015 will be nearly double. Whilst EXPO 2015 will be a great event for Milan, these figures show that spending almost half as much to protect Venice from high flooding should be affordable. Completing the payments for the dams would also be democratic. In January 2010, interviews were conducted by Milan's IPSO Institute with 1,500 people from Venice and the surrounding area, as well as 800 other Italians, and 77 percent of Italians were in favor of the dams, while 63 percent of Venetians were in favor and 30 percent of Venetians were against (*Il Gazzettino di Nord Est*, 12 January 2010). Environmentalists' campaigns against the dams are not representative of majority Venetian or Italian opinion and should not guide decision-making. By a similar token, UNESCO should be ejected from the Venice Safeguarding Commission. As explained in the Introduction, UNESCO has been the leading organization that has politicized the Venice problem, and it continues to play an active role through the Safeguarding Commission. It is undemocratic for an international organization to meddle in national decision-making about environmental protection.

Further important decisions about Venice's environment remain to be made. Policy is in place for the dams, yet they will not solve Venice's problems forever. The city's longer-term future and immediate decisions about subsidence need attention. There is a pressing debate about restarting natural gas extraction below the Adriatic Sea near Venice, which was proposed by a member of the government in 2009. Experience from the past and at Ravenna indicates a strong link between gas extraction and subsidence, so the extraction of natural gas near Venice should not be restarted (see Bertoni, Elmi and Marabini 2005, 23). This proposal needs to be distinguished from the re-gasification terminal opened in 2009 near Venice which does not increase subsidence. This terminal converts liquid gas delivered by ship from other countries into useable gas, which is then pumped into the Italian gas network. Italy has significant energy needs, which can be met using gas imports, restarting domestic nuclear energy production and other methods that do not exaggerate subsidence in Venice.

Rather than proposals that could increase subsidence in Venice, ideas to reverse subsidence deserve serious investigation. In 2005, Giuseppe Gambolati, Professor of Engineering at Padua University, suggested that Venice could be raised by pumping water into the soil under the lagoon. Gambolati recommended digging 12 holes with a 30cm (11.8 inch) diameter within a 30km (18.6 mile) distance around Venice, and pumping sea water into the ground to a depth of 700m (2297 feet) (*Associated Press* 2005). This could compensate for relative sea level rise (RSLR). "A pre-feasibility modelling study has offered encouraging results showing that the city may rise up to 30 cm over a 10 year period, thus significantly offsetting part of the expected RSLR," write Carbognin et al (2009, 8). Concern has been articulated about what would happen if the city rose unevenly: "Should parts of it be elevated in a different way, this would cause the city to crumble," argued Giovanni Mazzacurati, CVN President (*Associated Press* 2005). This concern is being investigated as part of a four-year pilot project announced in 2008 (Castelletto et al. 2008), and preliminary findings reported in 2010 indicate Venice would rise uniformly (*Il Gazzettino di Venezia*, 18 July 2010). CORILA identified the need for the research to progress to field experimentation, which required extra funding, and also rejected environmental campaigners' suggestion that raising Venice would make the mobile dams unnecessary (*Il Gazzettino di Venezia*, 18, 20 and 21 July 2010).

9) Maintenance Interventions and Restoration

Measures for reducing sinking and flooding for the whole lagoon need to be complemented by specific maintenance interventions. Internal improvements address the gradual deterioration of all the islands due to tide movement, waves, high levels of water against buildings, subsidence and degradation of the lagoon. Dredging canals regularly is one of the most important aspects of this work, which is concentrated in Venice's central islands. Restoring bridges and pavements, and renewing mains services (water, lighting, gas, fire prevention) also needs constant attention. The organization Insula was created in 1997 to oversee internal maintenance: by 2000, it had raised 30 percent (21,000m² - 22,6042 ft²) of the city's quays with paving stones lifted by between 20cm (7.87 inches) and 30cm (11.81 inches) (Davis and Marvin 2004, 196). A €60 million (US$78.5916m) project to raise the edge of the city in front of St Mark's Square began in 2003 and was completed, despite the diversion of funds from Insula to the mobile dams between 2003 and 2006. €28 million (US$36.6761m) for maintenance was provided in 2009 and regular financing should continue. Nonetheless, Panzeri (2009b) notes: "20 km of canals still need to be rehabilitated and 41 km of canal sides, 160 bridges and 340,000m² of pavement await restoration" (Panzeri 2009b, 96). Maintenance work for buildings is also required where the water level has gone above stone bases and is in contact with porous brickwork.

Renovations to the ancient sewage outlets have been recommended as part of internal maintenance work (Panzeri 2009b, 97). Venice's sewage outlets are wholly inadequate and many properties, especially hotels, have installed private septic tanks that are regularly emptied. The lagoon still functions as a flushing system for some sewage. A new sewage system needs to be installed for the whole city with a pump network to suck the waste away to be treated elsewhere, as has been constructed in the Netherlands. The installation of a new sewage system would mean that if the mobile dams are closed frequently pollution would be less of a problem, as discussed in Chapter Seven; and the cost has been estimated at €250 million (US$327.498). The absence of a sewage system demonstrates how Venice has failed to develop parts of its basic infrastructure beyond medieval expectations.

The restoration of ancient bridges, buildings, monuments and artwork is vital to maintain Venice's unique environment. Even so, restoration must not restrict new construction. "Of course the conservation of the city's architectural heritage should be a long-term aim, but this should not be allowed to stifle the creative energies of the architects of the future," writes Howard (2005, 301). Modernization is the priority, especially for buildings: as Rinaldo (2009) notes, "Venice is full of holes, of absurd architecture and infrastructure, of places to modify and things to do" (Rinaldo 2009, 215). The Constitution Bridge over the Grand Canal opened in 2008 and was berated as a modern eyesore. It is the fourth bridge over the Grand Canal and Venice's first new bridge for 70 years. Designed by the Spanish architect Santiago Calatrava, the 94 meter (308 feet) bridge linking Piazzale Roma with the railway station has a very modern design, with glass sides along a sleek arc of steel. Its placement between two modern areas on the edge of the historic city conforms to the marginalization of modernity from the city center, discussed in Chapter Three. Like the mobile dams, the bridge's completion was delayed (by two years) and was over-budget. The original cost of €5 million (US$6.54m) doubled to €10 million (US$13.099m), as estimated by Mara Rumiz, Venice City Council's former public works chief; according to some critics its final cost reached €20 million (US$26.234m) (Owen 2008). Plans to redesign the Accademia Bridge were announced by Venice City Council in 2009. The Council called for architects and building firms to provide sponsorship for the estimated €5 million (US$6.54m) reconstruction cost. No Council funds were offered (Spence 2009). The redesign of this bridge is well overdue. Numerous smaller bridges have been modernized, such as one to San Pietro Island. This type of restoration is positive and needs extending.

National and international non-profit organizations are now fundamental to restoration work in Venice, as outlined in Chapter Three. Without their support, raising the funds for restoration would be a considerable local burden. Restoration funding by American non-profit organizations has been impressive, assisted by the United States government granting income tax exemptions for donations by tax paying citizens contributing to American charities. Other governments should introduce comparable measures, especially in countries where substantial funds are raised for Venetian restoration. Nonetheless, international organizations should not overstep their roles in restoration and get involved in policy decisions about Venice's environment, which, as we have seen through the role played by UNESCO, would risk interfering in matters of broader political, infrastructural and democratic importance.

10) Funding

Most proposals put forward in this chapter to develop and modernize Venice need investment. The transformation of the port and the Marghera industrial area is particularly expensive, requiring €1.5 billion (US$1.96532bn) for reclamation of land contaminated by the petrochemical industry plus construction costs. A policy orientation based on public and private infrastructure investment is recommended. This includes changing the port and transforming Marghera, increasing container-handling and cruise ships, modernizing the Arsenal, providing new residential, tourist and student accommodation, building Tessera City, and the subway. Improving educational and cultural facilities will also help attract tourists, students and people who work in related fields. Venice City Council has been exploring policies to limit tourists entering the city and stigmatize them as environmental polluters: in fact, the expansion of tourism is crucial to help fund development.

In July 2008, it was reported that Venice's former Mayor Cacciari had requested permission from the national government to introduce a city tax for tourists arriving on cruise ships in Venice or staying overnight (Vanzan 2008b). "Yes. I would close Venice—or perhaps, on reflection, a little entrance examination and a little fee," he recommended (cited in Newman 2009). A tourist tax for Rome was legislated in 2010 to operate from 2011. Such schemes have been discussed for other places that are famous for being environmentally-threatened. President Mohammed Nasheed of the Maldives said he was planning a $3-a-day (€2.27) green tax for all tourists visiting local resorts to help fight climate change (*IHT,* 8 September 2009). Environmental tourist taxes should be rejected because tourists are not destroying Venice's environment. A flat rate entrance tax would also discriminate against less wealthy tourists. Although Cacciari touted an initial flat rate of €1 (US$1.32) per tourist, this would set a dangerous precedent: John Kay, a British economist, advises that Venice should charge tourists €50 (US$65.94) for admission (Kay 2008). Another idea floated in late 2009 by Venice Councilor Enrico Mingardi was that only tourists with overnight bookings should be allowed into the city (*ANSA*, 13 November 2009). It remains to be seen whether the Council administration elected in April 2010 will enforce schemes for tourist taxes or restricting entrance. These schemes would be consistent with the extra charges for tourists to use water buses and the segregated water bus services for tourists and residents. Even the city-wide wireless Internet network is divided between charging tourists and a free service for residents.

Venice City Council's treatment of tourists risks dissuading them from coming to the city, thus limiting the income of Venetian citizens and businesses. "[T]hose who are making the bucks are in fact very often the Venetians themselves. As tempting as many locals may find it to blame outsiders for exploiting their city, much if not most of the selling out of Venice that makes this possible is still being done by Venetians," note Davis and Marvin (2004, 297). Venice's tourist economy was estimated to be worth €1.5 billion (US$1.96532bn) annually in 2007 with the cruise ship market representing nearly 10 percent of this (Somers Cocks and Morel 2009, 45). Events like Carnival generate huge profits: one estimate of Carnival in 2000 calculated expenses of approximately $650,000 (€429,909), compared with an income just for the city center of over $80 million (€61.066m) (Davis and Marvin 2004, 256-7). Italy's leading financial newspaper reported that Venice is the richest city council area of Italy, along with Siena (*Il Sole 24 Ore* 2010). Both Venice and Siena have relatively small populations compared with other Italian cities and are firmly established on tourist itineraries. Economically, tourism is enriching Venice and its citizens rather than leading to the death of the city.

Investing in development that expands tourism will help Venice to prosper further. Despite the negative orientation of the previous Venice City Council administration towards tourism, it did endorse the construction of Tessera City outside Venice's center. Investment of €600 million (US$786.012m) for Tessera City has already been earmarked, plus €800 million (US$1048m) for extending Venice Gateway, the airport terminal and the second runway. This investment has been raised through a partnership between Venice City Council and SAVE, which is investing in the airports, shops, catering and tourist services that it manages. SAVE contributes to Venice City Council through taxes, licensing and rents, and it has a common interest with the Council in constructing Tessera City. This is a useful model relationship that can improve Venice. The City Council and other state bodies should support the development projects outlined in this chapter to secure private sector investment. Companies can be attracted by potential return on investment; in turn, investment encourages more businesses and tourists to come to Venice and this generates revenue for state institutions.

Venice has the potential to raise substantial funding to meet the recommendations in this chapter. This was indicated by a proposal in 2010 to invest €25 billion (US$32.75bn) over 10 years, outlined by Renato Brunetta, the losing candidate for the April 2010 Venice Mayoral election. Brunetta calculated that Venice needs €15 billion (US$19.65bn) of infrastructure spending on the port, airport, subway and roads near the city. An extra €10 billion (US$13.099bn) was identified by Brunetta for urban maintenance and new public housing (Vanzan 2010). He also supported Venice's bid for the 2020 Olympics, which would have expanded the sporting facilities of Tessera City: in May 2010, Rome was selected over Venice for Italy's bid.

Brunetta's electoral failure was not because his investment package was unrealistic or poorly-planned. Rather, it indicated that anti-developmental sentiments have the upper hand in the city. This is unsurprising given that conservationism became dominant in Venice, environmentalism is pervasive in Italy, and there is a general cultural unease about modernization. Even so, Brunetta lost the election very narrowly, and he responded to his defeat by promising to support a developmental agenda for Venice from Rome through his government job as Minister of Public Administration and Innovation (*Il Gazzettino di Venezia*, 11 April 2010). Efforts to pursue such a developmental agenda, including the creation of a new special law for Venice, should be supported.

THE CASE FOR DEVELOPMENT OVER 'SUSTAINABLE DEVELOPMENT'

This chapter has demonstrated the inaccuracy of the claim that tourism is leading to the death of Venice. The city is coping better now than it has historically. The problem is that sustainable policy-making in Venice is based on myths about tourism and inaccurate population claims. In reality, Venice is being held back due to policy-making guided by sustainable development.

Leading policy reports endorse sustainable development criteria for managing many aspects of life in Venice. The OECD Territorial Review of Venice advocates sustainable solutions for the city and the surrounding area (OECD 2010, 29 and 133). The COSES Report commissioned in 2008 by Venice City Council discusses sustainability of access, sustainability of accommodation, sustainability of circulation (on foot and by water transport), sustainability of cultural activities, sustainability of services and sustainability of the use of the city by different categories of people. Somers Cocks and Morel (2009) highlight that the principal objective of COSES is merely redirecting tourist flows: "Its objective is to end up with a digital platform capable of giving near real-time information on how full the city is, that will encourage visitors to come in the less crowded periods, offer price incentives and give advice to tour operators of the sort: 'Next week, predicted conditions would make it advantageous for you to stay on the Brenta canal and bring people in by the Burchiello boat'" (Somers Cocks and Morel 2009, 33).

This advice will not make a noticeable difference when Venice is faced with thousands of people at the height of the tourist season. COSES also recommended extending the 'Venice Connected' scheme, offering price incentives to come to Venice in quieter periods with various packages combining transport discounts and museum entry. It is fine to use technology and price incentives in this way; but such a 'sustainable' approach to tourism is insufficient because it is based on managing tourist flows only using the current infrastructure. There is no imagination about improving tourism using the subway or by building new hotels. Attempting to manage tourism at Venice's current level of development with sustainable criteria results in more regulation, surveillance, and 'education' in conservationist etiquette. Thus Francesco Bandarin, Director of UNESCO's World Heritage Centre, stated in 2009: "We need to use a multiple approach: physical ceilings if possible, bookings, pricing mechanisms, education, surveillance, diffusion of the flows. To increase positive impacts, we need to make the tourists more aware of the challenges of conserving a World Heritage site: leaflets, messaging, education of tour operators, etc" (UNESCO 2009). Tourists are told that if they are more 'aware' of Venice's problems, the city's environment will improve - and if people do not get the message that conservation comes first, this can now be legally enforced. Referring to a new legislative environmental code (Decrees 152/2006 and 4/2008), Fabrizio Fracchia, Roberto Agnoletto and Francesca Mattassoglio (2009, 101) point out that "[c]itizens may call for this principle to be applied, including before the courts, and if it is given its due weight it could provide new impetus for the protection of Venice." The code insists that the environment and heritage are to be prioritized over human interests: "[I]n comparative choices involving public and private interests over which there is a degree of discretion, priority must be given to the interests of protecting the environment and the cultural heritage" (Fracchia, Agnoletto and Mattassoglio 2009, 100-1).

Ultimately, the choice is a political one between development and sustaining the environment. To take one example: the dredging of the navigation channels in Venice for petrol tankers had a significant impact on the lagoon, but it also boosted the city's economy and benefited Venetians (Rinaldo 2009, 204). Today, we face similar choices about dredging channels to allow bigger cruise ships to dock, the transformation of Marghera, building the subway and creating Tessera City. These changes are likely to have some negative impacts on the natural environment, and serious attempts should be made to minimize these impacts. Yet it is necessary to decide whether to develop a modern Venice for the twenty-first century. "It must be accepted . . . that stopping and inverting of the process of degradation are connected to the choices related to the characteristics of the actions of man and his economic activity," stresses Rinaldo (2009). He added that preventing the recent unsustainable causes of lagoon degradation would require addressing the use of navigation channels by fishing and tourist vessels, which would reduce boat waves and improve the condition of the lagoon (Rinaldo 2009, 200-1). But prohibiting fishing and tourist boats would have a hugely detrimental economic and social impact on Venice. Constructing the subway, on the other hand, would reduce the use of boats and their waves, while not damaging Venice's economy.

Innovations like the subway, the mobile dams and creating a modern sewage system for Venice do put human interests before nature. All these innovations include short-term, highly localized, environmental damage. In the long term, these innovations improve conditions both for people and for Venice's environment. When environmentalists counterpose human and natural interests, the latter are narrowly defined around short-term impacts. There is a consistent inability to appreciate that human intervention in nature has done much to create the 'natural' environment we inhabit now, and can shape and improve that environment. Another disturbing aspect of the sustainable development orthodoxy is the presumption that current generations must make sacrifices for future generations. This is illustrated by Venice's environmental code. Article 3 of Legislative Decree 152/2006 (amended by Legislative Decree 4/2008), postulates that "every human activity legally relevant under the present code must conform to the principle of sustainable development in order to ensure that satisfying the needs of the present generation does not compromise the quality of life and opportunities of future generations" (cited in Fracchia, Agnoletto and Mattassoglio 2009, 100). In this way, the interests of both current and future generations are curtailed in the name of conservation.

Applying the limited imagination of 'sustainable development' is already reducing the quality of life of the present generation, by restricting tourism, cruise ships, container traffic and economic growth from these activities. To improve the lives of future generations, economic growth should be maximized. As the financial journalist Daniel Ben-Ami (2010) argues, "[r]estricting growth, as the advocates of sustainable development propose, means leaving future generations with many of the problems that exist today" (Ben-Ami 2010, 137). In Venice, conservationism and sustainable development have led to insufficient accommodation, and the delaying of key development projects including the mobile dams, subway and Tessera City. These projects have not been realized since they were conceived between 30 and 50 years ago. A generation of people has already lost most of the benefits they could have enjoyed from these projects.

Sustainable development is premised on the belief that putting nature and conservation first will benefit the environment. This book's developmental approach demonstrates that putting people first benefits humans, enables conservation of heritage, and improves the environment in the long term. Sustainable development is an oxymoron. To "develop" is defined as "to grow into a more mature or advanced state," according to the dictionary, while to "sustain" is defined as "to keep in existence; maintain" (Random House Dictionary 2010). To develop is to bring about progressive change, whereas to sustain is to prevent change. Unfortunately, sustainability has prevailed over development in Venice, both in practice and in contemporary metaphorical representations of the city.

NOTE

1. All currency conversions correct on 21 December 2010.

Conclusion: The Reconstitution of the Venetian Metaphor

"Never like today has Venice spoken to historians and men in general in a language more real and universal, offering experience of a microcosm that goes along a path full of dilemmas, of a relationship that is risky and perennially precarious with nature."

Piero Bevilacqua, Italian historian, La Sapienza University of Rome (2000, 20).

Bevilacqua's book *Venezia e le acque. Una metafora planetaria (Venice and the Waters. A Planetary Metaphor)* (2000) explores how the city's risky environment speaks to us all in modern times. Here, we discuss the contemporary international meaning of the Venetian metaphor: Venice as an environmentally-threatened retreat from modernity.

Chapter One confirmed Eglin's (2001, 8) conclusion that the Venetian political metaphor died with the fall of the Republic in 1797, while Chapter Two set out how the Venetian metaphor subsequently became more salient culturally than politically. As the twentieth century progressed, intellectuals frequently declared that Venice had lost any metaphorical significance: as indicated by Simmel's (1922) conviction that Venice could only regurgitate motifs and was no longer a creator of meaning. For Manfredo Tafuri (1978), an historian of Venice, the city lingered as "an allegory of a general condition" in which the metaphorical mask must be worn to save one's soul (Tafuri 1978, 299). Former Venice Mayor Massimo Cacciari draws on Simmel and Nietzsche to argue that Venice has lost its metaphorical signification, with the Carnival mask emblematic of the loss of direction: "All appearance exists in itself and for itself—a perfect mask that hides being, or rather, reveals the loss, the absence of being" (Cacciari 1995, 95). Similarly, the French journalist and political commentator Régis Debray (2002) signals his belief that

Venice no longer provides vision, offering little more than narcissistic "confirmations"; though he held out the possibility that the city could continue its tradition of anticipating the future:

> It seems to me that the relic is not sufficiently out of fashion to take a holiday; the graceful, the delightful carries too much weight. Perhaps this egocentric microcosm, which has always been a few centuries ahead of the rest—which invented the ghetto long before the camps, a department for monitoring correspondence long before telephone tapping and the letter of credit long before cashflow—is in the process of inventing before our unseeing eyes the insular Europe of tomorrow, reduced to picturesque features like half-timbering, wrought iron and inns but dead to space exploration, the planet and its century: a monocultural peninsula set in its lagoon, forgetting the open sea, suffocated by memory, and in which the tertiary sector will have eclipsed the primary and the secondary. (Debray 2002, 70-71)

We differ with Debray's presentation of Venice as a post-industrial model for an insular Europe. While it is true that Venice's petrochemical industry has declined and tourism is growing, the expansion of the port for cruise ships and goods-handling indicates that it is still open to the sea. But the suggestions that Venice has lost its meaning as a cultural center do seem to be describing real changes. Undoubtedly, the cultural importance of Venice at the beginning of the twenty-first century has diminished in comparison with the nineteenth century and the first 60 years of the twentieth century. Since the 1970s, the Lido has only attracted members of the international cultural elite occasionally, for specific exhibitions or the film festival. There have been recent attempts to revive Venice as a cultural center: we have seen how the Lido is being reconstructed as a prime vacation destination, and new cultural establishments have opened. In 2009, the renovated Punta della Dogana, the former customs house, was opened to display art from the collection of French billionaire François Pinault for 30 years, and a renovated warehouse in Venice's salt docks was opened as a museum dedicated to the post-war artist Vedova. Now Venice needs an exhibition and storage facility on the mainland to support museums and libraries within the city. A bid to be the European Capital of Culture in 2019 is an opportunity to extend Venice's cultural facilities. These initiatives are welcome, although Venice has not yet regenerated its past cultural magnetism.

We concur with Debray that Venice has not yet retired as a creator of meaning. Although death became the dominant Venetian motif in the twentieth century, the city has retained a potent impact. "The *ruin* of Venice was a given for the Romantics, but it was the death of Venice that was perpetrated in the twentieth century. Venice had become the very allegory for history, decay and the threat of the elements, powerful still at the onset of the twenty-first century," writes Plant (2002, 233. Emphasis in original). These comments identify environmental dangers lurking within the Venetian metaphor. Likewise, Eglin acknowledges the role of environmental issues for interpretations of Venice, and recognizes the potential for the reconstitution of the Venetian metaphor through environmental criticism: "The Venetian metaphor—the transfiguration of the myth of Venice—occasionally manifests itself even today. The Prince of Wales's 1989 jeremiad against modern and post-modern architecture approvingly cited John Simpson's 'Venetian' design for London Bridge City, and nostalgically evoked the Canalettian London, 'one of the architectural wonders of the world, a city built on the water like the centre of another great trading empire, Venice'" (Eglin 2001, 203-4).

We have registered how environmental activists have assisted the reconstitution of the Venetian political metaphor, especially through the campaign against the mobile dams that reached its height in 2005. Such campaigns have helped Venice to become synonymous with an environmentally-threatened retreat from modernity.

This book has identified two long-term themes that have contributed to the symbolic revival of the city: Venice as the expression of decadence, death and degradation of humanity; and Venice as representing the dominance of conservationism over modernization. In the twenty-first century, these themes have become central to Venice's association with a threatened watery environment. Venice's environment is now a reference point for cities worldwide: other cities are often referred to as "the Venice of the North," of the East, of Asia, of America. As Davis and Marvin (2004, 134) remark, nobody refers to Venice as "the Amsterdam of the Adriatic." Robert Browning coined the term 'Little Venice' for the network of canals in the London Paddington area, which is currently being redeveloped. Venice in California has its own replica canal and neo-Classical buildings. How many other cities have been reproduced as a hotel? The Venetian Hotel and Casino in Las Vegas, with its gondola rides along clean canals, may be a perverse reproduction of Venice, but it demonstrates how the Venetian metaphor has been stretched. "Venice's image has been appropriated and replicated by cities from Las Vegas to Macau," notes the OECD (2010, 14). Venice expresses international metaphorical meaning for watery environments and increasingly for threatened environments. Rinaldo (2009) describes Venice as a "planetary metaphor" with "global resonance" (2009, 136-138). Why is the contemporary Venetian metaphor so influential?

VENICE AS THE CRISIS OF HUMAN CIVILIZATION

André Chastel, the French art historian, refers to "the Venetian challenge: the central episode of the crisis of modern civilization" (cited in Rinaldo 2001, 61). Venice has come to represent an escape from the human-created modern world. This is largely due to the historical legacy of the Venetian Republic. Venice has a long history as a symbol of Western civilization, dating back to its days as the gateway to the Orient, and the pessimism that followed the fall of the Venetian Republic in 1797 weighed heavily on the minds of influential thinkers about the city. As the nineteenth century progressed, human intervention in nature was frequently interpreted as problematic. Culturally, Venice was depicted as a haven for conservationism against modernity. During the twentieth century, cultural associations with Venice as death, decadence and human degradation were consolidated into Venice's representation of the fall of man, as depicted by Berendt's (2005) book *The City of Falling Angels*. By the early twenty-first century, the sinking of Venice came to be seen as an allegory on human failure. "It will be submerged. It will descend into the water silently and permanently. It is the image of the city as the final end of all human achievement and aspiration," writes Ackroyd (2010, 403). Such profound pessimism about human intervention in nature became a vital component of conservationism, as Rinaldo (2009) describes:

> Venice, as the epitome of the global environmental challenge between natural environment and constructed environment and the central episode of modernity, gives authority to worries and hopes. For this subject, pessimism, indeed, has been the norm for centuries. Lord Byron already heard the screaming of nations about the sinking palaces, however from Ruskin until Braudel, Chastel and Indro Montanelli a lot has been done for the conservation of the city and its environment. (Rinaldo 2009, 207)

We have discussed how the transformation of conservationism into environmentalism required overcoming the modernist dichotomy between man and nature, which was central to conservationist protection of the natural world. Dissolving this modernist dichotomy has been perceived as an essential part of rediscovering a past equilibrium in Venice. "Establishing a modern environmental culture must begin with a change in direction which leads from the current anthropic conception of nature to a biocentric attitude, by recovering the concept that man is part of nature. Furthermore, man is a conscious part of nature, and therefore one which must become active in order to recover the equilibriums lost," write Mauro Bon et al (2001, 28). These observations depend more on Venice's ancient myths about political equilibrium and harmony than on environmental reality. The existence of a past natural equilibrium in Venice or anywhere has been questioned:

> The concept of natural equilibrium, in the static sense of maintenance which the notion suggests, does not feature at all in any natural evolutionary phenomena. However, if such a statement is true in general it is even more evident in the case of the Venice environment (its lagoon and drainage basin), which has been

> the object of much intervention for centuries. It has been maintained completely artificially only at the price of decisive transformations carried out by man. (Rinaldo 2001, 64)

For Rinaldo, "Venice, like Gaia, teaches that there cannot exist, in the life of complex systems like lagoons, a general notion of equilibrium" (Rinaldo 2009, 218). Notwithstanding the popularity of Gaia theory in green thinking, the re-establishment of equilibrium in Venice has been a key demand made by Venetian environmentalists. As reported in Chapters Five and Six, campaigners have claimed that the mobile dam project is violating natural equilibrium, while UNESCO has suggested that it will re-establish equilibrium: "MOSE will also help restore the equilibrium of the ecosystem of the lagoon; it will protect the surrounding marshes and help keep the canal banks in Venice's inner city and on the nearby islands from further harm" (UNESCO 2009).

The mobile dam project is only one point of tension in many drawn-out battles between modernizers and conservationists in Venice. The city built on the architectural and engineering pioneering that developed during the Venetian Republic to experiment with many innovative projects. Yet conservationism prevailed by the early twentieth century, and by that century's end Venice had become an emblem of the struggle between environmentalism and modernity. As discussed in Chapter Eight, environmental claims drew on a pervasive cultural unease with modernization and reticence about human intervention in nature. Defending Venice from sinking has become a specific focus for debating the perceived crisis of modern civilization.

For some, the creation and maintenance of Venice represents the triumph of human civilization. The International Council on Monuments and Sites (Icomos) describes how Venice "symbolizes the victorious struggle of mankind against the elements, and the mastery men and women have imposed upon hostile nature" (UNESCO 2009). While this was true during the Venetian Republic, the observation feels dated when every attempt to construct in Venice is subject to conservationist resistance. Numerous contributors to an article in the daily newspaper *Il Gazzettino di Venezia* (27 June 2009) bemoaned Venice's condition as a conservationist city. "To conserve can also mean to leave degrading, and the Costa-Cacciari 'reign' has registered a worsening of Venice," wrote Riccardo Calimani. For Antonio Alberto Semi, President of the scientific and cultural institution *Ateneo Veneto*, Venetian conservationism is merely representative of a wider climate: "It is true that there is resistance to change, but everywhere is like this". Venice has become synonymous with conserving the past against modern change: as Norwich (2003) comments, "[n]o city in the world has changed less over the last 200 years" (Norwich 2003, 259). However, the contemporary Venetian metaphor is not free of contradictions.

CURRENT VENETIAN PARADOXES

The belief that human intervention should be restrained against nature has created a sense of vulnerability. In Venice, this has become focused on the sea as a threat, which is fundamental to the contemporary Venetian metaphor. But the first Venetian paradox is that the sea is interpreted as threatening the city, when in fact it is a source of wealth and attraction. We have recorded how the ancient Venetians celebrated their relationship with the seas, especially during the Marriage to the Sea ceremony, which expressed their dominance over the sea and wider sense of power. Dominating the seas enabled Venice to become Europe's financial and trading center. By the end of the fourteenth century there was scarcely a single major commodity that was not largely transported in Venetian ships (Norwich 1983, 270-1). Venice became necessary to all European lands active in trade during the fifteenth century. The city was at the heart of Christian Europe, with its Rialto commercial district relying heavily on overseas connections. "The Venetians had made themselves masters of 'the gold of the Christians' because all of Europe was directly or indirectly supplied from there," writes Crouzet-Pavan (2002, 154).

Given the importance of the sea for ancient Venetian power, the overriding concern was its retreat. Fears that rivers into the lagoon would fill it with sediments plagued the ancient Venetians, and huge projects to divert the rivers Piave, Sile and Brenta away from the lagoon began in 1324, to be largely concluded by 1683. The loss of the lagoon would have made the city highly vulnerable to direct attack from the sea. Despite these mammoth efforts to maintain Venice's lagoon, Venetians in the seventeenth and eighteenth centuries feared that the sea would retreat as it had at Pisa (Eglin 2001, 9), and that this would unite the city with the mainland, threatening its island identity.

Since the fall of the Venetian Republic, fears about sinking and flooding have proliferated. Chapter Two discussed how nineteenth century thinkers including Ruskin, Byron, Shelley, Dickens, Rogers and Moore popularized the idea that Venice was heading for a watery grave: a fear that seemed to be confirmed by the collapse of the bell tower on St Mark's Square in 1902, and the high floods of November 1966. As Cesare Scarpa discerned, these floods were caused by a very rare combination of factors: "What happened in 1966 was extraordinary and will not happen again. Then there were two big sea tides that went on top of each other and they were pushed by the wind. Together with this, it rained a lot and the water inside the lagoon increased. So there were these extraordinary conditions in 1966. We have never had another condition like that. It was atypical" (Scarpa 2005).

Despite these well-documented "atypical" causes, Keahey (2002) writes that the events of November 1966 were "inevitable" (Keahey 2002, 93), and predicts that they will be repeated: "While subsequent storm surges have not yet equaled that 1966 ferocity, scientists believe it is only a matter of time" (Keahey 2002, 12). Keahey's negative assessments are not caused by consulting poor scientific forecasts, but by a more general environmental pessimism. Flooding in Venice has indeed increased over the past century, but its impact was much worse during the Venetian Republic than it has been over the past 200 years. Whereas the ancient Venetians celebrated their dominance over the sea in defiance of devastating floods, the contemporary tendency is to cower to the threatening waters. The French philosopher Jean-Paul Sartre visited Venice and reflected on the decline of the city "castrated" by the Adriatic Sea (Sartre 1991, 8). Fragile, vulnerable humanity is well-established within the modern Venetian metaphor, as indicated by the titles and principal themes of recent books: Keahey's (2002) *Venice Against the Sea: A City Besieged* and Plant's (2002) *Venice: Fragile City, 1797-1997.* Modern depictions of Venice tend to perceive the sea and lagoon as problems to be avoided: "[T]he vast, shallow and placid water surface of the lagoon is today considered more as an obstacle, that can be criss-crossed by fast, and often huge, destructive motorboats racing from land to sea and back. Or maybe bypassed with rapid underground connections, as some local planners propose" (Caniato 2005, 7). The attitude of the ancient Venetians was that the sea and lagoon were resources to be celebrated. Waters were a means of transport to trade, to prosper and for exploration, as well as a source of salt, fish and protection.

Today, the sea and lagoon continue to function as resources providing wealth for Venice, as container traffic connecting with China, India and Eastern Europe has supported the port's economy. Nonetheless, celebrations of Venice's revival as a gateway to the East have been conspicuous by their absence since writings by the celebrated Italian author Italo Calvino (1974a, 1974b). A similar pessimism is attached to tourism. The sea, lagoon and its canals provide Venice with its unique environment, attracting numerous tourists: they are now fundamental for the city's economy. Venice is one of the busiest ports and tourist destinations in the world. Yet the second paradox in the contemporary Venetian metaphor is that tourists have been depicted as causing the death of Venice.

The current paradoxes are not due to fundamental changes between the city and its physical environment; nor are they caused by the city being swamped by tourists. These paradoxes have been created because human resilience has been replaced by vulnerability. When the ancient Venetians were faced with environmental hazards, they typically responded with resilience. Likewise, the ancient Venetians realized that visitors to the city were resources and devised festivals to attract them. The spirit of resilience that prevailed during the Venetian Republic until the seventeenth century contrasts with the contemporary stress on human vulnerability. To address Venice's contemporary environmental challenges, we need to revive this spirit of resilience. For Venice and the surrounding territory, the OECD (2010, 90, referring to Bruneau and Reinhorn 2006) also recommends resilience, which it defines by the following features:

- Robustness: strength, or the ability of systems to withstand a given level of stress or demand without suffering degradation or loss of function . . .
- Redundancy: the extent to which substitutable systems exist that can satisfy functional requirements in the event of disruption, degradation or a loss of functionality . . .
- Resourcefulness: the capacity to identify problems, establish priorities and mobilise resources when conditions exist that threaten to disrupt some element . . .
- Rapidity: the capacity to meet priorities and achieve goals in a timely manner in order to contain losses, recover functionality and avoid future disruption.

Venice needs to embrace tourists and its surrounding waters as resources rather than threats. It must urgently move away from a redundant model of development on its outskirts and conservation in the city center. To achieve this end requires a critical challenge to the three key components of the contemporary Venetian metaphor: sustainability, climate change and the risks of tourism.

THE CONTEMPORARY VENETIAN METAPHOR

Venice as a Metaphor of Sustainability

Venice and its environment have become a metaphor of what sustainability means, as characterized in a contribution to a WWF publication by Davide Dal Maso (2001):

> Venice and its lagoon contain, express, are, an absolutely unique, unrepeatable cultural, environmental and social value. A Heritage of Mankind, they represent an asset to be safeguarded for our present and for the generations to come. As well as being, and perhaps more than, an 'asset', they give body to an idea, to a dream which is part of the collective image. But Venice and its lagoon are also more than this, they are a synthesis, a metaphor of what sustainability means. Venice is the physical place materializing the examples proposed in scientific literature to explain what is the problem of seeking a lasting balance among environmental, social and economic variables. For this reason, Venice is much more than itself:

it is also the image of a utopia. If the battle is won in Venice, if the possible is achieved here, then there is hope for the Alps, the tropical forests, the Antarctic, the planet. Because Venice is the synthesis of all this. (Dal Maso 2001, 4)

Presenting Venice as the synthesis of international sustainability followed the evolution of the sustainable development agenda. The UN 1992 Rio Conference propelled the sustainable development agenda to a new level and addressed Venice as a special case. In the influential book *Sustainable Venice*, Musu (2001) explains that debates about Venice have moved away from maintaining the city and its lagoon in relation to civil, military and economic needs, while other contributors discuss Venice's role as an international consideration for sustainability:

Venice is, in itself, a global problem and is threatened by global changes. Therefore, rather than asking what the Venetians can do to improve the planet's environment, it appears necessary to ask what can and must be done (in Venice, but not only in Venice) in order to pass on to future generations a Venice which is probably different from today's and yesterday's Venice but one which is, possibly, no less able to reproduce its order (social, economic and environmental). (Dente et al. 2001, 235)

In Venice in 2009, Prince Charles of Wales met a delegation from the Italian Green Party and was given a letter criticizing the MOSE dams by the party's Beppe Caccia. Prince Charles has become widely identified with environmental causes, and in 2007 received the Global Environmental Award in New York. Addressing a symposium on the regeneration of the lagoon, Prince Charles stressed that Venice must not become a "historical theme park", and emphasized the need for sustainability: "We need to recognise that sustainability is equivalent to a process of long-term construction; 100 years, not 20" (*ANSA,* 28 April 2009). Venetian environmentalists have similarly learned to frame Venice's protection in terms of sustainability. Dal Maso (2001) of the WWF argues that if the logic of sustainability was fully applied in Venice then "no such bizarre idea as MOSE would come into anyone's mind" (Dal Maso 2001, 6). As noted in Chapter Five, identifying risks and framing claims through sustainable development has enabled activists to redefine environmentalism (Jamison 2001, 152), and to present Venice as an international concern. In November 2009, the WWF and LIPU organized participative events around the Venetian lagoon in support of UNESCO's week dedicated to sustainable development (*Il Gazzettino di Venezia,* 14 November 2009).

Although environmentalists have framed their claims against the mobile dams in terms of sustainable development, others have presented the dams as an emblem of sustainability. Shortly after his election as Mayor of Venice in April 2010, Giorgio Orsoni said: "it is truly thanks to MOSE that Venice is a candidate to be a champion city of sustainable development" (Antonini 2010). This interpretation of the MOSE dams was supported by the application of sustainable criteria. When the EC decided to close the case against MOSE regarding the damage to bird habitats in April 2009, sustainable conditions were made mandatory. Employees of the CVN "must check every type of emission: dust, waste and noise"; and "to not disturb the island populations, the building work near houses will be suspended until the early morning and noise barriers will be installed" (*Il Gazzettino di Venezia,* 15 April 2009). To protect birdlife, it was stipulated that work be stopped during May 2009 at the Alberoni and Ca' Roman sites; from mid-November 2009 until the end of January 2010, construction work was ordered to be suspended at Bacan where many species of migratory birds visit.

It is difficult to determine if these sustainable conditions applied to the mobile dams were influenced by environmentalists' campaigning or whether the momentum behind EC environmental policy would have been sufficient to establish them anyway. It seems likely that the repeated appeals to the EC by various green organizations from Venice since 1998 and their delivery of the petition against the dams in 2006 would have helped exert some pressure on the EC. After the EC imposed the sustainable criteria for the dams, *Il Gazzettino di Venezia* (15 April 2009) ran a headline proclaiming, "Here is the green MOSE." Whether the EC regulations mean that the dams can now be considered sustainable will undoubtedly continue to be debated. But regardless of whether one believes that the dam project is sustainable or violates sustainability, there is no doubt that sustainable development permeates discussion about Venice.

Chapter Seven revealed how sustainability has been deployed in the construction of an environmental myth about the Venetian Republic, and Chapter Nine examined how it has become a guide to managing many real aspects of the city. However, as Rinaldo (2001) notes, the city's past traditions run contrary to the principles of sustainable development: "The sustainability of development implies the attainment and maintenance of a model of human life which is organised in harmony with its environment. Instead, the history of Venice teaches us the opposite; here, there is the resolute definition of an environment which moulds itself to the city's needs for living and development, with nothing or nearly nothing left to natural evolution" (Rinaldo 2001, 62-3). Nevertheless, he acknowledges that Venice has now become a case study in sustainable development that attracts international attention.

Venice as Climate Change Danger

In the early twenty-first century, Venice has been frequently referred to as a key location threatened by climate change. By 2004, the campaign against the mobile dams was at a low ebb. Reframing claims against the dams by linking them to climate change provided the campaign with urgency and new vitality. Claims-makers suggested that rising sea levels would mean the dams closing the lagoon from the sea so often that pollution could be exaggerated. When activists launched the No MOSE assembly in 2005, claims against the dams were regularly connected to the greenhouse effect. A combination of claims by environmentalists, journalists, institutional experts and academic scholars established Venice as an international emblem of climate change danger. This was achieved through the efforts of claims-makers in Venice, but also because climate change was rising to the top of political agendas in many countries. So Venice as a metaphor for the threat of global warming found a wider resonance. When politically prominent people, such as the Prince of Wales, want to issue warnings about rising sea levels, they habitually refer to Venice (*Daily Express*, 28 April 2009).

Although Venice's association with climate change is shaped by the political elevation of global warming, it is also due to the politicization of protecting Venice. Flood protection for Venice has not been treated as a technical issue, as it was for the Thames Barrier in London: it was an issue defined and politicized in the aftermath of the 1966 high floods. The intervention of various players, including successive Italian governments and UNESCO was decisive. More recently, UNESCO has played a pivotal role in promoting Venice as facing the perils of climate change. A UNESCO report released in April 2007, *Case Studies on Climate Change and World Heritage*, specifies how Venice is endangered by projected sea level rises due to climate change (*Environmental News Service* 2007). Venice is designated by UNESCO as a World Heritage Site threatened by climate change (*AFP,* 1 December 2008). "For its size and population, Venice probably suffers from more physical problems than any other city: this is one of the few world heritage sites that perpetually runs the risk of vanishing completely," warn Davis and Marvin (2004, 211). Other important locations are also associated with danger from climate change, including the Galapagos Islands in Ecuador and the Great Barrier Reef in Australia. Yet the well-known environmental critic Bjorn Lomborg writes that Venice is the world's "most famous" example of subsiding places threatened by rising sea levels (Lomborg 2007, 87).

Following the release of the 2007 Fourth Assessment Report by the IPCC, Osvaldo Canziani, an IPCC climatologist, stated that if current climate trends continue Venice "is destined to disappear" (*United Press International* 2007). The influential Worldwatch Institute report *State of the World 2007: Our Urban Future* referred to Venice's mobile dams as an example of how humans are adapting to climate change: Gianfranco Bologna, the Scientific Director of WWF Italy, responded that the dams will cause environmental damage and will not resolve the problems of rising sea levels (*Il Gazzettino di Venezia*, 12 January 2007). Environmentalists have fed off international debate about climate change to press their claims against the mobile dams: during the 2009 climate conference in Copenhagen, activists protested against the MOSE dams and global warming in St. Mark's Square, displaying a banner that read "Let's stop the planet's fever". "The Mose won't work because it has been designed to deal with exceptionally high tides that are temporary, and not for a permanent and progressive rise in sea levels," commented Gianfranco Bettin of Venice's Green Party, who joined the demonstration (Mayer 2010).

Environmentalists' campaigns about the dams and global warming have combined with political debate on climate change to generate international interest in the dams. A delegation from Tokyo visited Venice in February 2009 and examined the mobile dams as a model of protection. Comparisons have been made between the Venice dams and the construction of other flood barriers that have become highly politicized, such as the St Petersburg flood protection barrier (Gerritsen, Vis, Mikhailenko and Hiltunen 2005, 344). In September 2009, Venice's dam project was presented to a conference of approximately 400 scientists and others in New York by the Italian Language Inter-Cultural Alliance (*Il Gazzettino di Venezia*, 22 September 2009). *The Boston Globe* ran a feature about Venice's dams in 2008 (Kiefer and Murray 2008) as a contribution to the debate about whether to place barriers from Deer Island to Long Island to close Boston Harbor from rising sea levels. Since Hurricane Katrina devastated New Orleans in 2005, the dangers of climate change and inadequate flood protection have been frequently linked.

"Venice could be well under water in future centuries," predicted American journalist H.D.S. Greenway in 2008: "Nowhere are the problems more obvious—the fragile lagoon against the background of chimneys spewing greenhouse gases on the mainland" (Greenway 2008). We are left in no doubt that climate change is pre-eminent in the contemporary Venetian metaphor. As the dams are completed, Venice is likely to feature even more prominently in international debates about protection against global warming.

Venice Represents the Risks of Tourism

Although associations of Venice with global warming are formidable, tourism is being interpreted as an environmental risk comparable to the sinking threat. Norwich (2003, 260) highlights "the almost constant flooding—by tourists as well as tides", while Berendt (2005, 30) warns that rising sea levels and tourism are "the two major evils confronting Venice."

Tourism, along with many of the environmental themes examined in this book, enables the contemporary Venetian metaphor to find resonance with our contemporary cultural suspicion of human activity. "The ethos of sustainability, the dogma of the precautionary principle, the idealization of nature, of the 'organic', all express a misanthropic mistrust of human ambition and experimentation," explains Furedi (2005, 11). Concerns about sustainability in Venice present human development as being out of balance with nature. In the narrative of Venice as threatened by global warming, human-generated greenhouse gases have led to temperature increases, expanding oceans and sea level rises. According to this logic, increased human consumption and industry have exaggerating flooding of the city. Human desire and innovation are depicted as the root causes of an unsustainable and flooded Venice. With tourism, the problem is less subtle: it is just far too many people coming to the city. As intimated in Chapter Nine, it is particularly the mass character of tourism that is criticized.

Tourism illustrates the contemporary preoccupation with Venice's vulnerability to human impact. "Venice appeared vulnerable from all sides: from the sea and the lagoon, the environmental and mechanical projects to control them, from its success as a leading world site of tourism and from its own citizens, particularly those in power," remarks Plant (2002, 422), with reference to the corruption scandals among officials in the 1970s. In this view, both citizens and tourists are to blame for Venice's problems. Tourism has piggybacked on previous understandings of Venice's problems, especially climate change. "What we have found especially interesting in examining how the Lagoon and its waters impinge on Venice, however, are the ways in which the problems presented by such long-term changes as global warming, rising sea level, and the shifting Lagoon are all made much worse for the city by tourism itself," surmise Davis and Marvin (2004, 206). Given that tourism directly communicates a sentiment that the problem is people, it can be expected to become a more conspicuous component of the Venetian metaphor. It seemed appropriate that one of the leading movies released for Christmas 2010 was *The Tourist*, starring Johnny Depp and Angelina Jolie and set in Venice.

THE POLITICIZATION OF VENICE'S ENVIRONMENT

Venice's environment has been politicized through two combining trends; the politicization of the Venice problem and the politicization of environmentalism. Both these trends elevated consciousness about risk, which became an element within the contemporary Venetian metaphor. The character of risk consciousness has been shaped by how the relationship between nature and society has been redefined. Examining these trends will also help us appreciate that Venice's environment is, in fact, a political problem.

Although flooding had been tormenting Venice for centuries, it only emerged as a political problem after the 1966 floods across northern and central Italy. The 1966 floods were much worse in Florence than Venice, yet Venice attracted greater attention due to the perception of Venice's fragile relationship with the sea. The definition of the Venice sinking problem was also linked to the fall of the Venetian Republic. This insight explains Berendt's (2005) description of the shift in attention away from Florence to Venice after 1966:

> As for Venice, although no one had died and very little art had been damaged, it soon became evident that the situation was fundamentally worse than in Florence...When experts took a closer look at Venice, they discovered that most of its buildings and almost all of its works of art were in desperate condition, owing to two centuries of neglect following the city's defeat by Napoleon...After a part of a marble angel fell from a parapet of the ornate but sadly dilapidated Santa Maria della Salute Church, Arrigo Cipriani, the owner of Harry's Bar, posted a sign outside the church warning, 'Beware of Falling Angels.' (Berendt 2005, 260-1)

In the aftermath of the 1966 floods, claims-makers persuaded audiences to accept that Venice was a uniquely threatened city deserving special attention. Among the early transmitters of this claim were UNESCO, Italian governments, and preservationist organizations. The decisive intervention by UNESCO after 1966 was driven by a sense of moral obligation rather than the objective character of that year's flooding. It was believed that the Venice problem needed to be addressed by international political institutions and committees, as Italian bodies were not considered to be up to the job of safeguarding such a vital part of the world's cultural heritage. The perception of intensifying pollution in the Venetian lagoon during the 1970s, and the failure of the Italian authorities to react effectively, confirmed suspicions about poor Italian environmental management. Fay and Knightley (1976) depict corruption and lethargy in Italian politics as tainting the responses to the Venice problem. The roles of UNESCO and the private committees internationalized the Venice problem, and the international diffusion of the Venice problem has been highly politicized.

During the 1980s, sustainable development appeared to offer a coherent way to understand the Venice problem and guide institutional responses. Venice became an international case study in the politics of sustainability, rather than a question of local flood control. Scholars Dente et al (2001), researching sustainable governance of the city, assert: "Venice is, in itself, a global problem and is threatened by global changes" (Dente et al. 2001, 235). By the late 1990s, the Venice problem was linked to international fears of global warming. The redefinition of the Venice problem as an issue of climate change led to a new level of political attention from international institutions including the IPCC and the Worldwatch Institute. This attention reinvigorated UNESCO's intervention.

Venetian environmentalists have also helped to internationalize and politicize the Venice problem. Beppe Caccia (2005) explained the crucial need to develop the campaign against the MOSE dams at an international level[1]; Stefano Boato spoke at a landmark conference on the Venice problem in Cambridge (UK) in 2003; and Cristiano Gasparetto has been a member of several delegations for the No MOSE assembly that have appealed to the EC in Brussels. However, it was the wider politicization of environmental issues that helped to shape the Venice problem and, in turn, contributed to the reconstitution of the Venetian metaphor. Politicization was an outcome of the transformation of conservationism into environmentalism discussed in Chapters Three and Four. Radical environmentalists stressed political protests in contrast to conservationists, who prioritized educational change and restoration projects. In the 1980s, conservationists and radical environmentalists came together through the anti-nuclear issue and set the Italian political agenda for the first time. Green campaign issues had expanded beyond conservation to address broader social, economic and political concerns. This domain expansion also drew on the political discourse of sustainable development, which emphasized that environmentalism should not be limited to protecting nature.

Nationally, environmental organizations became more involved in political institutions during the 1990s, and Italia Nostra, Legambiente, Greenpeace and the WWF were partly institutionalized during this decade (Diani and Forno 2003, 164). The Green Party became highly bureaucratic through its roles in local government and by joining a national government for the first time in 1996. Political elites encouraged the incorporation of environmental agendas into state institutions (Melucci 1996, 164). In the early 1990s, Italian state institutions were suffering from weak legitimacy in the face of the Tangentopoli scandals, and environmental issues and organizations offered new points of public contact and popularity. Internationally, elites found that adopting environmentalism as a guide to policy provided them with direction and greater legitimacy (Eder 1998, 205). Throughout Europe, the EC exerted pressure on states to incorporate environmental policies, and environmentalism was transformed from a counterculture to mainstream political discourse (Szerszynski 1998, 105). With environmentalism absorbed into the dominant political culture, environmental groups adopted the role of putting pressure on state institutions to abide by green policies. Statham (1995) refers to the environmental movement as a "cultural pressure group", reflecting the close relationship between this movement and the dominant political culture (cited in Eder 1998, 204).

Since the late 1990s, Venice's environmental associations could also be characterized as pressure groups, concentrating on appeals to institutions. The principal institutional focus for campaigning has been Venice City Council, even though there were periods (between 2000 and 2005) when the Council was predominantly in favor of the mobile dams. Yet campaigners lobbied a wide range of institutions to block the dam project. As institutional appeals faltered, some activists turned to organizing more protests, which were increasingly independent from environmental associations. The citizens' committee The Damages of MOSE held its first protests in 2005, assisted by the Refounded Communist Party. The 2005 creation of the No MOSE assembly and the partial institutionalization of the campaign through the assembly followed patterns established in the campaigns against the Marghera industrial complex and EXPO 2000.

Della Porta and Diani (2004, 185) raise the question of whether Italian citizens' committees would fill the public space previously occupied by environmental associations, which had become largely institutionalized. We are now in a position to assess the relevance of this question for Venetian organizations. Despite the early 2005 protests of The Damages of MOSE citizens' committee, No Global Disobedients dominated the protests against the MOSE Project from mid-2005 into 2007. Research by Della Porta et al (2006, 4) documents the spread of demonstrations involving No Global Disobedients after the 2001 G8 Summit in the Italian city Genoa. This was a component of what Della Porta and Tarrow (2005, 15) interpreted as a new cycle of protest through the transnational movement against globalization, following the "low ebb" of mobilization during the 1990s. From the evidence in Chapter Five, the No Global Disobedients, citizens' committees and the RC have dominated the protests against the dams since 2005. But this did not alter the institutional direction of protests.

These shifts demonstrate the highly political character of environmental campaigning compared with the conservationist focus on restoration in the 1960s. They also reveal how the politicization of the Venice problem was shaped by the red subculture in the city, most graphically through the intervention of the RC. As described in Chapter Four, the red subculture has had a significant role in the local framing of environmental claims. The electoral victory of leftist Orsoni as the Mayor of Venice in April 2010 suggested that the city's red subculture remains a political undercurrent in the city, despite the erosion of left and right ideologies in Western politics; and this subculture meant that radical environmentalism had a politicizing impact on green campaigning in the city.

THE POLITICIZATION OF RISK

One of the findings of this book is that the intervention of claims-makers influences whether an issue becomes politically important and the meaning we attach to it. The campaign against the mobile dams by environmentalists helped to construct environmental risk as a component of the Venice problem and politicize it internationally. Perceived risk is another element in the contemporary Venetian metaphor to be considered. Environmental risk became a national priority for environmental organizations like Legambiente, whose campaigns against factories during the 1990s focused on claims about the risks to workers' health (Della Porta and Diani 2004, 35). Similarly, in Venice, campaigning about Marghera's industries concentrated on risks to workers from chemicals. Yet in the claims made about the mobile dams, risk was not raised in comparable terms.

When environmentalists interviewed for this book were asked about the risks connected with the mobile dams, claims about risks for the workers constructing the dams were entirely absent. This was striking given the difficulties with placing the casings and barriers on the sea bed bordering the lagoon. Rather, their claims developed a very specific, local definition of risk. Although campaigners referred to the internationally defined precautionary principle when discussing risks, they did not rely on the precautionary principle according to Principle 15 of the UN Rio Declaration or the Wingspread Declaration, but derived their definitions from Venice's safeguarding legislation. The precautionary principle was loosely deployed to back up claims about the project's lack of reversibility, experimentation and graduality, expressing a more general desire for restraint. Despite this specific definition of risk, environmentalists were still able to connect their claims with the pervasive politicization of risk in Western societies. The public's heightened risk consciousness is often sufficient for the manufacture and acceptance of risk.

Beck (1992) describes how the environmental movement gradually focused less on concrete cases and more on intangible threats that may only become visible for subsequent generations (Beck 1992, 162). When we analyzed risks with the mobile dams identified by environmentalists and others in Chapters Six and Seven, it was very difficult to prove if these risks were real or mythical. This is because these risks are typically derived from future possible scenarios, rather than scientific evidence based on past and present experience. Political claims about Venice's environment using future risks were most clearly illustrated by the discussion of global warming. Claims based on future scenarios introduce too much uncertainty for effective policy-making. Using past and present data about Venice's environment, we explained why the dams are a positive intervention to reduce the highest floods over the next 100 years. As circumstances change and human understanding and knowledge of the climate advances, Venetians will be in a better position to make decisions about continuing with the dams or alternatives. This does not mean we believe the claims about the dams and climate change are without basis: these claims broaden the objective risks of flooding and sinking, rather than constructing fictitious risks.

Risk is a theme running through the three components of the Venetian metaphor of sustainability, climate change and tourism. Redefinitions of nature, society and science have shaped the politicization of environmental risk in Venice and elsewhere.

REDEFINING NATURE, SOCIETY AND SCIENCE

The popularizing of environmental risks which inspired the environmental movement is identified with thinkers including Carson, Commoner, Ehrlich and Hardin, who built on earlier doubts about the use of science to change nature through modernization. Scientific and political thought during the eighteenth and nineteenth centuries was dominated by the belief that nature was to be used for human benefit and industry. Doubts were raised by nineteenth century romantics, including Byron, Ruskin and Wordsworth. In the early twentieth century, Mumford (1922, 1934) created a "human ecological" perspective, which challenged traditional natural science by integrating natural and social knowledge. After the Second World War, the traditions and applications of science were questioned more and more. The reaction against the human application of science in nature was illustrated by the popularity of Carson's (1962) book *Silent Spring*, which is widely regarded as having raised consciousness about environmental risks and inspired environmental movements.

Environmental thinkers of the 1960s developed the integration of natural and social knowledge to construct a holistic natural science, as discussed in Chapter Eight. Bramwell (1994, 16) argues that ecology does contradict the principles of modernist science, but is not a rejection of science itself because a new holistic version of science has been constructed. Yet holism has also been dismissed as alien to the scientific methods of the Enlightenment created by Descartes, Newton, Bacon and others (Merchant 1980; Popper 1982). By seeking to transcend the distinction between nature and society, holism elevates nature and belittles society, denying humanity's unique ability to transform nature. Holism thus signifies the end of Enlightenment privileging of the human subject over nature. Szerszynski (1998, 104) notes that nature has returned as our primary source of meaning.

Holistic environmentalism integrates nature and society. This integration was codified through the implementation of sustainable development policies from the 1980s. The principles of sustainability and the integration of nature and society are fundamental to the claims made by Venetian environmentalists about the mobile dams. Many of the comments by Venetian environmentalists contained objections to the use of science to change nature through modernization. Despite these objections, their remarks betrayed no hostility towards science. Rather, they argued for science-based interventions to be integrated with nature. While some environmentalists described unease with past interventions in the lagoon, other statements acknowledged the need for intervention to protect the city in the aftermath of the 1966 high floods. There is, however, one aspect of the modernist mentality behind the mobile dam project that environmentalists particularly reject: its engineering approach to problem-solving. Environmentalists in Venice have been able to make use of the emphasis on integration with nature to present modernist interventions like Project MOSE as using an outdated, engineering approach. This objection draws on the cultural reaction against modernization, identified by Giddens (1990, 2) as one component in the popularization of post-modernism.

The contemporary unease with modernization means that Venetian environmentalists' cultural narratives and myths have found wider cultural resonance. Yet unease with modernization is not universal, as was evident in the competing cultural narratives about the Venetian Republic sketched out in Chapter Seven. The myth about the Venetian Republic as environmentally-friendly was constructed through cultural narratives that presented significant interventions during the Venetian Republic, especially river diversions and the building of dam walls, as guiding nature according to the principles of experimentation, reversibility and graduality. This myth was contrasted with risky modern interventions, including the mobile dams. These cultural narratives about the Venetian Republic and their deployment as an argument against the mobile dams were referred to as Venecoism. However, Venecoism has been subject to counter-claims advocating anthropic intervention in the lagoon. Positive interpretations of significant interventions during the Venetian Republic have been built on to support the case for contemporary projects, such as the mobile dams. This outlook was referred to as Venanthropocentrism. Venecoism is based more on myth than Venanthropocentrism, but Venecoism is more in tune with current sentiments than Venanthropocentrism. This assessment was made because Venecoism is able to draw on the cultural resource of the contemporary unease with modernization.

In debates about Venice's environment, attitudes towards the use of science for modernization and intervention to transform nature have become increasingly contested, and the question of whether to intervene in nature is typically framed with reference to risk. Appreciating this contestation is fundamental to understanding Venice today.

UNDERSTANDING CONTEMPORARY VENICE: MYTH AND REALITY

Too much analysis of the Venice problem concentrates on clarifying physical changes to the relationship between the city, lagoon and sea. Not enough attention is given to cultural variations. It is important to take physical changes into account, yet these are insufficient to understand the relationship between Venice and its environment. Venice is not merely a physical entity: it is a human society and is shaped by human perceptions. Therefore, we need to investigate human perceptions to comprehend the relationship between the city and the sea.

For instance, when the ancient Venetians were physically more vulnerable to flooding, they believed that they could dominate the seas. Now we have much better technological ability to defend Venice, yet we fear the sea will engulf the city. To put matters bluntly, the flooding devastation and numerous deaths suffered by the ancient Venetians make it impossible to appreciate why they celebrated their relationship with the sea. Likewise, when we have technologies to respond to environmental hazards today, subsidence and rising sea levels cannot explain the fear of the sea engulfing Venice. It is only by examining ancient and contemporary myths that we can understand these interpretations, and objectively assess the extent to which myths contradict or correspond to reality. The resulting objectivity allows the creation of appropriate policy responses.

We agree with Rinaldo (2009) that the Venice problem "is dominated by politics that regulate the strategic choices for the city", and that "[t]he problem is not therefore of Science and of phenomena that can be predicted or explained" (Rinaldo 2009, 193 and 225). In an interview with the author, Jane Da Mosto explained how science has become tainted by politics in Venice: "It's important that the politicians are well informed about the underlying science to understand the trends in the system. But the problems can begin when the politics get into the science and I think in Venice it's impossible to ignore the cases where the science has been affected by politics" (Da Mosto 2010). The mobile dam project is the best illustration of the politicization of the Venice problem. Rinaldo (2001) explains: "[T]he mobile barriers' use and function can be justified in relation to any lagoon morphology. Therefore, from this point onwards the problem cannot be considered a technical one, but falls within the domain of the politics of strategic decisions concerning the city" (Rinaldo 2001, 71). Finding solutions for Venice's other problems are also primarily political, rather than solely questions of engineering or technology.

When we say that politics is the decisive realm for Venice's future, it is important not to frame politics in the past language of left and right. Rinaldo (2009, 209) describes the dilemma of considering himself on the left regarding questions of health, education and energy, yet finding his support for the dams results in him being associated with the right. This perceptual shift demonstrates how the political concepts of left and right and their ideological associations have become inadequate as political guides in the twenty-first century. Chapter Nine outlined that the key contemporary choices for Venice are between development and sustainability. Choosing polices based on development or sustainability depends on political priorities, especially whether we prioritize humanity or nature in the short-term. Of course, we want as much scientific research and analysis of Venice's problems as possible. But while science provides important context and data, ultimately we have to make political decisions about our priorities.

Due to the way that myth and reality have been intertwined in debates which have politicized the Venice problem, it is necessary to dissect the relationship between them to make clear political decisions. Mary McCarthy (1963) warns that, "The rationalist mind has always had its doubts about Venice. The watery city receives a dry inspection, as though it were a myth for the credulous—poets and honeymooners" (McCarthy 1963, 1). Although we have challenged many of the myths about Venice, we do not regard them as a distraction for the credulous. On the contrary, myths are fundamental to understanding the evolution and significance of Venice: as Lane (1973) observes, "Some myths have even been makers of reality and moulded Venice's history" (Lane 1973, 87). Chapter One registered how the fourteenth-century chronicler Andrea Dandolo constructed a mythical narrative about liberty and the founding of Venice. During the flourishing of Renaissance humanism in fifteenth and sixteenth century Venice, historians sought true explanations of reality. Lane offers one example:

> On the one hand they learned to stress a search for the truth of what happened. On the other hand, they regarded history as a branch of literature, to be written in flowing emotion-stirring rhetoric, while also giving a rational explanation of events. Bernardo Giustiniani applied these views to the early history of Venice, comparing chronicles and documents one with the other, and—more surprisingly—comparing them with archaeological findings and with geographical study to decide which of the traditions seemed most probably true. (Lane 1973, 220)

The quest for the truth by Venetian humanist Giustiniani provides a valuable lesson for our treatment of today's environmental myths. However, Giustiniani was careful not to question the core myths of Venice, including the belief that the city had been free since its foundation (Lane 1973, 220). We need to go one step further than Giustiniani and not shy away from challenging deeply-held myths, even when this risks offending dominant sensibilities. Questioning environmental myths is not easy or popular in today's Western culture when environmentalism has become so powerful. But to save and improve an important city like Venice, it is vital.

ARRIVING IN DEVELOPED VENICE

We conclude our critique by considering how one arrives in Venice, as an invitation to develop the city. Arriving in Venice is an indicator of how the relationship between humanity and nature is perceived. Pre-modern Venice appeared to dominate the seas and the ancient Venetians celebrated this relationship in their boats on the surrounding waters. Grand Tourists usually traveled to the city across the lagoon by ferry from the mainland at Fusina or Chioggia. The perspective from these journeys gave the impression that Venice was hovering in a commanding position above the water. "The prospect of this city, from the first entrance into the sea, is the most wonderful and extraordinary in the whole world, for the situation is such, that at a distance, which is a full five miles from the nearest land, it appears to the eye, as if floating on the waves," observed Nugent in 1749 (Nugent 1749, 44-45, cited in Davis and Marvin 2004, 33). Christopher Hervey's 1785 letters described the approach to Venice from Fusina: the city appeared "to exult over the subjected waves" (cited in Eglin 2001, 9).

Arriving in Venice changed with the onset of modernity, and so did perspectives on the city. Modernity brought the bridge between Venice's islands and the mainland, initially with the railway link in 1846 and followed by the addition of the road in 1933. Richard Wagner traveled across the newly-built railway bridge and remarked on "looking down from the causeway at the image of Venice rising reflected from the waters beneath" (cited in Ackroyd 2010, 35). In the nineteenth century, people arrived across the bridge while looking down on a Venice that appeared to be protruding from the waters as it subsided. Yet the city that produced the first commercial mirrors continued to provide a metaphorical reflection for humanity. Nineteenth-century intellectuals who felt repelled by modern development found solace in campaigning for Venice's conservation. The fall of the Venetian Republic was connected to the dying city and the degraded impact of man on nature. Venice's reputation for death, degradation and fallen humanity became entrenched in the twentieth century, and the modern Venice problem was constructed in the aftermath of the 1966 floods.

In the early twenty-first century, the Venice problem of flooding and sinking has been redefined as a myth of global warming. Contemporary environmental myths depict Venice as sinking and dying due to problems created by humans, including rising sea levels and tourism. Environmental claims about global warming, tourism and unsustainable management have helped reconstitute the Venetian metaphor. Environmentalists' campaigns have connected with the ascendancy of environmental politics internationally. Venice now represents an environmentally-threatened retreat from modernity. In reality, Venice is now better protected and flooding has less impact than during the Venetian Republic. But the ancient Venetians embraced the sea. In addition, when Venice was much more crowded and chaotic than it is today, the authorities encouraged visitors to flock to the city through the invention of festivals. Now we have a less populous and more orderly city, but claims about tourists leading to the death of Venice and discriminatory policies against them. Welcoming more tourists and others to Venice using the subway under the sea would be a positive step towards re-conquering the seas, by indicating how many other means we have of navigating them. With modernized transport links, people could fly to Venice and make the final stage of arrival from below the lagoon's surface through the subway. Such an arrival would illustrate our rediscovery of human dominance over the waters.

The mobile dams are a similar step in the right direction. Venice needs development on its outskirts and within the city center, as outlined in the ten proposals in Chapter Nine. All these improvements can be funded, particularly by expanding tourism. The real challenge is breaking free from the sustainable development agenda and contesting environmental myths. Then tourists, students, commuters and others will be able to join residents in a city that combines modern development with maintaining ancient heritage.

Venice is not in peril from sinking, rising sea levels or tourism. These are myths constructed by environmentalists and others through political campaigning that express contemporary cultural preoccupations. These myths can be compared with the ancient myth of St Mark, which was created to provide the city with wider meaning. There is reality to rising sea levels and increased flooding. But these problems are being addressed with the mobile dams and can be solved with other technologies. The real danger for Venice, and unfortunately for many parts of the world, is the sinking of human ambition, courage and resilience. Robustness, resourcefulness, redundancy and rapidity need to replace vulnerability in responses to Venice's challenges. Reviving ambition and resilience in combination with development are the pre-conditions for improving Venice.

NOTE

1. See Chapter Five.

Appendix A

Profiles of Venetian Environmentalists Interviewed

Please note the interviews for this book were conducted between 2005 and 2010 with the full consent of all interviewees. Interview dates are listed in the Bibliography. The interviews were digitally recorded for accuracy with the agreement of all interviewees. The profiles below were correct on the date of interview, although may have since changed:

- Orazio Alberti. Member of VAS. Former Green Party Councilor in Venice.
- Federico Antinori. Responsible for the lagoon environment for the Venice chapter of LIPU.
- Alvise Benedetti. President of Italia Nostra's Venice chapter.
- Michele Boato. Director of the Eco Institute. Stood in the 2005 and 2010 Venice Mayoral elections.
- Stefano Boato. Member of Venice Safeguarding Commission. Member of Italia Nostra. IUAV Professor of Urban Planning. Former Green Party Venice City Councilor.
- Beppe Caccia. Former Green Party Venice City Councilor.
- Fabio Cavolo. Director of the environmental citizens' committee *Associazione Rocchetta e Dintorni.*
- Flavio Cogo. Member of Legambiente. Director of the Committee from Certosa and Sant' Andrea. Member of the Refounded Communist Party. Active in the No MOSE permanent assembly.
- Marina Corrier. Founding member of the Association of Friends of the Sanpierota Boats. Active in the No MOSE permanent assembly.
- Gustavo De Filippo. Member of Legambiente. Works in The Natural Observatory of the Lagoon, Venice.
- Marco Favaro. Member of WWF. Venice Provincial Councilor for the Green Party. Works in The Natural Observatory of the Lagoon, Venice.

- Cristiano Gasparetto. Director on the Venice board of Italia Nostra.
- Giannandrea Mencini. National Vice-President of VAS and former VAS Venice President. Author of several books on Venice and its environment. Green Party member.
- Paolo Perlasca. Director of WWF Italy for the Venice lagoon and the upper Adriatic coast.
- Fabrizio Reberschegg. Vice-President of the Green Party in the Venice municipality.
- Edoardo Salzano. Member of Italia Nostra. Former member of Venice Safeguarding Commission. Previously IUAV Dean and Professor of Urban Planning. Former Deputy Mayor for Urban Affairs, Venice Council. Runs a leading environmental website, *Eddyburg*.
- Enrico Sambo. Board member of VAS Venice chapter. Active in the No MOSE permanent assembly.
- Giorgio Sarto. Previously a Senator in the Italian Senate for the Green Party. Former member of Venice Safeguarding Commission.
- Cesare Scarpa. Co-ordinator of the Committee from Certosa and Sant' Andrea. Active in the No MOSE permanent assembly. Former Director of Legambiente in Venice.
- Maurizio Zanetto. Secretary and former President of Italia Nostra's Venice chapter.

Appendix B

List of Non-Governmental Organizations

This is a list of the organizations referred to in the text rather than being comprehensive.

ENVIRONMENTAL ASSOCIATIONS

- *Ecoistituto del Veneto* (The Eco Institute of the Veneto Region)
- Friends of the Earth - FoE
- *Italia Nostra* (Our Italy)
- League against Hunting - LIC
- League against Vivisection - LAV
- *Lega Italiana per la Protezione degli Uccelli* - LIPU (The Italian League for Bird Protection, which is the Italian national bird association)
- *Legambiente* (The Environmental League)
- The National Forum of Associations for the Environment
- *Verdi Ambiente e Società* - VAS (Green Environment and Society)
- World Wide Fund For Nature - WWF (Europe, including Italy)
- World Wildlife Fund - WWF (United States and Canada)

CITIZENS' COMMITTEES

- *Amici di Venezia* (Friends of Venice)
- *Associazione Amici della Sanpierota* (Association of Friends of the Sanpierota boats)
- *Associazione Gabriele Bortolozzo* (Association in memory of Gabriele Bortolozzo, former worker at Marghera petrochemical plant)

- *Associazione per la Difesa dei Murazzi* (Association for the Defense of Sea Wall Dams)
- *Associazione Rocchetta e Dintorni* (Association of Rocchetta and Surroundings)
- *Comitato Certosa e Sant' Andrea* (Committee from Certosa and Sant' Andrea)
- *I Danni del MOSE* (The Damages of MOSE)
- *Medicina Democratica* (Democratic Medics)
- *Nosublagunare* (No subway)
- *Venessia*

UMBRELLA COMMITTEES/NETWORKS

- ARCI (Italian Recreative and Cultural Association)
- No Global *Disobbedienti* (Disobedients)
- No MOSE permanent assembly (also referred to as No MOSE 'forum')
- Save Venice and the Lagoon (also referred to as 'Save Venice with the Lagoon')

PRIVATE COMMITTEES SUPPORTING THE SAFEGUARDING OF VENICE

- *Accademia delle Belle Arti,* Italy
- American Committee to Rescue Italian Art, USA
- American Kress Foundation, USA
- *Arbeitskreis Venedig der Deutschen Unesco-Kommission,* Germany
- *Comitato Italiano per Venezia,* Italy
- *Comité des amis de Saint Marc,* France
- *Comité Francais pour la Sauvegarde de Venise,* France
- *Pro Venezia,* Sweden
- *Pro Venezia Foundation,* Switzerland
- *Nederlands Comtié Geeisterde Kunstaden Italie,* the Netherlands
- Save Venice, USA
- The Venice in Peril Fund, UK

Appendix C

List of Abbreviations of Political Parties

AO	*Avanguardia Operaia* (Workers' Vanguard)
DC	Christian Democratic Party
DP	*Democrazia Proletaria* (Democratic Proletarian)
DS	Party of the Democratic Left
FI	*Forza Italia* (Let's Go Italy)
Lega	*Lega Nord* (Northern League)
PCI	Italian Communist Party
PD	Democratic Party
PDL	People of Freedom
PSI	Italian Socialist Party
RC	Refounded Communist Party
RP	Radical Party

Appendix D

Glossary of Terminology

- Environmental Impact Assessment (EIA) (*Valutazione d'Impatto Ambientale) (VIA)*: An examination of a project's environmental impact.
- Environmental Impact Study (EIS): A specific study conducted as part of an EIA. Note that the terms EIS and EIA *(VIA)* are used interchangeably and imprecisely in some primary and secondary sources.
- Eustasy (eustacy): Changes in sea level on a global scale due to expansion in ocean volume.
- Fish farms: Shallow basins of water equipped for breeding and sometimes hunting fish.
- Groundwater: This is water held within interconnected openings of saturated rock beneath the land surface and can be used to supply wells.
- Habitat: The place where a plant or animal species lives.
- Inlet: A gap in the coastal strip between the sea and the lagoon.
- Insula: The Latin word *insula* means island. The plural *insulae* refers to the forty sectors into which the historic center of Venice has been divided. The term is now also used to refer to local flood defense strategies, which include raising the perimeter of individual islets and waterproofing measures.
- Jetty: A protective structure of stone and/or concrete extending into the sea to influence the current or tide and for protection.
- Lagoon: A shallow area of water separated from the sea by sandbanks or strips of low land and usually connected to the sea by one or more inlets.
- Morphology: The scientific study of the forms of things.
- Mud flat: Lagoon area characterized by soft soil. Normally submerged and emerging only during low tides.
- *Murazzi:* Sea walls built along the shoreline to prevent erosion and damage by wave action. Venice's *murazzi* were initially completed in the eighteenth century and constitute the first line of flood defense.

- Saltmarshes *(barene)*: A coastal wetland of plants that is periodically inundated by seawater. Saltmarshes support physical and biological processes of the lagoons in which they are situated.
- Sediment: Deposits of mud, silt or sand suspended in or settled out of water. In a lagoon, sediments continually shift within the lagoon or go out to sea.
- Storm surge: A rise in sea level caused by wind and pressure systems.
- Subsidence: Decrease in the elevation of land surface due to tectonic, seismic or artificial forces (such as the loss of underground water support).

Chronology: Venice and Its Protection

- 402. Some mainlanders begin their flight into the lagoons along the northeastern shores of modern Italy after pillaging by Alaric the Visigoth.
- 421. In popular mythology, Venice is founded at 12 noon on Friday, 25 March.
- 452-4. Attila the Hun begins his scourging of northern Italy and the lagoon again becomes a temporary sanctuary for fleeing mainlanders.
- 466. Representatives from the growing island communities within the various areas of the lagoon meet at Grado.
- 568. The Lombards invade Italy. Their campaigns during the final decades of the sixth and early decades of the seventh centuries mark a massive and definitive exodus to the islands of the lagoon.
- 726. Orso Ipato becomes the first Doge of Venice.
- 810. The Doge flees from Malamocco to Rialto after an attack by Pepin.
- 811. *Pax Nicephori* between Charlemagne and the Byzantine Empire confirms Venice as a semi-independent province of Byzantium.
- 820. Construction of Doge's Palace.
- 828. A body is taken from Alexandria to Venice and is rumored to be the body of St Mark.
- 1082. Venice is no longer a subject city of the Byzantine Empire and establishes itself as an autonomous state.
- Late eleventh century. Carnival begins as a regular event.
- Early twelfth century. The Arsenal is built.
- 1324. Diversions begin of the Brenta, Piave and Sile Rivers to protect the lagoon.
- 1348-9. Plague epidemic.
- 1405. Venice enters its "Golden Age" (Norwich 1983, 279).
- 1422. The old Doge's Palace is replaced by a Renaissance palace.

- 1453. The Turks conquer Constantinople, ending the Byzantine Empire and weakening the Venetian Republic's hold over the eastern Mediterranean.
- 1509. Venice is defeated in a battle by the League of Cambrai.
- 1571. Venice loses Cyprus to the Turks.
- 1576. Plague epidemic.
- 1606. The Pope places Venice under its fourth and most controversial interdict.
- 1669. Venice loses Candia (Crete) to the Turks.
- 1744. Construction begins of the defensive walls, known as *murazzi*, along the littoral islands that protect the lagoon from the Adriatic Sea. This takes 38 years.
- 1797. Napoleon conquers Venice, ending the Venetian Republic.
- 1798. Napoleon exchanges Venice with Austria to take control of Lombardy.
- 1806. The French retake control of Venice.
- 1807. Napoleon visits Venice and orders the dredging of major canals allowing larger ships to sail to the Arsenal. He also initiates repairs to the damaged sea walls and construction work in the city, especially in and around St Mark's Square.
- 1814. The Austrians recover Venice.
- 1846. The Austrians finish building the railway bridge between the historic center of Venice and the mainland.
- 1848-9. Revolution in Venice led by Daniele Manin and defeated by the Austrians.
- 1866. Venice joins the unified Kingdom of Italy.
- 1902. The bell tower on St Mark's Square collapses, but is rebuilt by 1912.
- 1903. Creation of the first Marghera industrial zone on the mainland bordering Venice.
- 1917. Marghera port is operational.
- 1925. Industrial pumping of groundwater from beneath Venice begins on a major scale and significantly adds to subsidence.
- 1933. The completion of the road link between Venice and the mainland.
- 1966. After storm surges on 3 and 4 November, very high floods swamp Venice. There is also severe flooding in other parts of Italy.
- 1969. UNESCO brings scientists together from around the world to discuss the 1966 Venice floods.
- 1970. The Italian government opens a 'competition of ideas' for a flood control system that would allow the lagoon to be closed when necessary.
- 1971. The 'competition of ideas' is won by a design for an underwater mobile dam system to close off the three lagoon inlets at times of storm surges. The government begins to shut down the industrial pumping of groundwater beneath Venice, slowing the accelerated subsidence this caused since 1925.

- 1973. Special Law 171 introduces legislation to safeguard the city of Venice and the lagoon.
- 1974. UNESCO's General Convention calls on the Italian state to guarantee the interventions specified in Special Law 171.
- 1975. The Senior Council of the Public Works Ministry calls for international bids for work to defend the city from high waters using a fixed, not mobile, dam system.
- 1978. A special adjudication panel decides that none of the proposed projects submitted in the international competition is suitable for flood control.
- 1980. The Public Works Ministry appropriates all plans submitted in the international competition and sets up a technical team to develop a project to protect the lagoon from flooding.
- 1981. The technical team comes up with the 'feasibility study and principal plan' for a major flood control project, known as the *Progettone* (big Project).
- 1982. Venice City Council approves the Progettone, specifying that it must be part of a general effort to restore hydro-geological and ecological balance to the lagoon.
- 1984. A Special Law establishes the New Venice Consortium (*Consorzio Venezia Nuova* - CVN) of engineering and construction companies to plan and implement measures to safeguard Venice and its lagoon. This law also creates an inter-ministerial council, known as the *Comitatone* (large committee) to decide on strategy to safeguard Venice and the lagoon, especially budgeting.
- 1988. At the Arsenal, the CVN rolls out an experimental model for the mobile dam project, which reproduces a single barrier in one to one scale. It is called the Electromechanical Experimental Model, *Modulo Sperimentale Elettromeccanico*, abbreviated to *MoSE*. Subsequently, the mobile dam project is commonly referred to as Project 'MOSE'.
- 1989. The conceptual design for the mobile dams is completed and considerably reduces the fixed parts of the barriers proposed in the Progettone.
- 1991. The Organisation for Economic Co-operation and Development (OECD) meets in Venice and appeals to the Italian government to take immediate action to protect Venice.
- 1992. The MoSE experimental model placed at the Arsenal is declared a success by the CVN and is removed.

  A new Special Law makes it mandatory to obtain the opinions of the Veneto Regional Council and Venice and Chioggia City Councils for safeguarding. This law also outlines specific safeguarding measures.
- 1993. Massimo Cacciari is elected as the Mayor of Venice and serves until 2000.

- 1994. The Public Works Ministry calls for detailed designs of the dams in preparation for construction.
  The Comitatone agrees to subject the mobile dam project to an Environmental Impact Study (EIS) following a request from Venice City Council.
- 1996. The Green Party becomes involved in national government for the first time and assumes control of the Environment Ministry.
- Venice's Fenice Theater burns down and is rebuilt in a similar style by 2003.
- 1997. The EIS is delivered to the national government. A Commission of International Experts is employed by the CVN to review the study. The Commission endorses the CVN's plans.
- 1998. The Environment Ministry's own commission produces a largely negative report on the mobile dam project. On the basis of this report, the Environment Minister, Edo Ronchi of the Green Party, issues a decree blocking progress of the project with support from the Cultural Affairs Minister Giovanna Melandri. This is known as the 'Ronchi-Melandri' decree.
  Italia Nostra launches an appeal to the European Commission (EC) against the CVN's 'monopoly'.
- 1999. The European Parliament asks the Italian government to quickly make a definitive decision about the mobile dams.
- 2000. The Veneto TAR tribunal cancels the 1998 'Ronchi-Melandri' decree.
- 2001. Silvio Berlusconi is elected Prime Minister in May and vows to start constructing the mobile dams, funding them directly through 'Strategic Objectives'.
  Italia Nostra and WWF Italy launch an appeal to the Veneto TAR tribunal against the approval of the dams' final design stage.
- 2002. The EC rejects the appeal launched by Italia Nostra against the CVN's 'monopoly' of the MOSE work.
  The CVN delivers the final design for the mobile dams with the introduction of locks for vessels to pass between the sea and lagoon when the barriers are raised.
  Italia Nostra and the WWF launch an appeal to the Veneto TAR tribunal against the inclusion of Project MOSE in the Strategic Objectives funding of infrastructure projects.
- 2003. The Italian government is asked to present its case to the EC within ninety days following an appeal by Legambiente. Legambiente appealed to the EC against the 2001 Strategic Objectives used to approve the final preparation and funding for Project MOSE.
  An appeal is launched in February at the Veneto TAR tribunal by the associations in the Save Venice and the Lagoon committee, led by the WWF and Italia Nostra. The appeal claims that the Executive Phase of Project MOSE needed an Environmental Impact Assessment (EIA) *(VIA)*.

Qualified approval is given by the Comitatone for implementation of the mobile dam system, while agreeing to consider eleven conditions related to environmental impact put forward by Venice City Council.
Prime Minister Berlusconi attends a high-profile inauguration of the start of construction of the mobile dams in Venice in May. Protests are held with participation by environmental organizations.

- 2004. The Green Party launches an appeal to the EC against the approval of the mobile dams without an EIA.
  A letter signed in February by 140 Italian MPs is sent to the EC Environment Commissioner highlighting the lack of an EIA for Project MOSE. A few days later another letter signed by 300 Italian MPs is sent to the Commissioner supporting Project MOSE.
  The Veneto TAR tribunal rejects all eight appeals against Project MOSE, placed by environmental associations, Venice City Council and the Venice Provincial Council. These organizations appeal to the Judicial State Council against this decision by the Veneto TAR tribunal. The Judicial State Council rejects these appeals and upholds the decision by the Veneto TAR tribunal.
- 2005. Massimo Cacciari is elected as the Mayor of Venice for the second time. Cacciari puts together a group of experts to evaluate alternatives to Project MOSE for closing the three inlets to the lagoon. The Minister of Infrastructure, Pietro Lunardi, insists that no alternatives are under consideration.
  Venice City Council orders the local police to block the construction work on Project MOSE and to check its authorization documents.
  Minister Lunardi visits the MOSE works in Venice. A protest of about 200 people is held on the Rialto Bridge including many environmental activists. Protestors occupy the offices of the Venice Water Authority and scuffles occur as police evict them.
  The No MOSE permanent assembly is created on 22 June.
  The Legambiente national campaign boat *La Goletta Verde* makes a highly publicized arrival in Venice in July to point out the illegality of Project MOSE.
  Several environmental organizations appeal to the EC that the construction work for Project MOSE is violating EC protection criteria for the areas of Ca' Roman, Alberoni, San Nicolò and Punta Sabbioni.
  An idea to raise Venice by injecting water under the lagoon is proposed by engineers and geologists from the University of Padua.
  The No MOSE permanent assembly presents a petition of 11,000 signatures against the mobile dams at a press conference in Rome.
  A debate is held in December at the Venice City Council building for invited attendees to view presentations on the alternatives to Project MOSE for closing the lagoon inlets.

- 2006. The Director General of the Environment for the EC rules that the construction work for Project MOSE has not respected European directives on protecting bird life.
  A delegation from the No MOSE permanent assembly goes to Brussels in March to present a petition against the mobile dams to the EC Commission investigating the legitimacy of the project.
  In May, 300 people in boats protest against Project MOSE outside the Venice City Council building before a Council vote on a document calling for verification and review of the Project. The vote is delayed until June.
  In December, the Minister of Infrastructure, Antonio Di Pietro, rejects alternative proposals to MOSE for blocking the lagoon inlets.
- 2007. A group of European politicians visits the MOSE construction site in February to inspect it following the appeal to the EC by the No MOSE permanent assembly.
  The Environment Minister and former Green Party leader, Alfonso Pecoraro Scanio, writes a letter to the Venice Water Authority calling for the MOSE works to be stopped due to the absence of an environmental impact assessment.
- 2008. The Constitution Bridge over the Grand Canal opens. It is the fourth bridge over the Grand Canal and Venice's first new bridge for 70 years.
  Venice faces its worst floods for 22 years on 1 December.
- 2009. The EC decides to 'archive' the charge that Italy was guilty of breaking a 2000 EC nature directive by failing to set aside protected areas for birds around the lagoon.
  Venice's People Mover train begins service.
- 2010. Reconstruction begins of parts of the Lido Littoral Island as luxury vacation destinations, despite protests.
  Venice City Council publicizes plans for a tram service to the outskirts of the city.
  Proposals are announced for an offshore docking platform to reduce the need for large ships to enter the lagoon.
  By early December, the number of flooding events in Venice during 2010 of 80cm and 90cm above the tide meter surpasses the number of these events in 2009.
  The mobile dam system is 63 percent constructed with completion predicted for 2014, depending on funding and political support.

Bibliography

NEWSPAPER AND NEWS AGENCY REFERENCES

AFP. 2008, December 1. "Venice suffers worst flooding in 22 years." Hosted by Google News. Accessed November 28, 2010. No author listed. http://www.google.com/hostednews/afp/article/ALeqM5iHOqqJsyjILTJBTjhOSVeyN2wfeQ.

AGI. 2009, April 14. "MOSE: Ronchi, UE ha archiviato la procedura d'infrazione." Accessed April 14, 2009. No author listed. http://www.agi.it/venezia/notizie/200904141321-cro-rt11077-mose_ronchi_ue_ha_archiviato_la_procedura_d_infrazione.

ANSA. 2009, April 28. "Britain's Prince Charles in Venice." Accessed April 28, 2009. No author listed. http://www.ansa.it/site/notizie/awnplus/english/news/2009-04-28_128346531.html.

ANSA. 2009, November 13. "Venice to stage its own funeral." No author listed.

ANSA. 2010, July 1. "Italian police foil Mafia bid to muscle in on EXPO 2015." Accessed July 1, 2010. No author listed. http://www.ansa.it/web/notizie/rubriche/english/2010/07/01/visualizza_new.html_1848487646.html.

Antonini, Alessio. 2010. "Orsoni benedice il Mose. Gli esperti lanciano l'allarme acqua salata." *Corriere della Sera - Corriere del Veneto*, May 11.

Associated Press. 2005, November 22. "Venice could be lifted by 30cm." *News24*. Accessed November 23, 2010. No author listed. http://www.news24.com/SciTech/News/Venice-could-be-lifted-by-30cm-20051121.

Bertasi, Gloria. 2008. "Marea eccezionale e improvvisa Venezia diventa una città sommersa." *Corriere della Sera - Corriere del Veneto,* December 2.

———. 2010a. "Bufera politca sull'inceneritore." *Corriere della Sera - Corriere del Veneto,* March 4.

———. 2010b. "'Report' boccia Tessera e Veneto city." *Corriere della Sera - Corriere del Veneto,* May 11.

Betts, Paul. 1999. "Rising tide of defiance: The fate of Venice provokes heated, but often sterile debate." *The Financial Times Weekend,* January 30.

Boato, Michele. 2010. "Noi alternativi a Orsoni-Brunetta." *Il Gazzettino di Venezia,* March 2.

Brown, Paul. 2002. "Flooded, stinking and sinking: Venice calls in British experts to rescue city." *The Guardian,* June 10. Accessed June 10, 2002 http://www.guardian.co.uk/world/2002/jun/10/arts.artsnews.

Brunetti, Roberta. 2010. No title. *Il Gazzettino di Venezia*, April 25.

Campostrini, Pierpaolo. 2010. No title. *Il Gazzettino di Venezia*, January 31.

Conor, Steve. 2002. "Venice flood barrier scheme 'will soon be obsolete.'" *The Independent*, May 13. Accessed May 20, 2002.

Corriere della Sera. 2006, November 29. "La Rossanda: sì al Mose, scusate ma difendo Venezia." No author listed.

Corriere della Sera - Corriere del Veneto. 2010, July 13. "Inaugurata a Fusina la prima centrale ad idrogeno del mondo." No author listed.

Crema, Maurizio. 2009. "Porti, Venezia ringrazia le crociere." *Il Gazzettino di Venezia*, August 29.

Daily Express. 2009, November 28. "Venice 'example of global warming'." Accessed November 28, 2010. No author listed. http://www.express.co.uk/posts/view/97575/Venice-example-of-global-warming.

De Rossi, Roberta. 2010. "People Mover a regime ridotto." *La Nuova Venezia,* April 22.

Di Marino, Jean. 2010. "Venice continues work on massive flood protection project." *Deutsche Welle*, October 6. Accessed November 30, 2010. http://www.dw-world.de/dw/article/0,,6082550,00.html.

Dina, Paolo Navarro. 2009a. No title. *Il Gazzettino di Venezia*, January 8.

———. 2009b. "Cacciari: 'Un superComune puo' salvare Venezia'." *Il Gazzettino di Venezia*, January 22.

Donadio, Rachel. 2009. "Locals fear Venice is becoming 'a city of ghosts'." *International Herald Tribune*, November 16.

Emsden, Christopher. 2000. "Venice Court Ruling Puts Plans for the Anti-Flood Dikes Back on Agenda." *International Herald Tribune: Italy Daily* section, July 15-16.

Environmental News Service. 2002. "UNESCO Asked to Save Venice and Lagoon from Dikes." November 14. Accessed November 30, 2010. http://www.ens-newswire.com/ens/nov2002/2002-11-14-01.asp.

———. 2007. "UNESCO World Heritage Sites in Danger of Global Warming." April 10. Accessed November 30, 2010. http://www.ens-newswire.com/ens/apr2007/2007-04-10-02.asp.

Favarato, Gianni. 2010. "Fincantieri, cassa integrazione vicina." *La Nuova Venezia*, September 7.

Favaro, Massimo. 2010. "Un grattacielo a Porto Marghera per i crocieristi." *Corriere della Sera - Corriere del Veneto*, April 23.

Fornasier, Claudia. 2003. "Prima pietra? Si, al collo di Venezia." *Corriere della Sera - Corriere del Veneto*, May 15.

Fullin, Michele. 2009. "Abbracci per la nautica, veleni per la chimica." *Il Gazzettino di Venezia*, March 15.

———. 2010. "Un futuro 'turistico' per Arsenale, forti e caserme." *Il Gazzettino di Venezia*, April 13.

Fumagalli, Marisa. 2006. "Cacciari frena sul Mose: ecco i progetti alternativi." *Corriere della Sera*, January 25.

Gasparetto, Cristiano. 2010. "Zaia in passerella al MOSE." *Eddyburg*, June 16. Accessed November 23, 2010. http://eddyburg.it/article/articleview/15283/0/178/.

Giannini, Claudia. 2003. "Venezia, a febbraio la prima pietra del MOSE." *Il Gazzettino di Venezia*, January 15.

Greenway, H.D.S. 2008. "Facing climate truths." *International Herald Tribune*, October 15.

Il Gazzettino di Nord Est. 2010, January 12. "In Italia otto su dieci sono favorevoli al Mose." No author listed.

Il Gazzettino di Venezia. 2007, January 12. "E per il Worldwatch Institute il Mose diventa l'esempio di come combattere i mutamenti del clima." No author listed.

Il Gazzettino di Venezia. 2008, October 9. "Mose, fondi dall'Europa ma slitta la fine dei lavori." No author listed.

Il Gazzettino di Venezia. 2009, April 15. "Barene, velme e pochi rumori: ecco il Mose 'verde'." No author listed.

Il Gazzettino di Venezia. 2009, June 27. "La città 'conservatrice' dà segnali di reazione." No author listed.

Il Gazzettino di Venezia. 2009, September 22. "Delegazione a New York." No author listed.

Il Gazzettino di Venezia. 2009, November 14. "Immergersi nella natura." No author listed.

Il Gazzettino di Venezia. 2009, December 18. "Il Pd propone di utilizzare la graduatoria già esistente favorendo l'impiego di imprese locali." No author listed.

Il Gazzettino di Venezia. 2010, January 3. "MOSE, con Cacciari persi sette anni." No author listed.

Il Gazzettino di Venezia. 2010, January 10. "Giusta l'azione per chiedere i danni." No author listed.

Il Gazzettino di Venezia. 2010, April 11. No title or author listed.

Il Gazzettino di Venezia. 2010, July 18. "'Punturoni' di acqua salata per sollevare Venezia." No author listed.

Il Gazzettino di Venezia. 2010, July 20. No title or author listed.

Il Gazzettino di Venezia. 2010, July 21. "Alzare Venezia, il Comune vuole lo studio." No author listed.

Il Sole 24 Ore. 2010, April 26. "Venezia e Siena i comuni più ricchi." No author listed.

International Herald Tribune. 2009, September 8. "Maldives Plans Environmental Tax." No author listed.

James, Barry. 2002. "A global threat laps at the gates of Venice." *International Herald Tribune*, March 22.

Kay, John. 2008. "Welcome to Venice, the theme park." *The Times* (London), March 1.

Kiefer, Matthew and Hubert Murray. 2008. "For Boston, the lessons from Venice." *The Boston Globe*, December 11. Accessed November 27, 2010. http://www.boston.

com/bostonglobe/editorial_opinion/oped/articles/2008/12/11/for_boston_the_lessons_from_venice/.

La Padania. 2001, December 7. "Via libera alle grandi opere." No author listed.

La Repubblica. 2009, January 9. "Scoppia la rivolta bipartisan contro le estrazioni in Adriatico." No author listed.

Mayer, Peter. 2010. "Italians trying to keep above water." *DPA Press Agency, The China Post*, January 4. Accessed November 27, 2010. http://www.chinapost.com.tw/life/environment/2010/01/04/239239/p1/Italians-trying.htm.

Morris, Jim. 1998. "High Level Crime: Italy develops a case for manslaughter because workers breathed vinyl chloride." *Houston Chronicle,* June 28. Reproduced by *Gabriele Bortolozzo Association.* Accessed November 12, 2010. http://agb.provincia.venezia.it/RaStampa/StampaEstera/Usa/HoustonChronicle.html.

Morris, Roderick Conway. 2000. "The Rise and Fall of Venice." *Geographical,* July. Reproduced by *CBS bnet.* Accessed December 2, 2010. http://findarticles.com/p/articles/mi_hb3120/is_7_72/ai_n28779589/.

Nadeau, Barbie. 2009. "Ciao, Ciao, Venice." *Newsweek*, November 13.

Newman, Cathy. 2009. "Vanishing Venice." *National Geographic Magazine*, August 3.

Owen, Richard. 1998. "Greens Prove a Barrier to Saving Venice." *The Times* (London), December 4.

———. 2008. "Venice cancels opening ceremony for hated Santiago Calatrava bridge." *The Times* (London), August 28.

Palazzo, Francesco. 2009. "Social housing in Venice." *myvenice.org*, January 5. Accessed November 20, 2010. http://www.myvenice.org/Social-Housing-in-Venice.html.

Physorg.com. 2010, November 24. "New wave of planning for coastal zones." Accessed December 8, 2010. http://www.physorg.com/news/2010-11-coastal-zones.html.

Pietrobelli, Giuseppe. 2010. "Venezia, il porto sfida l'Europa." *Il Gazzettino di Venezia,* May 11.

Piran, Giacomo. 2010. "Italia Nostra: 'No allo stravolgimento della Riviera.'" *La Nuova Venezia*, March 28.

Pirazzoli, Paolo. 2002. "Did the Italian Government Approve an Obsolete Project to Save Venice?" *EOS - Transactions* 83, 20:217-223.

———. 2009. "For a revision to the Italian project to save Venice." *Eddyburg*, December 19. Accessed November 27, 2010. http://eddyburg.it/article/articleview/14384/0/178/.

Possamai, Paolo. 1999. "Azzerati tutti i veti, il Mose riparte." *La Tribuna di Treviso*, February 27.

Poveledo, Elisabetta. 2006. "Death of Venice? Tourists pour in as residents head out." *International Herald Tribune*, September 29.

———. 2010. "Putting the polish back on a historic Venetian gem." *International Herald Tribune*, October 1.

Raccanelli, Giulietta and Silvio Testa. 2000. "Mose, accolto il ricorso della Regione." *Il Gazzettino di Venezia,* July 15.

Scalzotto, Davide. 2008. "Venezia 'per carità, 'pentimento'…". *Il Gazzettino di Venezia*, April 4.

Scano, Luigi. 2004. "The Lagoon protection (1973-1985)." *Eddyburg*, July 6. Accessed November 28, 2010. http://www.eddyburg.it/article/articleview/1346/0/122/.

Spence, Rachel. 2009. "Venice's Accademia Bridge to be redesigned." *The Art Newspaper*, July/August. Accessed November 28, 2010. http://www.theartnewspaper.com/articles/Venice-s-Accademia-Bridge-to-be-redesigned.

Squires, Nick. 2009. "Venice tries to turn back the tides." *The National*, March 28. Accessed March 30, 2009. http://www.thenational.ae/article/20090328/FOREIGN/205163796/1036.

Standish, Dominic. 2002. "Will sinking Venice raise our ambitions?" *International Herald Tribune: Italy Daily* section, December 3.

———. 2004. "Attempts to sink Venice dam project flood into EU." *European Voice* (*The Economist Group*), 10, 11:25-31.

———. 2008. "The Death of Venice is Greatly Exaggerated." *Spiked*, December 8. Accessed December 8, 2008 http://www.spiked-online.com/index.php/site/article/6004/2004.

Tamburrini, Pierluigi. 2009. "Con il Mose disponibili 4.700 alloggi al pianterreno." *Il Gazzettino di Venezia*, February 2.

Tantucci, Enrico. 2010. "Sul progetto di Tessera City la Regione non si esprime e la Salvaguardia lo boccia." *La Nuova Venezia*, May 12.

Testa, Silvio. 1999. "'Mose, se ne riparlerà a fine anno." *Il Gazzettino di Venezia,* March 9.

———. 2005. "Non ha ancora nome, ma intanto…." *Il Gazzettino di Venezia,* June 24.

———. 2010. "I lavori del Mose estranei all'alta marea." *Il Gazzettino di Venezia*, January 3.

Tsai, June. 2009. "Venice trip teaches local students about home." *Taiwan Review*, October 4. Accessed November 15, 2010. http://taiwanreview.nat.gov.tw/fp.asp?xItem=49569&CtNode=205.

UNESCO (United Nations Educational, Scientific and Cultural Organisation). 2009. "Tides of Time." *New York Times*, Advertising Supplement, September 2.

United Press International (UPI). 2007. "Report: Future is grim for Venice." Reproduced by *Earth Times*, April 6. Accessed November 28, 2010. http://www.earthtimes.org/articles/news/48838.html.

Vanzan, Alda. 2008a. "Sublagunare, pronta la delibera da spedire a Roma." *Il Gazzettino di Venezia*, December 30.

———. 2008b. "Cacciari: 'Tassa di un euro per chi dorme a Venezia.'" *Il Gazzettino di Venezia*, July 27.

———. 2010. No title, *Il Gazzettino di Venezia*, January 24.

Vianello, Alessio. 2009. "Un nuovo ruolo per Porto Marghera." *Il Gazzettino di Venezia*, March 17.

Vitucci, Alberto. 2010. "'IUAV non spinge la sublagunare adesso il convegno sulle alternative'." *La Nuova Venezia*, May 9.

Zamparutti, Tony. 2003. "Venice in Peril." *The Ecologist,* March 22. Accessed December 3, 2010. http://www.theecologist.org/investigations/natural_world/268502/death_of_venice.html.
Zanon, Gigio. 2009. "La subluganare e l'opinione di Cacciari." *Il Gazzettino di Venezia* (letters), March 4.

PRIMARY AND SECONDARY SOURCES

Abby, Edward. 1975. *The Monkey-Wrench Gang*. Philadelphia: Lippincott.
Ackroyd, Peter. 2010. *Venice: Pure City*. London: Vintage.
Agnoletto, Roberto, Jane Da Mosto, Fabrizio Fracchia, Renato Gibin, Francesca Mattassoglio, Thierry Morel, Lidia Panzeri, Anna Somers Cocks, Tom Spencer, Enrico Tantucci, and Stefania Tonin. 2009. *The Venice Report*. The Venice in Peril Fund. Cambridge: Cambridge University Press.
Alberti, Orazio. 2005. Interview with Dominic Standish, November 10.
Ammerman, Albert, and Charles McClennen. 2000. "Saving Venice." *Science* 289, 5483: 1301-1302. doi: 10.1126/science.289.5483.1301.
ANSA. 1997. *Venice 1966-1996: 30 Years of Protection as Covered by the Press.* Rome: ANSA Dossier.
Antinori, Federico. 2005. Interview with Dominic Standish, July 16.
Assemblea Permanente No MOSE. 2005. "Nasce l'Assemblea Permanente NoMOSE di Chioggia." November 29. Accessed December 5, 2010. http://www.nomose.info/doc/Chioggia291105.pdf.
Barrès, Maurice. 1990 [1903]. *La Mort de Venise suivie de carnets de voyage inédits et de documents*. Edited by Marie Odile Germain. Paris: Christian Pirot.
Beck, Ulrich. 1992. *Risk Society.* London: Sage.
———. 1995. *Ecological Politics in an Age of Risk.* Cambridge: Polity Press.
———. 1999. "What is 'Risk (Society)'?" *Prometheus* 1:74-79.
———. 2000. *The Brave New World of Work.* Cambridge: Polity Press.
Ben-Ami, Daniel. 2010. *Ferraris For All: In Defence of Economic Progress*. Bristol: Policy Press.
Benedetti, Alvise. 2005. Interview with Dominic Standish, October 10.
Benford, Robert. 2002. "Controlling Narratives and Narratives as Control within Social Movements." In *Stories of Change: Narrative and Social Movements*, edited by Joseph Davis, 53-75. New York: State University of New York Press.
Berendt, John. 2005. *The City of Falling Angels*. London: Hodder and Stoughton.
Bertoni, Werter, Carlo Elmi, and Francesco Marabini. 2005. "The Subsidence of Ravenna." *Giornale di Geologia Applicata* 1:23-32. doi: 10.1474/GGA.2005-01.0-03.0003.
Best, Joel. 1999. *Random Violence: How We Talk about New Crimes and New Victims*. California: University of California Press.
———. 2001a. "Introduction: The Diffusion of Social Problems." In *How Claims Spread. Cross-National Diffusion of Social Problems,* edited by Joel Best, 1-18. New York: Aldine de Gruyter.

———. 2001b. *Damned Lies and Statistics: Untangling Numbers from the Media, Politicians, and Activists*. Berkeley: University of California Press.

Bettin, Gianfranco. 1991. *Dove volano i leoni: fine secolo a Venezia*. Milan: Garzanti.

———. 1997. *Laguna Mondo*. Venice: Ediciclo Editore.

Bevilacqua, Piero. 2000. *Venezia e le acque. Una metafora planetaria*. Rome: Donzelli Editore.

Biorcio, Roberto. 1992. "Il movimento verde in Italia." Istituto Superiore di Sociologia di Milano Working Paper No. 46, presented at Institut de Ciències Politiques i Socials, Barcelona, 1992.

———. 2002. "Green Parties in National Governments: Italy." *Environmental Politics* 11, 1: 39-62.

Biorcio, Roberto and Giovanni Lodi, eds. 1988. *La sfida verde*. Padua: Liviana.

Boato, Michele. 2005. Interview with Dominic Standish, December 16.

Boato, Stefano. 2005. Interview with Dominic Standish, October 13.

Bon, Mauro, Danilo Mainardi, Luca Mizzan, and Patrizia Torricelli. 2001. "The biodiversity in the Venice lagoon as the basis of a sustainability project." In *Sustainable Venice: Suggestions for the Future*, edited by Ignazio Musu, 27-60. Dordrecht, Boston, London: Kluwer Academic Publishers.

Bookchin, Murray (pseudonym: Lewis Herber). 1963. *Our Synthetic Environment*. New York: Knopf.

van der Borg, Jan and Antonio Paolo Russo. 2001. "Towards sustainable tourism in Venice." In *Sustainable Venice: Suggestions for the Future*, edited by Ignazio Musu, 159-193. Dordrecht, Boston, London: Kluwer Academic Publishers.

Bramwell, Anna. 1989. *Ecology in the 20th Century. A History*. New Haven and London: Yale University Press.

———. 1994. *The Fading of the Greens*. New Haven and London: Yale University Press.

Brundtland, Gro. 1987. *Our Common Future*. World Commission on Environment and Development. Oxford: Oxford University Press.

Bruneau, Michel, and Andrei Reinhorn. 2006. "Overview of the Resilience Concept." Paper No. 2040, presented at Proceedings of the Eighth US National Conference on Earthquake Engineering, San Francisco, April 18-22.

Bruzelius, Nils, Bent Flyvbjerg, and Werner Rothengatter. 2005. *Megaprojects and Risk*. Cambridge: Cambridge University Press.

Bull, George. 1981. *Venice: The Most Triumphant City*. London: Michael Joseph.

Butcher, Jim. 2003. *The Moralisation of Tourism: Sun, Sand... and Saving the World?* London: Routledge.

Byron, Lord George Gordon. 2008. *The Major Works*. Oxford and New York: Oxford University Press.

Caccia, Beppe. 2005. Interview with Dominic Standish, July 13.

Cacciari, Massimo. 1995. *Architecture and Nihilism: On the Philosophy of Modern Architecture*. Yale: Yale University Press.

Caciagli, Mario. 1988. "Quante Italie? Persistenza e trasformazione delle culture politiche subnazionali." *Polis* II:429-457.

Callenbach, Ernest. 1975. *Ecotopia*. Berkeley: Heyday Books.

Calvino, Italo. 1974a. *Invisible Cities*. Translated by William Weaver. London: Picador.

———. 1974b. "Venezia: archetipo e utopia della città acquatica." In *Italo Calvino: saggi, 1945-1985, vol. 2*, edited by Italo Calvino, 2688-92. Milan: Arnoldo Mondadori.

Caniato, Giovanni. 2005. "Between salt and fresh waters." In *Flooding and Environmental Challenges for Venice and its Lagoon: State of Knowledge,* edited by Caroline Fletcher and Tom Spencer, 7-14. Cambridge and New York: Cambridge University Press.

Carbognin, Laura, Pietro Teatini, Alberto Tomasin, and Luigi Tosi. 2009. "Global change and relative sea level rise at Venice: what impact in term of flooding." *Climate Dynamics* 35, 6:1039-1047. doi: 10.1007/s00382-009-0617-5.

Carson, Rachel. 1962. *Silent Spring*. Boston: Houghton Mifflin.

Castelletto, Nicola, Massimiliano Ferronato, Giuseppe Gambolati, Mario Putti, and Pietro Teatini. 2008. "Can Venice be saved by pumping water underground? A pilot project to help decide." *Water Resources Research* 44: W01408. doi: 10.1029/2007WR006177.

Cavolo, Fabio. 2005. Interview with Dominic Standish, July 6.

Cecconi, Giovanni. 1997a. "The Venice Lagoon Mobile Barriers. Sea Level Rise and Impact of Barrier Closures." *Italian Days of Coastal Engineering*. Reproduced by Consorzio Venezia Nuova. Accessed November 30, 2010. http://www.salve.it/banchedati/Letteratura/testi/89.pdf

———. 1997b. "Real time storm and watershed inflow forecasting in the Venice lagoon." Paper presented at RIBAMOD Workshop, Monselice, September 25-26. Reproduced by Consorzio Venezia Nuova. Accessed November 30, 2010. http://www.salve.it/banchedati/Letteratura/testi/87.pdf.

Cederna, Antonio. 1975. *La distruzione della natura in Italia*. Turin: Einaudi.

Cogo, Flavio. 2005. Interview with Dominic Standish, November 5.

———. 2010. Interview with Dominic Standish, February 24.

Commission of the European Communities. 2000. "Communication from the Commission on the precautionary principle." EC, Brussels. Accessed November 30, 2010. http://ec.europa.eu/dgs/health_consumer/library/pub/pub07_en.pdf.

Commission (*Collegio*) of International Experts. 1998. "Report on the mobile gates project for the tidal flow regulation at the Venice lagoon inlets." *Numero speciale dei Quaderni Trimestrali* Special Edition: 57-106.

Commoner, Barry. 1971. *The Closing Circle: Nature, Man, and Technology*. New York: Knopf.

Consorzio Venezia Nuova (CVN). 2009a. "Defence from high waters. MOSE system-mobile barriers at the inlets. Studies and Experimentations." Accessed September 23, 2009. http://www.salve.it/uk/soluzioni/f_acquealte.htm.

———. 2009b. "Design development and alternatives examined." Accessed July 9, 2009. http://www.salve.it/uk/soluzioni/f_acquealte.htm.

———. 2009c. "Frequently asked questions. Is it true that the gates would become rapidly useless as they would be unable to cope with the rapid rise in sea level?" Accessed November 28, 2010 http://www.salve.it/uk/domande/f-domande.htm.

———. 2009d. "Mose system: questions and answers. Mose is characterised by being an underwater structure. Is there not a risk that the metal structures of the gates and operating mechanisms might corrode in the marine environment to the point of jeopardising the operation?" Accessed September 29, 2009. http://www.salve.it/uk/domande_risposte.htm.

———. 2010a. "Bulletin. MOSE system. Work at the inlets." Accessed November 13, 2010. http://www.salve.it/uk/news/f_news.htm.

———. 2010b. "Venice – Lagoon: The MOSE System for the Defence against High Water." Accessed November 28, 2010. http://www.salve.it/uk/soluzioni/acque/mose_uk.htm.

Cooper, James. 1831. *The Bravo.* New York: G.P. Putnam & Sons, The Knickerbocker Press.

Corrier, Marina. 2005. Interview with Dominic Standish, November 5.

Costa, Paolo. 1993. *Venezia: economia e analisi urbana.* Milan: Etas Libri.

Cozzi, Gaetano. 1973. "Authority and the Law in Renaissance Venice." In *Renaissance Venice*, edited by John Rigby Hale, 293-345. London: Rowman and Littlefield.

Crouzet-Pavan, Elizabeth. 2002. *Venice Triumphant: The Horizons of a Myth.* Maryland: Johns Hopkins University Press.

Crovato, Giorgio, and Maurizio Crovato. 2008. *The Abandoned Islands of the Venetian Lagoon.* Teddington (UK): San Marco Press.

Da Mosto, Jane. 2009. "Venice: Its Demography." In *The Venice Report*, Agnoletto et al., 10-21. Cambridge: Cambridge University Press.

———. 2010. Interview with Dominic Standish, November 11.

Da Mosto, Jane, Tom Spencer, Caroline Fletcher, and Pierpaolo Campostrini. 2005. "Venice and the Venice Lagoon: creating a forum for international debate." In *Flooding and Environmental Challenges for Venice and its Lagoon: State of Knowledge*, edited by Caroline Fletcher and Tom Spencer, 3-5. Cambridge and New York: Cambridge University Press.

Dal Maso, Davide. 2001. "Venice, metaphor of the possible." In *Venice: A Future Capable System*, edited by Adriano Paolella and Paolo Perlasca, 4-7. WWF: Rome.

D'Annunzio, Gabriele. 1991 [1900]. *The Flame.* Translated by Susan Bassnett. London: Quartet Books.

Davis, Joseph. 2002. "Narrative and Social Movements: The Power of Stories." In *Stories of Change: Narrative and Social Movements*, edited by Joseph Davis, 3-29. New York: State University of New York Press.

Davis, Robert and Garry Marvin. 2004. *Venice, the Tourist Maze.* Berkeley and Los Angeles: University of California Press.

De Filippo, Gustavo. 2005. Interview with Dominic Standish, May 13.

Debray, Régis. 2002. *Against Venice.* Translated by John Howe. London: Pushkin Press.

Della Porta, Donatella. 1996. "The system of corrupt exchange in local government." In *The New Italian Republic: From the fall of the Berlin Wall to Berlusconi*, edited by Stephen Gundle and Simon Parker, 221-233. London: Routledge.

Della Porta, Donatella, and Massimilliano Andretta. 2000. "National Environmental Organizations in the Italian Political System. Lobbying, Concertation and Political Exchange." ECPR Joint Sessions, Copenhagen. Accessed November 30, 2010. http://www.essex.ac.uk/ECPR/events/jointsessions/paperarchive/copenhagen/ws5/porta_a.PDF.

Della Porta, Donatella, Massimilliano Andretta, Lorenzo Mosca, and Herbert Reiter. 2006. *Globalization from Below: Transnational Activists and Protest Networks*. Minneapolis and London: University of Minnesota Press.

Della Porta, Donatella and Mario Diani. 1999. *Social Movements*. Oxford: Blackwell Publishing.

Della Porta, Donatella and Mario Diani. 2004. *Movimenti senza protesta? L'ambientalismo in Italia*. Bologna: Il Mulino.

Della Porta, Donatella and Gianni Piazza. 2008. *Voices of the Valley, Voices of the Straits: How Protest Creates Communities*. New York and Oxford: Berghahn Books.

Della Porta, Donatella and Sidney Tarrow. "Transnational Processes and Social Activism: An Introduction." In *Transnational Protest and Global Activism*, edited by Donatella Della Porta and Sidney Tarrow, 1-17. Oxford and New York: Rowman and Littlefield.

Dente, Bruno, Cinzia Griggio, Andrea Mariotto, and Carolina Pacchi. 2001. "Governing the sustainable development of Venice: Elements of the institutional planning procedure." In *Sustainable Venice: Suggestions for the Future*, edited by Ignazio Musu, 227-261. Dordrecht, Boston, London: Kluwer Academic Publishers.

Diani, Mario. 1988. *Isole nell'arcipelago: Il movimento ecologista in Italia*. Bologna: Il Mulino.

———. 1995. *Green Networks: A Structural Analysis of the Italian Environmental Movement*. Edinburgh: Edinburgh University Press.

Diani, Mario and Francesca Forno. 2003. "Italy." In *Environmental Protest in Western Europe*, edited by Christopher Rootes, 135-165. Oxford: Oxford University Press.

Dorigo, Wladimiro. 1972. *Una laguna di chiacchiere*. Venice: Emiliana.

———. 1973. *Una Legge contro Venezia: natura storia interessi nella questione della città e della laguna*. Rome: Officina.

Dubos, René Jules and Barbara Ward (Lady Jackson). 1972. *Only One Earth: The Care and Maintenance of a Small Planet*. Stockholm: United Nations Conference on the Human Environment.

Duggan, Christopher. 1994. *A Concise History of Italy*. Cambridge: Cambridge University Press.

Dundes, Alan. 1976. "Myth." In *Encyclopaedia of Anthropology*, edited by David Hunter and Phillip Whitten, 279-281. New York: Harper and Row.

Eagleton, Terry. 1996. *The Illusions of Postmodernism*. Oxford: Blackwell Publishers.

Eder, Klaus. 1998. "The Institutionalisation of Environmentalism: Ecological Discourse and the Second Transformation of the Public Sphere." In *Risk, Environment and Modernity: Towards a New Ecology*, edited by Scott Lash, Bronislaw Szerszynski and Brian Wynne, 203-223. London: Sage.

Eglin, John. 2001. *Venice Transfigured: The Myth of Venice in British Culture, 1660-1797*. New York: Palgrave.

Ehrlich, Paul. 1971 [1968]. *The Population Bomb*. New York: Macmillan General Books.

European Parliament. 1999. "Resolution on the crisis situation in Venice." Reproduced by Consorzio Venezia Nuova. Accessed November 30, 2010. http://www.salve.it/uk/sezioni/itermose/allegati/parlameuropeo.html.

Farro, Antimo. 1990. *La lente verde: Cultura e politica ambientaliste*. Milan: Franco Angeli.

Favaro, Marco. 2005. Interview with Dominic Standish, July 5.

Fay, Stephen and Phillip Knightley. 1976. *The Death of Venice*. London: Andre Deutsch.

Fersuoch, Lidia. 2004. "The Serene Republic and the Maintenance of the Lagoon." Unpublished letter, May 15. Accessed November 30, 2010. http://www.italianostra-venezia.org/3laguna/fersuoch.htm.

Fletcher, Caroline, and Jane Da Mosto. 2004. *The Science of Saving Venice*. Turin, London, Venice and New York: Umberto Alemandi.

Fletcher, Caroline and Tom Spencer, eds. 2005. *Flooding and Environmental Challenges for Venice and its Lagoon: State of Knowledge*. Cambridge and New York: Cambridge University Press.

Folin, Marino and Mario Spinelli. 2005. "Local flood protection measures in Venice." In *Flooding and Environmental Challenges for Venice and its Lagoon: State of Knowledge*, edited by Caroline Fletcher and Tom Spencer, 147-158. Cambridge and New York: Cambridge University Press.

Fracchia, Fabrizio, Roberto Agnoletto, and Francesca Mattassoglio. 2009. "Public funding: Is the future of Venice sustainable?" In *The Venice Report*, Agnoletto et al., 86-95 and 98-101. Cambridge: Cambridge University Press.

Frassetto, Roberto. 2005. "The facts of relative sea-level rise in Venice." In *Flooding and Environmental Challenges for Venice and its Lagoon: State of Knowledge*, edited by Caroline Fletcher and Tom Spencer, 29-40. Cambridge and New York: Cambridge University Press.

Furedi, Frank. 2002. *Culture of Fear: Risk-taking and the morality of low expectation*. Revised edition. London and New York: Continuum.

———. 2005. *Politics of Fear: Beyond Left and Right*. London and New York: Continuum.

Gaeta, Franco. 1961. "Alcune Considerazione sul Mito di Venezia." *Bibliothèque d'Humanisme et Renaissance* 23:58-75.

Gasparetto, Cristiano. 2005a. Interview with Dominic Standish, July 27.

———. 2005b. "A chi serve il MoSE." *Il Manifesto*, July 28.

Gerritsen, Herman, Rinus Vis, Rosa Mikhailenko, and Marjukka Hiltunen. 2005. "Flood protection, environment and public participation – case study: St Petersburg Flood Protection Barrier." In *Flooding and Environmental Challenges for Venice and its Lagoon: State of Knowledge*, edited by Caroline Fletcher and Tom Spencer, 341-351. Cambridge and New York: Cambridge University Press.

Ghetti, Augusto. 1988. "Subsidence and Sea-level Fluctuations in the Territory of Venice." *Landscape and Urban Planning* 16:13-33.

Gibin, Renato and Stefania Tonin. 2009. "Change of use of buildings." In *The Venice Report*, Agnoletto et al., 46-55 and 59-67. Cambridge: Cambridge University Press.

Giddens, Anthony. 1990. *The Consequences of Modernity*. Cambridge: Polity.

———. 1991. *Modernity and Self-Identity: Self and Society in the Late Modern Age.* Cambridge: Polity.

———. 1999. *Runaway World.* London: Profile Books.

Ginsborg, Paul. 2001. *Italy and its Discontents 1980 -2001*. London: Allen Lane.

Goethe, Johann Wolfgang. 1962. *Italian Journey (1816-29).* Translated by Wystan Hugh Auden and Elizabeth Mayer. Harmondsworth: Penguin.

Goklany, Indur. 2001. *The Precautionary Principle: A Critical Appraisal of Environmental Risk Assessment.* Washington DC: The Cato Institute.

Greenpeace. 1999. "Greenpeace Activists face a rough time while occupying a chemical plant to protect Venice lagoon." September 15. Accessed December 2, 2010. http://archive.greenpeace.org/pressreleases/toxics/1999sep15.html.

Groppello, Andrea, and Paola Virgioli, eds. 2009. *Venezia sistema Mose: Studi di inserimento architettonico delle opere mobili alle bocche lagunari per la difesa dalle acque alte.* Venice: IUAV/Marsilio.

Grubb, James. 1986. "When Myths Lose Power: Four Decades of Venetian Historiography." *Journal of Modern History* 58:43-94.

Gundle, Stephen. 1996. "The rise and fall of Craxi's Socialist Party." In *The New Italian Republic: From the fall of the Berlin Wall to Berlusconi*, edited by Stephen Gundle and Simon Parker, 85-98. London: Routledge.

Hajer, Maarten. 1998. "Ecological Modernisation as Cultural Politics." In *Risk, Environment and Modernity: Towards a New Ecology*, edited by Scott Lash, Bronislaw Szerszynski and Brian Wynne, 246-268. London: Sage.

Hannigan, John. 2006. *Environmental Sociology.* London and New York: Routledge.

Hibbert, Christopher. 1988. *Venice: The Biography of a City.* London: Grafton Books.

Holgate, Simon. 2007. "On the decadal rates of sea level change during the twentieth century." *Geophysical Research Letters* 34:L01602, 4PP. doi:10.1029/2006GL028492.

Howard, Deborah. 2005. *The Architectural History of Venice.* New Haven and London: Yale University Press.

Illich, Ivan. 1971. *Deschooling Society*. London: Marion Boyars Publishers Limited.

———. 1973. *Tools for Conviviality*. London: Calder and Boyars.

Intergovernmental Panel on Climate Change (IPCC). 2007. "Climate Change 2007: Synthesis Report. Summary for Policymakers." Accessed December 2, 2010. http://www.ipcc.ch/pdf/assessment-report/ar4/syr/ar4_syr_spm.pdf.

Istituzione Centro Previsioni e Segnalazioni Maree. 2009. "La subsidenza e l'eustatismo." Venice City Council. Accessed December 4, 2010. http://www.comune.venezia.it/flex/cm/pages/ServeBLOB.php/L/IT/IDPagina/1844.

Italia Nostra. 2007a. "Bulletins of interventions in Venice, 1957-1999." Italia Nostra Venice Chapter. Accessed December 2, 2010. http://www.italianostra-venezia.org/4notizie/4news.htm.

———. 2007b. "Is Venice Sinking? Our Opposition to the Dam Project." Italia Nostra Venice Chapter. Accessed November 12, 2010. http://www.italianostra-venezia.org/3laguna/3no-en.htm.

———. 2007c. "Is Venice Sinking? The Alternatives to Protect Venice and its Lagoon." Italia Nostra Venice Chapter. Accessed December 2, 2010. http://www.italianostra-venezia.org/3laguna/3alt-en.htm.

———. 2010a. "Motor Boat Waves: The Impact on Venice and on its Lagoon." Italia Nostra Venice Chapter. Accessed November 12, 2010. http://www.italianostra-venezia.org/3campagn/SoloMO/MOndosoOKOK-en.htm.

———. 2010b. "Italia Nostra: Venezia, la città più commissariata d'Italia, ma sempre più in pericolo." National Press Conference Document, March 2. Accessed December 2, 2010. http://www.italianostra-venezia.org/.

James, Henry. 1909. *Italian Hours.* London: William Heinemann.

Jamison, Andrew. 1998. "The Shaping of the Global Environmental Agenda: The Role of Non-Governmental Organisations." In *Risk, Environment and Modernity: Towards a New Ecology*, edited by Scott Lash, Bronislaw Szerszynski and Brian Wynne, 224-245. London: Sage.

———. 2001. *The Making of Green Knowledge: Environmental Politics and Cultural Transformation.* Cambridge: Cambridge University Press.

Jamison, Andrew, and Ron Eyerman. 1994. *Seeds of the Sixties.* Berkeley: University of California Press.

Keahey, John. 2002. *Venice Against the Sea: A City Besieged.* New York: St. Martin's Press.

Kuhn, Thomas. 1962. *The Structure of Scientific Revolutions.* Chicago: University of Chicago Press.

Kumar, Krishan. 1999. *From Post-Industrial to Post-Modern Society.* Oxford: Blackwell.

Labalme, Patricia. 1969. *Bernado Giustiniani: A Venetian of the Quattrocento.* Edizioni di Storia e Letteratura: Rome.

Lane, Frederic. 1973. *Venice: A Maritime Republic.* Baltimore and London: The Johns Hopkins University Press.

Lash, Scott, Bronislaw Szerszynski, and Brian Wynne. 1998. "Introduction: Ecology, Realism and the Social Sciences." In *Risk, Environment and Modernity: Towards a New Ecology*, edited by Scott Lash, Bronislaw Szerszynski and Brian Wynne, 1-26. London: Sage.

Latour, Bruno. 1992. *We Have Never Been Modern.* Cambridge (MA): Harvard University Press.

Lauritzen, Peter. 1986. *Venice Preserved.* London: Michael Joseph.

Lewanski, Rudolf. 1997. *Governare l'ambiente. Attori e processi della politica ambientale: interessi in gioco, sfide, nuove strategie.* Bologna: Il Mulino.

Liberatore, Angela and Rudolf Lewanski. 1990. "The Evolution of Italian Environmental Policy." *Environment* 32, 5:10-40.

Lodi, Giovanni.1984. *Uniti e diversi. Le mobilitazioni per la pace nell'Italia degli anni Ottanta.* Milan: Unicopli.

———. 1988. "La Lista Verde di Milano tra ambientalismo e politica." In *La sfida verde*, edited by Roberto Biorcio and Giovanni Lodi, 147-160. Padua: Liviana.

Lomborg, Bjorn. 2007. *Cool It: The Skeptical Environmentalist's Guide to Global Warming.* London: Cyan/Marshall Cavendish.

Loseke, Donileen. 2003. *Thinking about Social Problems.* New York: Aldine de Gruyter.

Lowe, Philip, and Jane Goyder. 1983. *Environmental Groups in Politics.* London: Allen and Unwin.

Macdonald, Douglas. 1991. *The Politics of Pollution: Why Canadians are Failing their Environment.* Toronto: McClelland and Stewart.

Magnami, Carlo. 2009. "Infrastructure and landscape design at the inlets." In *Venezia sistema Mose: Studi di inserimento architettonico delle opere mobili alle bocche lagunari per la difesa dalle acque alte,* edited by Andrea Groppello and Paola Virgioli, 125. Venice: IUAV/Marsilio.

Mann, Thomas. 1998 [1912]. *Death in Venice and Other Stories.* Translated by David Luke. New York: Bantan Dell.

McCarthy, Mary. 1963. *Venice Observed.* Orlando: Harcourt Inc.

McCarthy, Patrick. 1997. *The Crisis of the Italian State.* London: Macmillan Press.

McGann, Jerome. 2008. Introduction to *The Major Works*, by Lord George Byron, xi-xxiii. Oxford and New York: Oxford University Press.

Melucci, Alberto. 1996. *Challenging Codes: Collective action in the information age.* Cambridge: Cambridge University Press.

Mencini, Giannandrea. 1996. *Venezia acqua e fuoco. La politica della salvaguardia dall'alluvione del 1966 al rogo della Fenice.* Venice: Il Cardo Editore.

———. 2003. Introduction to *Venezia. Ambiente Laguna,* edited by Giannandrea Mencini, 7-9. Venice: Supernova.

———. 2005a. Interview with Dominic Standish, October 5.

———. 2005b. *Il Fronte per la difesa di Venezia e della Laguna e le denunce di Indro Montanelli.* Venice: Supernova.

Merchant, Carolyn. 1980. *The Death of Nature: Women, Ecology and the Scientific Revolution.* New York: Harper Collins.

Ministry of Public Works (Italy), Venice Water Authority and Consorzio Venezia Nuova. 1997. "Environmental Impact Study (EIS) of the Preliminary Plan for mobile barriers at lagoon inlets for the defence against high water, EIS – Results." Accessed November 15, 2010. http://www.salve.it/uk/sia/sia-esiti.htm.

Ministry of the Environment (Italy) (Environmental Impact Assessments Commission). 1998. "Summary of the conclusions of the opinion of the Environmental Impact Assessments Commission on the preliminary project for tidal flow control measures at the lagoon inlets." Accessed November 15, 2010. http://www.salve.it/uk/sezioni/itermose/allegati/parere.html.

Moore, Howard. 2005. "Collaboration between the worlds of science and culture." Preface to *Flooding and Environmental Challenges for Venice and its Lagoon: State of Knowledge*, edited by Caroline Fletcher and Tom Spencer, xxiii-xxiv. Cambridge and New York: Cambridge University Press.

Muir, Edward. 1981. *Civic Ritual in Renaissance Venice*. Princeton: Princeton University Press.
Mulier, Eco Haitsma. 1980. *The Myth of Venice and Dutch Republican Thought in the Seventeenth Century*. Assen: Van Gorcum.
Mumford, Lewis. 1922. *The Story of Utopias*. New York: Boni and Liveright.
———. 1934. *Technics and Civilization*. New York: Harcourt, Brace and Company.
Musu, Ignazio. 2001. "Venice and its lagoon." In *Sustainable Venice: Suggestions for the Future*, edited by Ignazio Musu, 1-25. Dordrecht, Boston, London: Kluwer Academic Publishers.
Nietzsche, Friedrich. 1982. *Daybreak: Thoughts on the Prejudices of Morality*. Translated by Reginald John Hollingdale. Cambridge: Cambridge University Press.
Norwich, John Julius. 1983. *A History of Venice*. London: Penguin.
———. 2003. *Paradise of Cities. Venice and Its Nineteenth-century Visitors*. London: Viking (Penguin).
Nugent, Thomas. 1756. *The Grand Tour; or, A Journey through the Netherlands, Germany, Italy and France, First Volume*. London: Browne, Millar, Hawkins, Johnston, Daven and Law.
OECD (Organisation for Economic Co-operation and Development). 2010. *OECD Territorial Reviews: Venice, Italy 2010*. Available at http://www.oecd.org/document/28/0,3746,en_2649_33735_45370460_1_1_1_1,00.html.
Panzeri, Lidia. 2009a. "The Arsenale: a 50-year stalemate over its changing use." In *The Venice Report*, Agnoletto et al., 56-58. Cambridge: Cambridge University Press.
———. 2009b. "Venice needs constant, costly maintenance: the vital role of Insula." In *The Venice Report*, Agnoletto et al., 96-97. Cambridge: Cambridge University Press.
Pemble, John. 1996. *Venice Rediscovered*. Oxford: Oxford University Press.
Perlasca, Paolo. 2001. "Morphological Transformations." In *Venice: A Future Capable System*, edited by Adriano Paolella and Paolo Perlasca, 61-65. WWF: Rome.
———. 2005. Interview with Dominic Standish, July 15.
Pertot, Gianfranco. 2004. *Venice: Extraordinary Maintenance*. Venice: Paul Holberton Publishing.
Plant, Margaret. 2002. *Venice: Fragile City. 1797-1997*. New Haven and London: Yale University Press.
Poggi, Giancarlo. 1968. *L'organizzazione partitica del PCI e della DC*. Bologna: Il Mulino.
Polanyi, Michael. 1998 [1958]. *Personal Knowledge*. London: Routledge.
Popper, Karl. 1982. *Unended Quest: An Intellectual Biography*. London: Open Court Publishing Company.
Principia R.D. 2009. "Comparison of two barriers of gate systems. Hydrodynamic study." Reproduced by *Venice City Council*. Accessed November 27, 2010. http://www2.comune.venezia.it/mose-doc-prg/documenti/Venice_Report_V7.pdf.
Proust, Marcel. 1994 [1971]. *Against Sainte-Beuve and Other Essays*. Translated by John Sturrock. Harmondsworth: Penguin.
Putnam, Robert. 1993. *Making Democracy Work: Civic Traditions in Modern Italy*. Princeton: Princeton University Press.

Queller, Donald. 1973. "The Development of Ambassadorial Relazioni." In *Renaissance Venice*, edited by John Rigby Hale, 174-196. London: Rowman and Littlefield.
———. 1986. *The Venetian Patriciate: Reality versus Myth.* Urbana: University of Illinois Press.
Random House Dictionary. 2010. Accessed November 27, 2010. http://dictionary.reference.com/.
Reberschegg, Fabrizio. 2005. Interview with Dominic Standish, June 24.
Redclift, Michael and Graham Woodgate. 1997. "Sustainability and social construction." In *The International Book of Environmental Sociology*, edited by Michael Redclift and Graham Woodgate, 55-71. Cheltenham and Massachusetts: Edward Elgar.
Reich, Charles. 1971. *The Greening of America.* London: Allen Lane.
Rhodes, Martin. 1995. "Italy. Greens in an overcrowded political system." In *The Green Challenge: The development of Green parties in Europe*, edited by Dick Richardson and Christopher Rootes, 168-192. London: Routledge.
Rice, John Steadman. 2002. "Getting Our History Straight: Culture, Narrative, and Identity in the Self-Help Movement." In *Stories of Change. Narrative and Social Movements*, edited by Joseph Davis, 79-99. New York: State University of New York Press.
Rifkin, Jeremy. 1983. *Algeny.* Harmondsworth: Penguin.
Rinaldo, Andrea. 2001. "On the natural equilibrium of the Venice Lagoon (Will Venice survive?)." In *Sustainable Venice: Suggestions for the Future*, edited by Ignazio Musu, 61-94. Dordrecht, Boston, London: Kluwer Academic Publishers.
———. 2009. *Il governo dell'acqua. Ambiente naturale e Ambiente costruito.* Venice: Marsilio.
Rootes, Christopher. 2003. "The Transformation of Environmental Activism: An Introduction." In *Environmental Protest in Western Europe,* edited by Christopher Rootes, 1-19. Oxford: Oxford University Press.
———. 2005. "The British Environmental Movement." In *Transnational Protest and Global Activism*, edited by Donatella Della Porta and Sidney Tarrow, 21-43. Oxford and New York: Rowman and Littlefield.
Rosand, David. 2001. *Myths of Venice: The Figuration of a State.* Chapel Hill and London: University of North Carolina Press.
Rubin, Charles. 1994. *The Green Crusade: Rethinking the Roots of Environmentalism.* New York: The Free Press.
Runca, Eliodoro, Alberto Giulio Bernstein, Leo Postma, and Giampaolo Di Silvio. 1996. "The framework of analysis to evaluate environmental measures in the Venice lagoon." *Consorzio Venezia Nuova.* Accessed November 28, 2010. http://www.salve.it/banchedati/Letteratura/testi/46.pdf.
Ruskin, John. 1960 [1860]. *The Stones of Venice.* Edited by Joseph Gluckstein Links. New York: Da Capo Press.
Sachs, Wolfgang. 1999. *Planet Dialectics: Explorations in Environment and Development.* London: Zed Books.
Said, Edward. 1999. "Not all the Way to the Tigers: Britten's 'Death in Venice'." *Critical Quarterly* 41:46.
Salzano, Edoardo. 2005. Interview with Dominic Standish, December 20.

Sambo, Enrico. 2005. Interview with Dominic Standish, November 5.
Sarto, Giorgio. 2005. Interview with Dominic Standish, December 8.
Sartre, Jean-Paul. 1991. *La Reine Albemarle ou le dernier touriste, fragments*. Paris: Gallimard.
Scarpa, Cesare. 2005. Interview with Dominic Standish, November 24.
Schumacher, Ernst Friedrich. 1973. *Small is Beautiful: Economics As If People Mattered*. London: Blond and Briggs.
Scotti, Alberto. 2005. "Engineering interventions in Venice and in the Venice Lagoon." In *Flooding and Environmental Challenges for Venice and its Lagoon: State of Knowledge*, edited by Caroline Fletcher and Tom Spencer, 245-255. Cambridge and New York: Cambridge University Press.
Simmel, George. 1922. *'Venedig', Zur Philosophie der Kunst: Philosophische und Kunstphilosophische Aufsatze*. Potsdam: Gustav Kiepenheurer.
Snow, David and Robert Benford. 1992. "Master Frames and Cycles of Protest." In *Frontiers in Social Movement Theory*, edited by Aldon Morris and Carol McClurg Mueller, 133-155. New Haven and London: Yale University Press.
Snow, David. A, Burke E. Rochford, Steven K. Worden, and Robert D. Benford. 1986. "Frame alignment processes, micromobilization, and movement participation." *American Sociological Review* 51:464-81.
Somers Cocks, Anna and Thierry Morel. 2009. "Tourism." In *The Venice Report*, Agnoletto et al., 30-45. Cambridge: Cambridge University Press.
Sontag, Susan. 1983. *Unguided Tour*. Film produced by Giovanella Zannoni for Lunga Gittata Cooperative, RAI Rete 3.
Spencer, Tom. 2009. "Deepening the navigation channels is still a risk to the lagoon and therefore also Venice." In *The Venice Report*, Agnoletto et al., 25-26. Cambridge: Cambridge University Press.
Standish, Dominic. 2003a. "Barriers to barriers: why environmental precaution has delayed mobile floodgates to protect Venice." In *Adapt or Die*, edited by Kendra Okonski, 39-55. London: Profile Books.
———. 2003b. "Barriere alle barriere: perché il principio di precauzione ha ostacolato il progetto MOSE che salverà Venezia." In *Dall'effetto serra alla pianificazione economica*, edited by Kendra Okonski and Carlo Stagnaro, 43-78. Milan: Rubbettino/Leonardo Facco.
———. 2009. "Nuclear Power and Environmentalism in Italy." *Energy and Environment* 20, 6:949-960.
Statham, Paul. 1995. "Political pressure or cultural communication? An analysis of the significance of environmental action in public discourse." Unpublished manuscript. Florence: European University Institute.
Stefanizzi, Sonia. 1987. "Alle origini dei nuovi movimenti sociali: gli ecologisti e le donne in Italia (1966-1973)." *Quaderni di Sociologia* 34, 11:99-131.
Strassoldo, Raimondo. 1993. "The Greening of the Booth: Environmental Awareness, Movements and Policies in Italy." *The European Journal of Social Sciences* 6, 4:457-72.
Szerszynski, Bronislaw. 1998. "On Knowing What to Do: Environmentalism and the Modern Problematic." In *Risk, Environment and Modernity: Towards a New*

Ecology, edited by Scott Lash, Bronislaw Szerszynski and Brian Wynne, 104-137. London: Sage.

Tafuri, Manfredo. 1978. *The Sphere and the Labyrinth: Avant-Gardes and Architecture from Piranesi to the 1970s*. Massachusetts and London: Pellegrino d'Acierno and Robert Connolly.

Tantucci, Enrico. 2009. "Venice on the mainland." In *The Venice Report*, Agnoletto et al., 24 and 27-29. Cambridge: Cambridge University Press.

Tarrow, Sidney. 1989. *Democracy and Disorder: Protest and Politics in Italy 1965-1975.* Oxford: Oxford University Press.

———. 1994. *Power in Movement. Social Movements, Collective Action and Politics.* Cambridge: Cambridge University Press.

Tickell, Crispin. 2006. Foreword to *The Revenge of Gaia,* James Lovelock, xi-xiii. London: Allen Lane.

Trigila, Carlo. 1986. *Grandi partiti, piccole imprese.* Bologna: Il Mulino.

Trovò, Francesco. 2010. *Nuova Venezia Antica, 1984-2001.* Santarcangelo di Romagna: Maggioli Editore.

UNEP (United Nations Environment Programme). 1992. "Rio Declaration on Environment and Development." *UNEP*. Accessed November 28, 2010. http://www.unep.org/Documents.Multilingual/Default.asp?DocumentID=78&ArticleID=1163.

UNESCO (United Nations Educational, Scientific and Cultural Organisation). 1969. *Rapporto su Venezia*. Milan: Mondadori.

Urry, John. 1990. *The Tourist Gaze: Leisure and Travel in Contemporary Societies.* London: Sage.

Venice City Council. 2005. "Comparison of the Alternative Interventions at the Port Inlets." *Venice City Council,* Document No. 200458, November 15. Accessed November 28, 2010. http://www2.comune.venezia.it/mose-doc-prg/documenti/prg-mose/MOSE/relazione.pdf.

Vianello, Michele. 2004. *Un'isola del tesoro.* Venice: Marsilio.

Visconti, Luigi. 2004 [1971]. *Death in Venice*. United Kingdom: Warner Brothers (film).

Von Archenholtz, Johann Wilhelm. 1785. *England und Italien*. Translated by Joseph Trapp. In *A Picture of Italy*, 1791, Volume 1. London: G.C.J. and J. Robinson.

Wagner, Richard. 1983. *My Life.* Translated by Andrew Gray, edited by Mary Whittall. Cambridge: Cambridge University Press.

Wilkes, David and Sarah Lavery. 2005. "The Thames Barrier – now and in the future." In *Flooding and Environmental Challenges for Venice and its Lagoon: State of Knowledge*, edited by Caroline Fletcher and Tom Spencer, 287-294. Cambridge and New York: Cambridge University Press.

Wolters, Wolfgang. 2010. "Wolfgang Wolters' introduction to Francesco Trovò's book about building restoration in Venice." Reproduced by *The Venice in Peril Fund*, February 28. Accessed November 28, 2010. http://veniceinperil.org/newsroom/news/wolfgang-wolters-introduction-to-francesco-trovos-book-about-building-restoration-in-venice.

Wordsworth, William. 2000. *The Major Works*. Oxford and New York: OUP.

Worldwatch Institute. 2007. "State of the World 2007: Our Urban Future." Accessed November 28, 2010. http://www.worldwatch.org/taxonomy/term/467.

Worster, Donald. 1977. *Nature's Economy: The Roots of Ecology.* San Francisco: The Sierra Club.

Woudhuysen, James and Joe Kaplinsky. 2009. *Energise! A Future for Energy Innovation.* London: Beautiful Books.

Yates, Joshua J. and James J.D. Hunter. 2002. "Fundamentalism: When History Goes Awry." In *Stories of Change: Narrative and Social Movements*, edited by Joseph Davis, 123-148. New York: State University of New York Press.

Yearley, Steven. 1992. *The Green Case: A Sociology of Environmental Issues, Arguments and Politics.* London: Routledge.

Zanetto, Maurizio. 2005. Interview with Dominic Standish, October 21.

Zucchetta, Gianpietro. 2000. *Storia dell'acqua alta a Venezia.* Venice: Marsilio.

Index